Prompt Cartography

Prompt Cartography: Interactive Map Design with LLMs is a practical, research-informed guide to directing large language models and AI agents to design modern interactive maps across web, mobile, and emerging natural-language GIS platforms. It highlights how cartographic principles such as design, ethics, and spatial communication remain vital despite technological shifts and emphasizes the importance of understanding coding and data literacy to responsibly integrate AI tools like LLMs into workflows. It reframes cartographers as media directors who articulate intent, ethics, and design decisions in structured natural language that LLMs then execute, critique, and iterate at scale.

Features

- Demonstrates how to design complete prompt-to-map pipelines for web cartography, from semantic data discovery and cleaning through visual form, symbology, interactivity, and publication decisions.
- Connects classic cartographic principles like visual hierarchy, visual literacy, thematic map types, color, and typography rules directly to prompt structures and reusable natural-language patterns.
- Shows how to manage LLM defaults, guard against hallucinations, and encode ethical constraints around provenance, privacy, bias, and civic accountability directly into prompts and agent workflows.
- Explains techniques for orchestrating multiple LLM agents, such as data finders, critics, visualizers, and publishers, into robust map design pipelines that make expert judgment visible instead of implicit.
- Provides end-to-end seminar-style case studies, from exploratory analysis to civic-facing web maps, that expose real missteps, revisions, and decision points in prompt cartography practice.
- Includes conceptual exercises, applied labs, and companion website materials with evolving prompt templates, schemas, and implementation examples for a variety of web mapping platforms.
- Acts as a stand-alone text, but it can also serve as a bridge for readers of *Web Cartography* to bring their existing cartographic expertise into the era of AI-assisted, prompt-driven interactive mapping.

This book is aimed at undergraduate and graduate students in GIS, cartography, geography, and related fields, as well as working professionals who need to incorporate LLMs and natural-language workflows into their map design practice without surrendering aesthetic quality, ethical responsibility, or authorial judgment and control.

Ian Muehlenhaus, Ph.D., is a cartographer, author, and educator with more than 15 years of experience in higher-education teaching, research, and academic program leadership in mapping and GIS. He wrote the widely used textbook *Web Cartography* and has spent his career helping students and professionals design interactive maps that are both visually compelling and intellectually honest. Ian currently works as a principal product engineer at Esri, where he develops and tests visualization tools, prototypes new LLM-based mapping workflows, and explores how AI tools and large language models can be woven responsibly into the mapmaking user experience. His research and professional writing span persuasive and rhetorical cartography, flow and panscalar maps, and cartographic ethics, with a particular focus on how design choices shape public belief in the messages maps are communicating. Ian also serves as Chair of the International Cartographic Association's Commission on Map Design, where he collaborates with colleagues worldwide to advance contemporary map design practice for both print and digital media.

Prompt Cartography

Interactive Web Map Design with LLMs

Ian Muehlenhaus

CRC Press
Taylor & Francis Group
Boca Raton London New York

CRC Press is an imprint of the
Taylor & Francis Group, an **informa** business

Designed cover image: Ian Muehlenhaus

First edition published 2027
by CRC Press
2385 NW Executive Center Drive, Suite 320, Boca Raton FL 33431

and by CRC Press
4 Park Square, Milton Park, Abingdon, Oxon, OX14 4RN

CRC Press is an imprint of Taylor & Francis Group, LLC

ISBN: 978-1-041-35474-1 (hbk)
ISBN: 978-1-041-35475-8 (pbk)
ISBN: 978-1-003-79792-0 (ebk)

DOI: 10.1201/9781003797920

Typeset in Palatino LT Std
by SPi Technologies India Pvt Ltd (Straive)

Access the support material: www.routledge.com/9781041354741

To Svenja and Antja.

The future can be scary. But it's *your* journey. I hope you embrace it; roll with it; and pause long enough to find joy in both its ups and downs. No matter where it takes you, I'll be cheering you along. Always.

Contents

Preface ... xix
Acknowledgments ... xxii

Part I Prompt Cartography Fundamentals

1. From Web Cartographer to Prompt Cartography Director ... 5
What Has Changed (And What Never Does) ... 5
The Prompt Director Mindset ... 9
Maps Still Persuade – Even When Machines Draw Them ... 11
Why This Book Is Structured the Way It Is ... 12
Learning to Speak Prompt Design Fluently ... 13
Takeaways ... 14
Conceptual and Applied Exercises ... 15
Purpose First. ... 15
Audience Reality Check. ... 15
LLM Framing Exercise. ... 15
Micro-Prompt Exercise. ... 16

2. Learning to Speak via Prompt ... 17
Casual Prompting Versus Directive Prompting ... 17
Different Prompt Types for Different Roles ... 21
Prompts as Design Model Artifacts ... 22
Prompt Structure Beats Cleverness ... 23
The CRAFT Framework as a Prompt Aid ... 24
Prompt Literacy as a Professional Skill ... 25
Looking Ahead: From Literacy to Practice ... 25
Takeaways ... 26
Conceptual and Applied Exercises ... 27
Interactivity Audit ... 27
Control Versus Clarity ... 27
LLM Critique Prompt ... 27
Design Judgment Exercise ... 27

3. The Cartography Process as Knowledge Recall ... 28
From Tacit Knowledge to Explicit Recall ... 28
Retrieval-Augmented Generation ... 29
Knowledge Training Versus Retrieval Augmentation ... 30

Chunking as Editorial Decision ... 32
Balancing and Structuring Memories ... 35
FAISS, Briefly, and Then We Move On ... 36
Retrieval as Constraint ... 37
Why Forgetting Is a Useful Feature ... 37
What This Enables – And What It Doesn't ... 39
From Memory to Workflow ... 39
Takeaways ... 40
Conceptual and Applied Exercises ... 41
Tacit Versus Explicit Recall ... 41
Memory as Design Choice ... 41
LLM Memory Probe ... 42
Conceptual Prompt Exercise ... 42

4. **The Prompt Cartography Workflow and Pipelines** ... 43
From Improvisation to Repeatable Design Moves ... 43
A Conceptual Prompt-to-Map Workflow ... 44
Data Engineering: Determining What Your Map Can Show ... 46
Visual Planning: Designing Before Anything Is Drawn ... 48
Interface Planning: Anticipating Use and Saying "No" ... 49
Specifications as Contracts as Opposed to Commands ... 50
Review, Critique, and the Value of Slowing Down ... 52
Publishing as a Moment of Accountability ... 54
Pipelines Don't Replace Judgment: They Make It Visible ... 55
Takeaways ... 55
Conceptual and Applied Exercises ... 57
Improvisation Check ... 57
Judgment Visibility ... 57
LLM Framing Exercise ... 57
Pipeline Articulation Exercise ... 57

5. **Provenance, Ethics, and Privacy** ... 58
The Chain of Custody for Data ... 58
Prompt Metadata Fields ... 59
Derivative Works and Style Attribution ... 60
Attribution, Transparency, and Intent ... 60
Declaring Intent and Generating Prompt Metadata ... 61
Archiving Your Pipeline Prompts to Maintain Trust ... 61
Bias, Fairness, and Explainable Style Decisions ... 62
Privacy, Representation, and Sensitivity ... 65
Ethics and Archiving for Trust ... 69
Takeaways ... 70
Ready to Learn How to Design via Prompting? ... 71
Conceptual and Applied Exercises ... 71

Ordinary Ethics 71
Provenance Under Pressure 71
LLM Ethics Framing 72
Disclosure as Design Choice 72

Part II From Reorientation to Application

6. The (Constant) Critique Pipeline 75
From "It Works!" to "But… Is It Right?" 76
What We Mean by "Critique" (And What We Don't) 78
Design Drift: When Maps Slowly Become Something Else 79
Aesthetic Drift 80
Data Drift 81
Ethical Drift 81
Critique Prompts as Conversation 82
Ode to Keeper Prompts 84
Metrics Without the Panic 85
Contrast 86
Performance 86
Design Quality Assurance Process 88
Intent Alignment 88
Experiential Clarity 89
Credibility and Care 89
From Critique to Continuity 89
From Conversations to Pipelines 90
Takeaways 90
Conceptual and Applied Exercises 92
Critique Versus Confirmation 92
Drift Detection 92
LLM Critique Framing 92
Critique as Continuity 92

7. The Data Engineering Pipeline 93
How to Use LLMs to Search for Datasets 94
Semantic Search (A Primer) 95
Semantic Search for Datasets 95
Evaluating and Assessing Dataset Integrity for Your Map's Purpose 99
Evaluating Dataset Quality and Ethics 102
Cleaning and Engineering Spatial Datasets via LLM 104
Describing What You Need from Your Datasets 106
Ethical and Legal Considerations 106
Minority Perspectives and Representation 107
Map Ethics Always Starts with Your Dataset 107

The Data Engineering Pipeline ... 108
Looking Ahead ... 109
Takeaways ... 109
Conceptual and Applied Exercises ... 111
Truth Versus Fit ... 111
From Keywords to Constraints ... 112
Semantic Fingerprint Test ... 112
Dataset Accent Exercise ... 112
Ethical Clause Challenge ... 112
Mini Data Analysis ... 112

8. Learning the Language of Map Form and Style ... 114
Visual Literacy and the Dondis's Framework ... 115
Visual Literacy: Same Data, Different Form ... 116
Naming the Building Blocks of Visual Form ... 118
Dot, Line, Shape ... 118
Direction and Movement ... 118
Scale and Proportion ... 118
Texture and Pattern ... 118
Color ... 118
Visual Literacy: From Visual Dimension to Prompt Language ... 119
Aesthetic Styles in Cartography ... 123
Map Form and Function ... 130
From Style to Promptable Variables ... 131
Style Drift and Aesthetic Defaults in LLMs ... 132
Declare Style Early ... 132
Use Style Templates or Schema References ... 132
Chunk Your Prompts ... 133
Regularly Verify and Re-Specify Your Expectations ... 133
Self-Critique ... 133
On to Building Map Form Piece-by-Piece ... 133
Takeaways ... 134
Conceptual and Applied Exercises ... 135
Finding the Missing Words ... 135
Style as Constraint ... 135
LLM Vocabulary Test ... 135
Structured Prompt Exercise. ... 136

9. Map Elements, Layouts, and the Curse of LLM Defaults ... 137
Titles, Framing, and the Rhetoric of First Impressions ... 140
Title and Splash Screen Design Tips ... 142
Mapped Area, Extent, and the Myth of Infinite Zoom ... 143
Mapped Area and Extent Design Tips ... 146
What About Neatlines and Map Navigation Tools? ... 147

Pan and Zoom Controls Design Tips ... 147
Interface Elements via Language Prompts ... 148
Interface Element Design Tips ... 149
Visual Hierarchy Remains in Charge ... 150
Visual Hierarchy Design Tips ... 151
Layout Systems: From Screen Real Estate to Spatial Grammar ... 153
Layout Systems Design Tips ... 158
Layout and Balance ... 159
Legends, Insets, and the Danger of Letting the Model Decide ... 159
From Design Rules to Prompt Literacy ... 163
Takeaways ... 166
Conceptual and Applied Exercises ... 167
Default Autopsy ... 167
Extent as an Argument ... 167
LLM Defaults Stress Test ... 168
Structured Prompt Exercise ... 168

10. Color Theory and Palette Design ... 169
A Brief Color Literacy (So Prompting Isn't Guesswork) ... 171
Hue, Value, and Saturation ... 171
Warm and Cool Colors (And Why They Matter) ... 172
How Screens Actually Make Color ... 173
Why This Matters for Prompting ... 174
Choosing the Right Color Scheme (and Making the LLM Explain It) ... 174
Color by Prompt ... 175
From Palette Generator to Palette Critic ... 175
Semantic Color Prompts ... 177
Harmony and Data Type ... 177
Accessibility and Culture ... 179
Color as More than Looking Pretty ... 179
Takeaways ... 180
Conceptual Applied Exercises ... 181
Color as Persuasion ... 181
Scheme Mismatch ... 181
LLM Palette Critique ... 181
Constraint Exercise ... 182

11. Prompting for Typography ... 183
Don't Waste Your Time on the Typeface Versus Font Debate ... 184
Serif Versus Sans-Serif (Or: Why Web Maps Are Biased against Serifs) ... 184
Legibility Isn't Just Taste: A Few Rules That Still Stand ... 185

Label Placement Norms: The Quiet Rules Your Prompts Should Encode ... 186
Web Reality Check: You Don't Control Fonts Like You Can in Print ... 187
Prompting for Fonts and Font Styles ... 188
Type Sets the Tone of Your Map ... 189
Fallbacks and Robustness ... 189
Density and Collision Rules ... 191
Typography and Web Fonts as Key Component in Your Visual Design ... 191
Takeaways ... 192
Conceptual Applied Exercises ... 193
Typography Tone Audit ... 193
Fallback Failure ... 193
LLM Type Reasoning ... 193
Hierarchy Through Type ... 194

12. Map Symbology ... 195
Symbols Are Visual Arguments ... 195
Prompting for Symbology ... 197
Shape for Categorical Difference, Not Storytelling Flair ... 198
Size and Quantitative Logic (or Trust but Verify Your LLM's Output!) ... 200
Opacity: Hierarchy, Context, and Uncertainty (Without Lying) ... 201
Texture: Visual Redundancy Establishes Emphasis ... 201
Using Map Symbol Libraries and Prompting Your Own ... 204
PNG Symbol Prompting ... 204
SVG Symbol Prompting ... 205
Symbols as Structured Decisions ... 206
Takeaways ... 207
Conceptual Applied Exercises ... 208
Symbol Rhetoric Check ... 208
LLM Symbol Justification ... 208
Abstraction Versus Detail ... 208

13. Thematic Representation ... 209
Data and Phenomenon: The Fundamental Distinction ... 209
Choropleth Maps ... 211
When Choropleths Work ... 211
The Choropleth's Default Lies ... 211
Classification Decisions That Shape Meaning ... 211
Prompting for Choropleths ... 213
Prompt Variation Example ... 214

Proportional and Graduated Symbol Maps 214
The Critical Distinction 215
Why Proportional Symbols Excel 215
The Look-Right-Is-Wrong Error 216
The Drawbacks of Proportional Symbols 217
Prompting for Proportional Symbols 217
Prompt for Graduated Symbols 217
Dot Maps 218
When Dot Maps Shine 218
The Recurring Failure Mode 218
Design Considerations 218
Prompting for Dot Maps 220
Prompt Variation: When Point Data Overwhelms 220
Heat Maps and Density Surfaces 220
When Heat Maps Work 220
The Critical Drawbacks of Heat Maps 221
Binning for Human Legibility 222
Prompting for Heat Maps 223
Isarithmic Maps 223
Two Types Based on Data Collection 223
When Isarithmic Maps Excel 224
The Beautiful Lie 224
Design Best Practices 224
Prompting for Isarithmic Maps 225
Prompt Variation: Showing Data Collection Points 225
Flow Maps 225
What Makes Flow Maps Powerful? 225
The Aesthetic and Technical Challenge 226
Prompting for Flow Maps 226
Advanced Flow Map Techniques 228
Cartograms: The Rhetorical Sledgehammer 228
Two Approaches 229
When Cartograms Work 230
The Double-Edged Sword of Areal Distortion 230
Prompting for Cartograms 230
Prompt Variation: Maintaining Orientation 230
Multivariate Maps 231
When Multivariate Representation Succeeds 231
The Cognitive Load Problem 231
Thematic Combinations 231
Colored Dots for Multiple Categories 232
Chart Maps 233
When to Avoid Multivariate Mapping 233
Agent Prompt for Multivariate Restraint 233

Thematic Mapping via Prompting: Integrating Method Selection....... 234
Method Selection Framework 234
Iterative Refinement Pattern.......... 234
Prompt Chaining for Complex Decisions.......... 235
Guardrails as Structural Elements.......... 235
The Tip of the Iceberg.......... 235
Takeaways.......... 236
Conceptual and Applied Exercises.......... 237
Method Mismatch.......... 237
Binning Consequences 238
LLM Method Reasoning 238
Prompt Specification Exercise 238

14. Change over Time.......... 239
When Animation Earns Its Place 239
Prompting Temporal Controls 241
Temporal Legends: Make Time Visible.......... 242
Distinguishing Animation Types in Prompts 243
Motion Constraints: The Non-Negotiables.......... 243
When to Choose Stop-Frame Versus Smooth Transitions.......... 244
Stop-Frame Prompt.......... 244
Smooth-Transition Prompt 245
Simplicity and Duration.......... 245
Human in the Loop: Review Your Animation Output.......... 245
Use Animation for Visual Rhetoric and Temporal Explanation – Never Decoration.......... 246
Takeaways.......... 246
Conceptual and Applied Exercises.......... 247
Animation Justification 247
Motion Meaning.......... 248
LLM Temporal Design.......... 248
Constraint Specification 248

15. Bringing Your Visual Plan to Life via Interactivity.......... 249
From Wiring Events to Directing Experience 249
Interaction Grammars 252
Interactivity Versus Animation: A Quick Clarification.......... 253
Disclosure Systems: Legend + Filters + State 254
What Disclosure Controls 254
State as a Rhetorical Decision.......... 255
Common Disclosure Failures 256
Prompting Patterns.......... 257
Integration Reminder 257

Commitment Systems: Tooltips + Pop-Ups (i.e., Information Windows) 257
Temporal Commitment 258
Pacing as Meaning 258
Common Commitment Errors 261
Prompting Patterns 261
Memory and Persistence Systems: Panels + Dashboards + Persistent Context 262
Why Memory Matters 262
Behavior Questions 263
Prompting Patterns 263
Control Systems: Layer Toggles + Menus + Locator Maps 264
Architecting User Choice 264
Behavior Questions 265
Prompting Patterns 265
Menus: Hide Complexity Without Hiding Meaning 266
Locator Maps (Mini-Maps): Context + Navigation Alternative 267
Scope Creep and the Director's Job 267
Motion and Continuity Systems 267
Prompting Motion and Timing 268
When Abrupt Is More Honest 268
Behavioral Orchestration: Designing User Flows 269
Common Behavior-Based Prompting Mistakes 270
Under-Specification and Leaving Defaults Unspecified 271
Over-Specification: Adding Too Many Affordances Without Purpose 271
Inconsistent States Across Elements 272
Gratuitous Animation and Novelty-First Design 272
Validation Bias (Or: LLMs Are Paid to Praise You, Not Tear Your Hear Out) 272
Testing and Refining Interactive Prompts 273
Test Communicatively: Does It Clarify or Clutter? 273
Iterate by Subtraction: Constrain, Simplify, Remove 273
Variation Testing: Small Prompt Changes Reveal Sensitivity 274
Over Time, Prompts Become Shorter Because Thinking Becomes Clearer 274
Integrating Interactivity with Earlier Design Decisions 274
But Are Your Map Behaviors Working? 275
Takeaways 276
Conceptual and Applied Exercises 278
Interactivity Archaeology 278
State Specification Practice 278
LLM Behavioral Critique 278
Flow Versus Chaos 279

16. From Prompt Cartographer to Interactive Web Map Director 280
The Prompt Cartography Workflow 281
The Prompt Cartography Stack 282
What You Have Learned 283
A Training Arena, Not the Point 283
Framing the Dataset Like a Director 284
Style Emerges from Framing (Not the Other Way Around) 285
Ethical Dilemmas Can Show Up Early, Even in Fictional Places 286
Generalizing Beyond the Example 286
Visual Direction: Style, Symbology, and Hierarchy 287
Style Logically 287
Symbology Is About Meaning Before Beauty 288
Time for a Mid-Process Critique 288
Your Hierarchy as Silent Hero 289
Direction Before Composition 289
Layout and Composition as System Design 289
Layout Is an Issue of Constraint 291
Directing Layout with Intentional Prompts 291
Layout as the Bridge Between Visuals and Behavior 292
Direction Locked In 292
Interaction, Behavior, and User Experience 294
Interaction Is Animated Rhetoric 294
Designing for Confidence, Not Cleverness 294
A Mid-Journey Behavioral Critique 295
Accessibility Is Interaction Design 295
Letting the Map Breathe 295
Critique, Ethics, and Continuous Checking 296
Using the Critique Pipeline Throughout the Workflow 296
Ethics Lives in Ordinary Decisions 297
Critique Without Defensiveness 298
Know When to Stop or You Will Never Finish 298
Publishing, QA, and Readiness 299
Publishing Is a Design Decision 299
Quality Assurance Is About Fragility 300
Testing Without Overthinking 301
Readiness Is Not the Same as Perfection 301
Preparing for Reuse and Evolution 301
Peripheral Sparks: Seeing the Workflow Everywhere 302
Transfer Without Translation 302
You Have Directed Your First Feature Length Web Map 303
Takeaways 303
Conceptual and Applied Exercises 304
From Maker to Director 304
Hierarchy Check 304

LLM Direction Probe ... 304
Knowing When to Stop ... 305

17. Sustaining a Map That Has Left the Station: The Maintenance Pipeline ... 306
From Prompt Spec to Live Deployment ... 306
Prompt-Driven Automation Pipelines ... 307
Automation as Continuity as Opposed to Convenience ... 308
Designing Small, Boring Pipelines ... 308
Automation Without Intimidation ... 309
Monitoring Drift and Version Integrity ... 309
What Technical Drift Looks Like in Practice ... 310
Versioning for Map Integrity Over Perfection ... 310
Watching Entropy Happen Without Succumbing to Paranoia ... 310
Human-in-the-Loop QA and Accessibility ... 311
Accessibility as Ongoing Care ... 311
The Rhythm of Review ... 311
Maintenance as Creative Practice ... 312
Embracing the Change ... 312
Stewardship over Control ... 312
Why This Matters Beyond the Map ... 313
Takeaways ... 313
Conceptual and Applied Exercises ... 314
A Map's Second Life ... 314
Drift in the Wild ... 314
LLM Maintenance Probe ... 314
Knowing What to Preserve ... 315

Part III The Prompt Cartography Seminar

18. Mapping Community Data with Accountability ... 321
Civic Maps ... 321
Intent Under Scrutiny ... 322
Choosing and Shaping Civic Data ... 323
Design Decisions with Consequences ... 324
Setting Up Interpretation without Accusation ... 326
Publishing and Knowing When Not To ... 327
Revisiting What You Just Applied ... 327

19. Mapping Presence, Absence, and Uncertainty ... 329
What Observations Can and Cannot Say ... 330
Overtly Highlighting Dataset Bias ... 330
Choosing Observational Data ... 331

Designing for Uncertainty 332
Symbolizing Less to Show More 333
Contextual and Base Map Layers 333
Map Elements and Interactivity 333
Temporal Thinking Without Prediction 334
The Ethics of Mapping Uncertainty 335
What You Just Practiced 336

20. The Main Feature Presentation: Visualizing Complex Connections 338
From Prompting Maps to Managing Systems 338
Production Does Not Equal Optimization 339
Deciding What Belongs Together 340
Requirements 340
Designing for Separation Without Isolation 341
Responsible Map Design over Accurate Map Design 341
I Finished the Capstone… What Now? 342

Conclusion 343

Glossary of Terms 346

References and Further Readings 358

Index 363

Preface

When *Web Cartography* was first published, the web map itself still felt slightly rebellious. We were wrestling geographic information out of desktop silos and coaxing it into browsers with HTML, JavaScript, and a good bit of patience. At the time, simply making a map pan smoothly without crashing a lab computer felt like progress. The conceptual work of cartography – projection, symbolization, hierarchy, color, narrative – remained intact, but the technical scaffolding beneath it was shifting quickly. We were moving from command lines and static exports to graphical interfaces and dynamic tiles.

That shift has not stopped. It has accelerated.

This book exists because the ground moved again. Large language models emerged, natural language interfaces matured, and the once-rigid boundary between "the person who knows GIS" and "the person who asks for a map" began to blur. I did not set out to revise *Web Cartography*. In fact, I resisted the idea. A second edition, had it merely updated APIs (application programming interfaces) and screenshots, would have felt incremental. Technology, fortunately, refused to be incremental. It evolved fast enough to demand something new.

So this is not simply a revision. It is a progression.

Cartography, at its core, remains what it has always been: the careful, ethical, intentional design of spatial communication. The tools, however, continue their restless march. We have moved from pen and ink to command-line GIS; from graphical user interfaces to scripting; from scripting to automated pipelines; and now from pipelines to prompting. The hand that once drew contours now writes functions. The cartographer who once memorized labeling rules now composes prompts that orchestrate multi-agent systems.

And yet, the intellectual heart of the discipline remains steady.

Prompt Cartography does not replace cartographic design principles. It repositions them. When a map can be generated in seconds from a sentence, design judgment becomes more – not less – important. When an LLM can write the code for a choropleth, it becomes the cartographer's responsibility to ensure that the data are appropriate, the classification meaningful, the color ramp perceptually sound, and the narrative honest. Automation amplifies both excellence and error.

The shift toward natural language GIS (NLGIS) is not about avoiding rigor; it is about relocating it. (Note: NLGIS is not necessarily ubiquitous at the time of writing, but I would bet my retirement on them becoming ubiquitous. I define them as GISystems that use and promote natural language interfaces over those of command-line and graphical-user ones.)

There is also, inevitably, a personal progression embedded in these pages. When I wrote *Web Cartography*, I was a graduate student navigating academia and learning the rhythms of teaching and research. Since then, I have been a professor, a department chair, a university administrator, and now a product engineer at a technology company. I have gone from single to married, from husband to father. Through those transitions, I remain Ian. The questions that animate me – how maps persuade, how design shapes understanding, how technology mediates truth – are the same. But my vantage point has changed.

So has my writing.

Early in my career, I wrote with the earnestness of someone determined to prove he belonged. Now, I write with the calm realism of someone who has seen technologies crest and recede. The web mapping frameworks of 2010 are largely archaeological artifacts. The Python scripts that once felt cutting-edge have been abstracted into higher-level tools. What persists is not the syntax, but the sensibility.

This book reflects that evolution.

I hope it serves multiple audiences at once. For experienced cartographers, it offers a conceptual bridge into workflows that incorporate LLMs, prompt engineering, retrieval-augmented generation, and multi-agent orchestration. If you are integrating AI into production environments, you will find both enthusiasm and caution here. The goal is not to automate thinking away, but to augment it responsibly.

For students just beginning their journey, this text may represent an entirely different starting line. Some of you may never need to write a Python script from scratch. That would have sounded heretical a decade ago. Today, it is plausible. Natural language interfaces can scaffold complex geoprocessing tasks. They can draft code, design layouts, and critique symbolization.

Still, I will say this plainly: understanding how code works – even at a basic level – remains deeply valuable. Not because you must memorize syntax, but because literacy in the underlying mechanics strengthens your judgment. Prompting without comprehension risks becoming magical thinking. Prompting with comprehension becomes orchestration.

The arc from *Web Cartography* to *Prompt Cartography* mirrors the broader arc of digital cartography itself. Each era abstracts complexity. Each abstraction widens access. Each widening demands renewed attention to ethics and design. If GIS once required specialized hardware and expertise, and web mapping required JavaScript literacy, prompt cartography may require little more than a well-constructed sentence. That accessibility is powerful. It is also destabilizing.

Maps have always been persuasive artifacts. When their production becomes frictionless, persuasion scales. That reality demands a generation of cartographers who are fluent not only in visual variables and projections, but also in model limitations, data provenance, and algorithmic bias.

I did not want to write a simple update to a previous book. I wanted to write something that acknowledges where we are headed. The technologies described in these chapters will evolve. The APIs will change. The model names will be replaced. But the underlying orientation – designing maps thoughtfully within natural language systems – will remain relevant longer than any particular framework.

If this book succeeds, it will do two things. First, it will help practitioners transition confidently into workflows that include LLMs and NLGIS. Second, it will provide emerging cartographers with a foundation robust enough that they can navigate whatever comes next – whether that involves prompting, scripting, or something we have not yet imagined.

Cartography has always been adaptive. So have I.

We evolve. The discipline evolves. The medium evolves. But the responsibility endures: to design maps that inform, persuade responsibly, and invite critical engagement.

That continuity – across tools, across careers, across seasons of life – is the thread connecting *Web Cartography* to *Prompt Cartography*.

I am still Ian.

Cartography is still cartography.

The human-computer interface connecting the two, however, has changed.

Acknowledgments

Birgit, Svenja, and Antja. I'm so sorry that every 10 years or so I sign up for a project like this that nearly does me in emotionally. I appreciate how forgiving and patient you all are with my "side projects" and "hobbies." You three are my life. Without you, there is no book. There is no… anything. *Ich hab euch Lieb! Vielen Dank.*

Linda Beale. The most intelligent person I've ever met (across all areas of intelligence to boot). The best professional mentor I've ever had. Pragmatic. Inspiring. So *damn* smart! You are the best manager I've ever had. I can't begin to express what a positive impact you've had on my intellectual growth, professional development, and perspective on life in general. *Thank you.*

Guillaume Touya, Eric Losang, and Ken Field. My friendships with you provide me the energy and rebellious zest to keep pursuing new projects even during the darkest of geopolitical times. I'm grateful for our conversations, discussions, and shared vulnerabilities. *Thank you.*

The ArcGIS Enterprise AI Studio team I work with daily at Esri. All of you! Thanks for exploring LLMs and AI agents daily with me. I've learned more from working with you all than I ever possibly could on my own. I cherish it. *Thank you.*

Sam Bary (Blenheim, New Zealand, 1998), Josh Morrell (Green Valley, Wisconsin, 2019), and His Royal Highness, the Infant King of Greater Ruthenia, Herr Doktor Heiko Edgar Fuerst (Berlin, Germany, 2025). I think of you all daily. *Thank you* for having been a significant part of my life. I miss you.

The University of Wisconsin – River Falls Geography and Mapping Science Department (or whatever new incarnation it takes due to budget changes). Neither *Web Cartography* nor *Prompt Cartography* would exist if you hadn't given me my academic start back in 2008. *Thank you.*

All my former students. It was teaching that gave me the most joy during my career. I am forever grateful to you all. *Thank you.*

ChatGPT 4.1, ChatGPT 5.2, Claude 4.5, and Perplexity were used for feedback, editorial suggestion, and constructive criticism on my early chapter outlines and during revision. Several captions were rewritten using Claude 4.5 for clarity.

Prompting is always better with music! *Thank you* to all the musicians and bands who got me through this. Playlists available on the companion website, but special shout outs to: Oval, Low, The Smiths, Kruder Dorfmeister, Philip Glass, Tom Waits, Rage Against the Machine, Bad Bunny, Pavement, Thievery Corporation, Smashing Pumpkins, and the Lumineers.

Part I

Prompt Cartography Fundamentals

If you've been making maps for any length of time, this moment probably feels familiar.

Tools have changed on you before, and workflows have shifted. GIS software replaced darkrooms. GUI-based GIS replaced command-line. Interactive web maps quickly displaced paper maps in use and ubiquity earlier this century. Mapmaking skills that once felt essential have slowly receded into the background, replaced by new abstractions and technologies that promised speed, access, or flexibility. Every time, for me at least, the adjustment felt bigger while going through it than it does now in retrospect.

Google Maps was revolutionary. It brought tears to a fellow cartographer's eyes when we met for coffee in 2007, and he was explaining to me that he could design by scripting in something called KML instead of using Adobe Flash. He's no longer alive today, sadly, but I think back to that day at Mapps Café (yeah, it was really called that) whenever my daughter flippantly opens Google Maps and takes for granted the speed and accuracy of the results delivered to her in seconds.

The emergence of large language models (LLMs) and their use for making web maps carries the same mix of curiosity and unease. Something is clearly changing, in our work in cartography and the world at large, but it's not particularly clear how the world of cartography is changing or which tools can make your mapmaking life easier. What makes the current situation so confounding compared to previous ones?

DOI: 10.1201/9781003797920-1

Some of the discomfort may come from automation itself. But more likely, I imagine, it's coming from something quieter. Scripts have been automating parts of our work for years now. It's not likely about delegation either; we already depend on software daily to execute decisions and tasks we don't want to perform manually – projections, dataset classification, symbolization (i.e., things I was still teaching students to do manually as recently as 2008). So our unease must be coming from some quieter place. Perhaps the sense, or is it a fear, that cartographic design decisions are going to be made somewhere else, in ways that are harder to see or explain.

It's reasonable to have such concern. Your sense is real; mapmaking is going through a change. In fact, it seems almost preordained that GIS and cartography processes will become largely natural language based in the future, not necessarily done using traditional GIS or cartography software at all, but rather LLM-agent assisted tools. Such a shift will reshape what we mean by cartographic work. A lot of things that we cartographers "do" today will not be done by us at all tomorrow.

Moments like this are usually less about abandoning what we know than about applying what we know in new ways. Mastering new, more efficient technologies to apply knowledge is how disciplines and crafts stay relevant. The guiding principles of stellar mapmaking – clarity, purpose, responsibility – aren't disappearing. What's changing is the way we will apply these principles. Basically, it's going to change most cartographers' day jobs.

This book starts from that recognition. Not from urgency, nor fear, but from the assumption that experienced cartographers already have most of what they need for the transition – knowledge from prior training and experience. Your intuition will pay dividends. If you are a young professional, current student, or novice cartographer, you are in luck, too. You don't have to unlearn or try to forget old techniques and their accompanying tasks that will no longer be as relevant.

The goal of this book is to help you learn and begin to master prompt cartography – creating maps using individual or systems of LLM and AI-based assistants and agents. So without further ado, let's get started!

A NOTE ON WHAT I MEAN BY LLMS, AGENTS, AND AGENTIC SYSTEMS

Before we begin, I want to make a note about the terms I use. This book's intended audience is cartographers, lifelong learners, and people interested in making interactive web maps using LLM prompts. This book is *not* aimed at software engineers. Thus, some liberties with nomenclature have been taken.

Namely, when I refer to LLM agents, I am using the term generically. In software development, agents are generally considered to be autonomous LLM-based AI tools that can act independently. I have no doubt that many readers of this book will be using and developing such agents in this sense. However, when I refer to "LLM agents" loosely, I'm speaking about specialized LLM chatbots or GPT-based assistants that the user has created using chunked knowledge documents and system prompts (to be explained in the book). I'm not necessarily expecting readers of this book to necessarily be building LLM agents from scratch. (Though, to any of you that are doing this, respect!) Thus, "LLM agent" in this book means specialized agents (typically created within major LLM interfaces), built around their own set of knowledge documents (retrieval-augmented generation [RAG]) and a user's system prompts to achieve specific cartographic pipeline goals.

1

From Web Cartographer to Prompt Cartography Director

Prompt cartography is the practice of designing interactive web maps through structured natural-language directives rather than direct manipulation of code or graphical interfaces. It does not replace cartographic judgment; it relocates it. The cartographer's role shifts *from manually executing* each step *to directing systems* that carry those steps out. The craft remains the same. The interface has changed. Prompt cartography is the newest iteration of cartography.

A prompt cartographer, then, is not someone who "lets AI make maps." It is a person who deliberately frames, constrains, critiques, and refines map outputs iteratively through language. The responsibility for clarity, accuracy, ethics, and all other facets of the cartographic discipline remains fully human.

I've taught mapmaking and design for nearly two decades. I've *never* been remotely as excited to teach and write about anything cartography-related as I am about this. My goal with this book is to help you become or evolve into a prompt cartographer.

What Has Changed (And What Never Does)

At a surface level, the shift to an LLM-power, *prompt-based workflow* may feel dramatic. Tasks that once required dozens of lines of code, or hours of interacting with esoteric tools found in a GUI, can now be described in a few *carefully* chosen sentences. Interfaces that demanded procedural thinking now respond to declarative requests. The mechanics of mapmaking have undeniably changed.

The deeper structure of the work, however, is far more stable than it first appears. LLMs aren't directly replacing human cartographers. LLMs can't parrot a mapmaker and spit out masterpieces—yet. Just as a novice can't sit down and create a cartographic masterpiece when presented with desktop GIS, such as Josef Minard's map of Napoleon's March as seen in Figure 1.1,

DOI: 10.1201/9781003797920-2

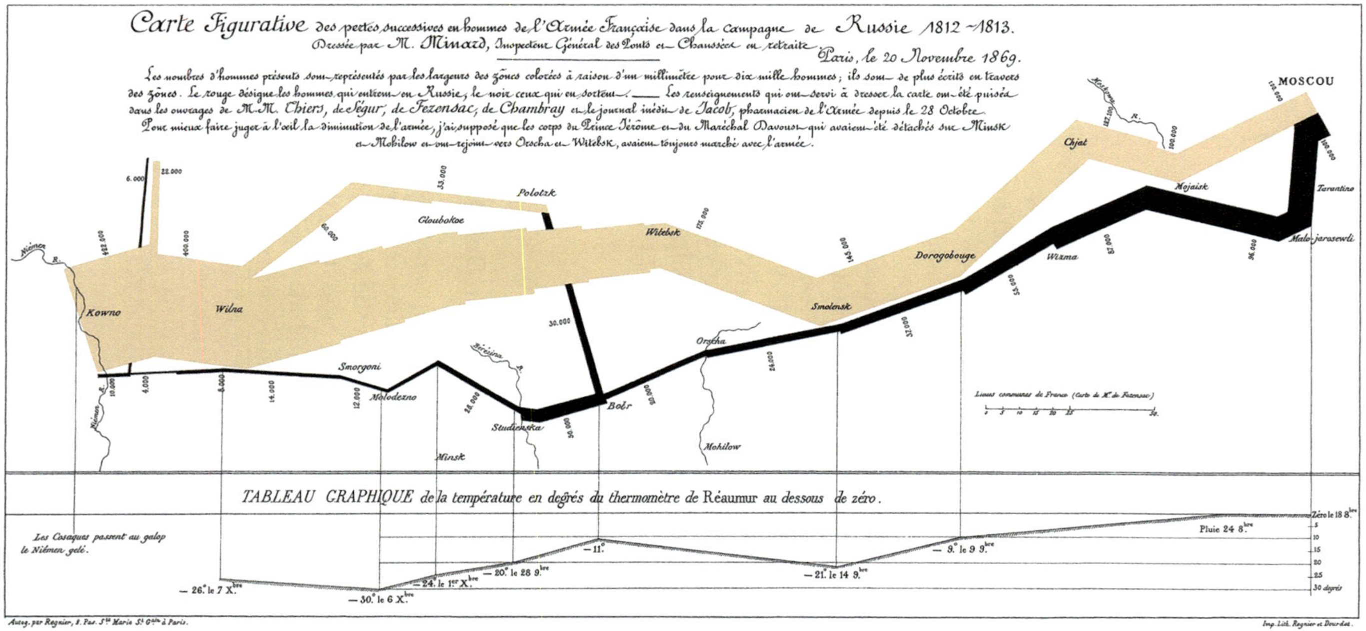

FIGURE 1.1
Minard's famous map of Napoleon's march to Russia. Whether drawn by hand (as the original was), coded (as innumerable versions have been), or prompted into existence, all outputs are manifestations of human intent and direction.

or a full-stack software developer won't be able to magically produce a stellar map design using their coding skills alone, prompt cartography is a skill that requires direction, intent, and practice.

Cartographers have always been guided by intent. We decide what a map should emphasize, what it should omit, how it should look, and whom it should target with its messaging. Traditionally in web cartography, those specifications were translated into code, configuration files, or design artifacts that were then fed to a computer and executed without too much more human involvement or consideration. The translation from human to computer – done via scripting or GUI clicks – was crucial, of course, but it often minimized the design thinking or human inputs (at least, as much of what we cartographers do becomes subconscious) that preceded it.

Large language models (LLMs) compress the translation step between humans and computers (as shown in Figure 1.2). Importantly, though, they don't remove it. Instead of writing and inputting clear, rule-based procedures in byzantine scripting languages, we now articulate needs and constraints in English (or Hungarian, Mandarin, German, French, Esperanto, pick your linguistic flavor). Instead of defining and calling every operation in an unnatural syntax, we describe preferred outcomes and priorities in natural languages. (Programming languages can be considered "unnatural" in the sense that they use a different part of the human brain than natural languages.)

Don't think of the shift from coding to natural language as one from rigor to looseness, however. Rather, it's a change from years of procedural precision to expressive clarity – something those trained in the liberal arts may excel at, whereas those with backgrounds in mathematics or engineering may find a bit more foreign. Mapmaking is again expressed creatively, albeit in everyday wording rather than via analog drawing as it was before the rise of GIS software, or pedantically, using complex syntax and exacting parameters

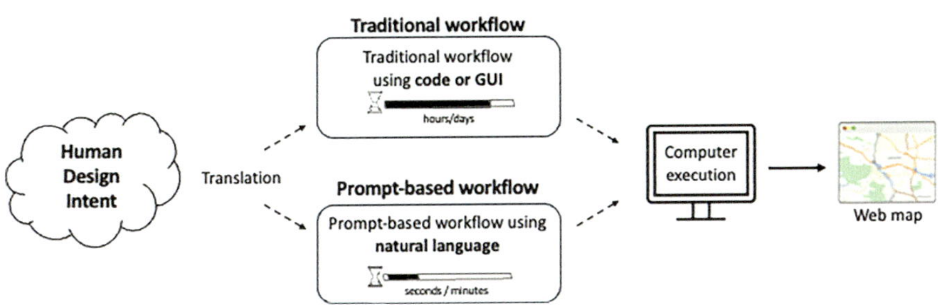

FIGURE 1.2
A conceptual contrast between procedural mapmaking workflows and prompt-based workflows. Same outcome, sped-up processes.

and numbers. Parameters and numbers are still very much required. It's just that they are taken care of behind the scenes by LLMs.

This is where confusion often sets in. Because prompts look conversational, it's easy to mistaken them for informal requests rather than design specifications. This doesn't only impact cartography, but everyday use of LLMs. Since we've never been able to communicate to computers via natural languages before, most of us (begin) as subpar prompters. Then we get frustrated when outputs don't match our immediate gratification needs and claim that the LLM is "hallucinating" or producing garbage. It's a reassuring assessment that typically doubles as self-serving when one investigates the prompts being fed to the machine.

Hallucinations – a term used to report nonsensical or completely erroneous answers from LLMs – are nothing of the sort, mind you. Such answers might better be thought of as "confabulations" – similar to humans misremembering and conflating facts (they know) incorrectly when attempting to impress someone with a quick answer. Other times, an LLM accused of hallucination is returning what the user *actually* asked for (if you reread the question and critique its grammar) or producing garbage because the prompter gave it an impossible task to begin with. Perhaps a plurality, if not majority, of subpar responses generated by LLMs I work on as an AI product engineer are caused by human communication error or how humans planned the multi-agentic system (behind the scenes). In sum, many of us aren't as great at communicating in natural language as we assume, i.e., we often aren't saying what we think we are.

But how can this be?! Isn't it true that we communicate with one another all the time. We are all expert natural language communicators, right? As they might say in Germany: "Jein" (or in New Zealand, perhaps, "Yeah, nah, yeah"). We do communicate with one another using natural languages, *but* most often in an unstructured, casual, open-ended manner.

Few of us plan or structure what we are saying before saying it. But if you want to make good-looking maps with minimal frustration, it's crucial to do so in a structured manner. *Structure, limits, and guardrails are what drive consistent LLM output.* It's what will separate you, a *prompt cartographer*, from my teen daughter casually asking an LLM to make a map.

PARENTING IS PROMPTING HUMANS

Metaphor of a subpar prompt to my teenage daughter:

```
You can hang out with your friends until midnight
if you promise to be good.
```

Example of a detailed prompt:

```
Yes, you may hang out with your friends tonight
until midnight, but you must adhere to the fol-
lowing rules:
  • maintain physical health and mental wellbeing
    at all times
  • You are **forbidden** from:
      • [insert every parent's worst fear here]
  • If any of the above requirements are about
    to be or are broken, immediately call or
    text me for further direction.
```

You see the difference? The first prompt has vibes; the second has specific guardrails. If she wasn't human and instead an LLM, the first would allow her to do anything, and the second would ensure she hangs out at with her friends *and* allow me to sleep soundly before she came home.

In other words, designed prompts allow cartographers to make the same kinds of decisions they have always made. Prompt cartographers just make these decisions using structured language. That change lowers the barrier to entry for many cartographic tasks, but it also raises the bar for articulating your intent clearly. Ambiguity that might have been absorbed silently by a script accepting concrete parameters now shows up directly in your LLM output.

Understanding this distinction is critical. As a prompt cartographer, you still make all of the cartographic decisions. Prompt cartography is about acknowledging what is genuinely new – language as the primary design interface. Your cartographic judgment isn't being swept away or made obsolete. If you make maps, you are a cartographer. To become a capable prompt cartographer, you must learn how to express that thinking in forms that agentic systems can reliably work with. Yes, there is a learning curve, but there is positive news too. You will *never* make a map alone again. By definition, prompt cartography is collaboration.

The Prompt Director Mindset

When people first encounter the prompt-based mapping workflow, it's common to frame the shift as one of skill: learning how to write better prompts, how to phrase requests more clearly, how to coax a system into producing the desired result. That framing isn't wrong, but it's incomplete. What's really changing is not just *how* instructions are written, but *who* is responsible for shaping them.

In traditional web cartography workflows, responsibility is often distributed across tools. Decisions are embedded in scripts, configuration files, and templates that execute faithfully but quietly. Once those systems are in place, much of the cartographer's role shifts toward maintenance and refinement rather than direct orchestration. Prompt-based workflows bring this orchestration back to the foreground.

Rather than acting as a technician who implements decisions through code, the cartographer increasingly acts as a director – someone who specifies intent, constraints, and priorities, then evaluates what comes back. The work becomes less about managing every operation and more about guiding a process that unfolds across multiple steps and iterations.

This does not make the cartographer less technical. It makes their technical judgment even more visible, which is why following the prompt cartography workflow (as shown in Figure 1.3) will pay dividends.

In film production, directors do not relinquish control simply because they are not operating the camera or the lighting rig themselves. They are responsible for coherence, pacing, and meaning across the whole production. In much the same way, a prompt director remains accountable for what a map communicates, even when execution is delegated to systems that operate at a distance.

What distinguishes this mindset is not authority, but attentiveness. Prompt directors pay close attention to how decisions are framed, how context is provided, and how constraints – or what are interchangeably called *guardrails* – shape outcomes. If you produce a map that uses red-green and is a hot mess for colorblind users, that's on you, not the AI. As a prompt director, it's

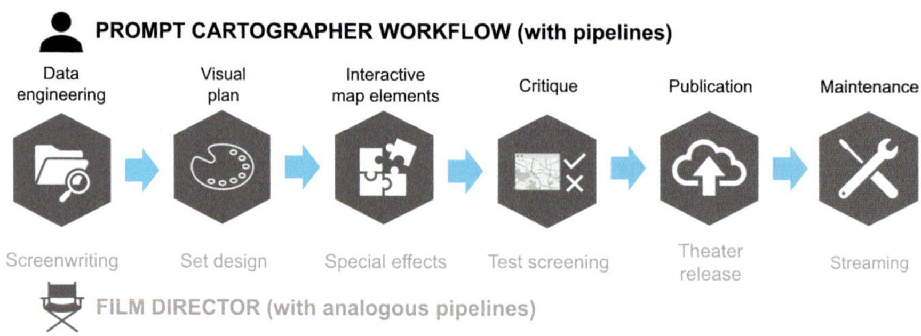

FIGURE 1.3
The prompt cartography workflow. Prompt cartographers remain directors throughout the production process, guiding specialized LLM agents through a series of conversational pipelines rather than coding or clicking through GUIs (graphical user interfaces). Each pipeline represents a stage.

up to you to recognize that small changes in language can produce meaningful shifts in LLM behavior. Your design choices are shaped by shifting your language.

This perspective also clarifies where responsibility lies when things go wrong. If a map misleads, confuses, or persuades in unintended ways, the cause is rarely that the system "failed." More often, the intent was underspecified, the constraints were unclear, or the context was poorly chosen. Seeing prompts as part of the design record makes those failures easier to diagnose and, importantly, easier to learn from.

The prompt director mindset prepares us for the chapters that follow. It sets the stage for thinking carefully about what gets remembered, how workflows are structured, and where critique belongs. Before any of that can happen, though, it helps to understand prompting itself – not as a bag of tricks, but as a form of design literacy.

Maps Still Persuade – Even When Machines Draw Them

The involvement of automated systems in mapmaking can create a subtle but persistent misconception: that persuasion has somehow been neutralized by delegation.

If a machine generates the map, the thinking goes, then the map must be more objective, or at least less entangled with individual intent. This assumption has a long history in cartography. It has surfaced before with new projections, new data sources, and new computational tools. Each time, it has proven unreliable.

Like all forms of media consumption and interaction, maps play a fundamental role in shaping user's beliefs and ideas whether we intend for them to or not.

Choices about scale, classification, emphasis, and omission shape how spatial information is interpreted. Those choices do not vanish when execution is automated. They are simply expressed earlier in the process, often through language rather than code. A prompt that specifies what should be highlighted is no less rhetorical than a script that implements the same decision procedurally (Figure 1.4).

What changes in an LLM-assisted workflow is the visibility of those choices. Because intent is articulated in words, it becomes easier to see how a particular outcome came to be. This can feel unsettling at first, especially when results diverge from expectations. *But that visibility is also an opportunity*. It allows cartographers to examine their assumptions more directly and to revise them with greater care. It creates transparency and allows for replicability.

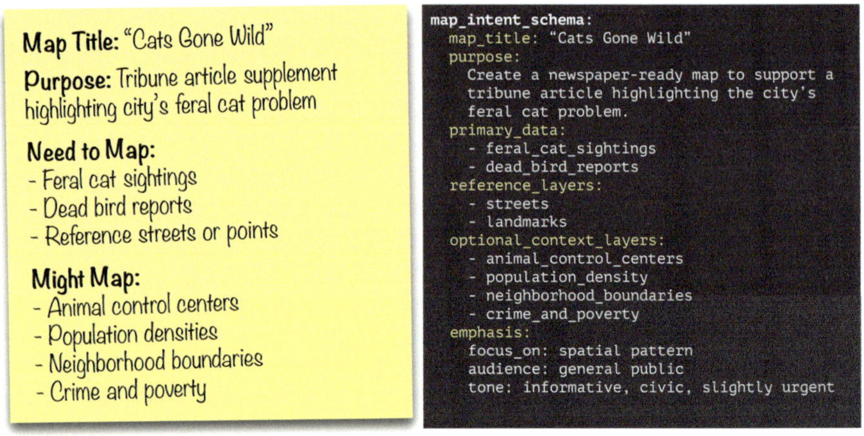

FIGURE 1.4
Intent and purpose as prompts. Writing down your map's purpose and intention before beginning is still important. What has changed in prompt cartography is that you're now writing this for you *and your LLM assistants* – not just yourself. Thus, the format has evolved from sticky note to a format you can share with your LLM.

Delegation does not absolve responsibility. It relocates it. The cartographer remains accountable for what a map communicates, regardless of who – or what – performed the final steps of execution. Recognizing this early helps prevent a common mistake: treating automated LLM outputs as brainless rather than as human-directed design artifacts. This book focuses on the cartographic workflow in prompt cartography and on techniques you can use to become a responsible prompt cartographer. It will require effort, and a change of mindset at times, but my hope it that you will see the power of this new technique and go on to design some amazing maps.

Why This Book Is Structured the Way It Is

By now, it should be clear that this book is not organized around tools. You won't find specific workflows in this book regarding particular LLMs, GIS applications, or code blocks. The end of most chapters will have applied exercises to help you reinforce what you have just learned through conceptual application. The companion website has labs, tutorials, schemas, and prompts that can be updated over time. I recommend it be used in conjunction with this book.

This structure is deliberate. Web technology and APIs – i.e., application programming interfaces, which allow software developers to access and other software features – are in constant flux, and workflows built around them tend to age poorly. (The original *Web Cartography* book proposal planned to have labs using Adobe Flash and ActionScript. I've learned my lesson!) What is timeless, thus far, are the decisions cartographers make – about purpose, audience, emphasis, and responsibility – and the ways those decisions are expressed and evaluated over time.

In that light, Part I of this book is devoted to orientation. Its chapters focus on how the mental models of cartography change with prompting, before worrying about how to execute the prompts themselves. The goal is not to delay hands-on work, which is coming, but to ensure that when execution begins, it does so with a shared vocabulary and a clear sense of what matters.

Part II shifts into design by prompt. That is where the prompting strategies, structured specifications, and concrete design decisions (that make up different pipelines) come into view. By this point, the conceptual groundwork has been laid, and the details about prompting should feel purposeful rather than overwhelming.

Part III is comprised of different prompt cartography scenarios, walking you through all the processes covered episodically in Part II real-life contexts. You will be asked to think systematically in these chapters to get through scenarios.

Throughout the book, the companion website (www.promptcartography.com) serves a supporting role. It hosts extended examples, labs, and resources that benefit from iteration and revision. Keeping those materials separate allows the book itself to remain focused on ideas and processes that will hold up even as implementations and AI tools evolve.

If any of my former students happen to be reading this book, they will recognize this pattern; it's how I taught them (just with GUI- and code-based technologies). Conceptual knowledge and procedural thinking are foundational. Mapmaking tools and applications are evolving and require lifelong retooling. So far, at least, this structure is timeless. Understanding comes first. Practice follows. Reflection and adaptation are how we grow.

Learning to Speak Prompt Design Fluently

The chapters that follow assume that prompting is not a novelty, but a language – one that cartographers can learn to use fluently by practicing prompt cartography. My hope is that by the end of this book you will be comfortable planning, designing, and implementing interactive web maps as competent prompt cartographer. The aim is not to master a particular system, but to become comfortable directing them. The work of cartography remains what it always has been: making careful, consequential choices about how space is represented and understood.

Fluency, here, does not mean speed or cleverness. It means being able to express intent clearly, to recognize when a system's response aligns with that intent, and to intervene thoughtfully when it does not. It means treating prompts, specifications, and workflows as design artifacts rather than as incidental instructions.

Chapter 2 begins by building that literacy at a conceptual level. It introduces prompting as a form of design communication, setting aside recipes and techniques in favor of understanding what structured language makes possible – and what it cannot do on its own.

From there, Part I turns to memory and workflow. Chapters 3 and 4 explore how design knowledge is retrieved, how decisions are sequenced, and how critique is reintegrated into processes that move quickly by default. We wrap up your introduction to prompt cartography fundamentals, with a discussion about integrating ethics within your work. Together, these chapters establish a way of working that keeps human judgment visible, even as execution becomes increasingly mediated.

Takeaways

- **The Medium Changed, Not the Craft**

  Web cartography reshuffled how maps are made and used, but it didn't erase cartographic fundamentals. Many long-standing principles still apply – and often matter more than ever.

- **Maps Make Arguments**

  Every map communicates claims through design choices like projection, color, scale, and interaction. Intentional or not, maps are rhetorical objects.

- **Purpose Comes First**

  Effective maps begin with a communicative goal. Data selection, design, and interactivity should all serve that purpose – not compete with it.

- **Audience Is Not Optional**

  Maps designed "for everyone" rarely work well for anyone. Clear audience definition simplifies design decisions and strengthens communication.

- **Web Maps Are Meant to Be Used**

  Unlike print maps, web maps invite manipulation. Designing for users – not passive readers – changes how map elements and interfaces must be conceived.

- **Plan for What Users Remember**

  Most users will retain only a few ideas from a map. Deciding in advance what truly matters helps focus visual hierarchy and design effort.

- **More Data ≠ Better Maps**

  The ease of web mapping makes over-inclusion tempting. Thoughtful restraint in data selection remains essential to clear communication.

- **Tools Shape Norms, Not Truth**

  Popular platforms have normalized certain styles and projections. Familiar design conventions should still be questioned for appropriateness and accuracy.

- **Maps Are Still About Communication**

  Regardless of platform or technology, maps exist to help people understand spatial relationships and patterns. That responsibility has not changed.

Conceptual and Applied Exercises

Purpose First

Think of a web map you regularly use. What do you believe its primary communicative purpose is, and where does that purpose begin to break down as features and interactivity accumulate?

Audience Reality Check

Identify a map you have created or encountered that tried to appeal to "everyone." What design compromises were made as a result, and who do you think the map actually served best? How might the map have been designed differently if it were targeting a more specific audience?

LLM Framing Exercise

Ask any generic LLM:

```
What is the communicative purpose of a map that shows
global earthquake activity?
```

Compare its response to how you or a professional cartographer might phrase the same purpose more precisely. You can also find a real web map, provide the URL in your prompt, and ask what it thinks its communicative purpose is.

Micro-Prompt Exercise

Write a *single sentence* that defines the purpose and intended audience of a map you might want to make someday. Do not include tools, colors, datasets, or technologies – only your purpose and intended audience.

For extended tutorials, hands-on labs, prompt templates, and updated examples related to this chapter, visit www.promptcartography.com.

2

Learning to Speak via Prompt

When people first encounter prompt-based systems, it's tempting to treat them as conversational partners. You ask the system a question, you get an answer. It sounds pretty good. It may even pass the proverbial Turing Test – i.e., if you didn't know it's a large language model (LLM), you may assume you are communicating with a human.

If you are a tinkerer, you may have discovered that when you refine your prompt's grammar and wording, the answers begin to improve. Like learning to communicate with a new colleague, as you start to figure out how to communicate well with your LLM tool or agent over time, your interactions begin to feel less like human–computer interactions and more like human-to-human conversations.

This conversational framing works – up to a point. The trouble is that conversational ease obscures what is really happening. Though it may not feel like it, you are still communicating with a computer. However, instead of clicking buttons and menu items with a mouse or finger (Figure 2.1) or commanding the computer via esoteric scripts and code that you spent months and years learning (Figure 2.2), you are using a natural language interface as shown in Figure 2.3. Prompts *are not* casual or neutral requests. They are information that is parsed and processed by a computer. Prompts replace graphical user interfaces (GUI) and scripting.

Casual Prompting Versus Directive Prompting

Then, what is a prompt? A prompt, at its most simple, is a natural language directive sent to an LLM. It's how humans communicate with LLMs – or if software has a built-in LLM component, how humans can communicate with the software.

Prompts represent your input specifications, directives, written in a language that feels natural to you that are parsed and interpreted using a massive probability matrix that consumes vast amounts of energy so that the computer can make sense of your request. Your prompts, and importantly how you write them, shape and steer what a system considers relevant, permissible, and important before any output is ever generated.

DOI: 10.1201/9781003797920-3

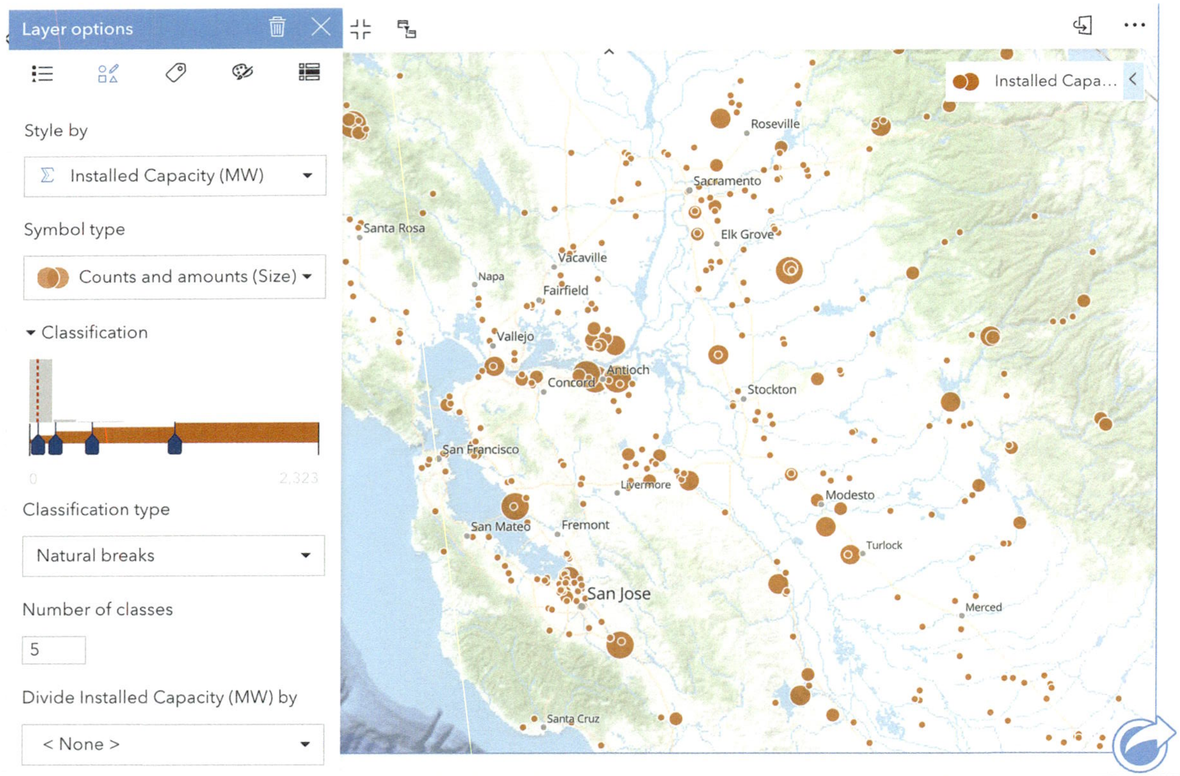

FIGURE 2.1
GUI-based map design required graphic interface controls to design a data visualization of California power plant capacities. All GUI-based software requires training or experience to master but is typically code free or code light. The software then translates from human interaction with the GUI to coded commands.

Source: Esri ArcGIS Insights, RIP.

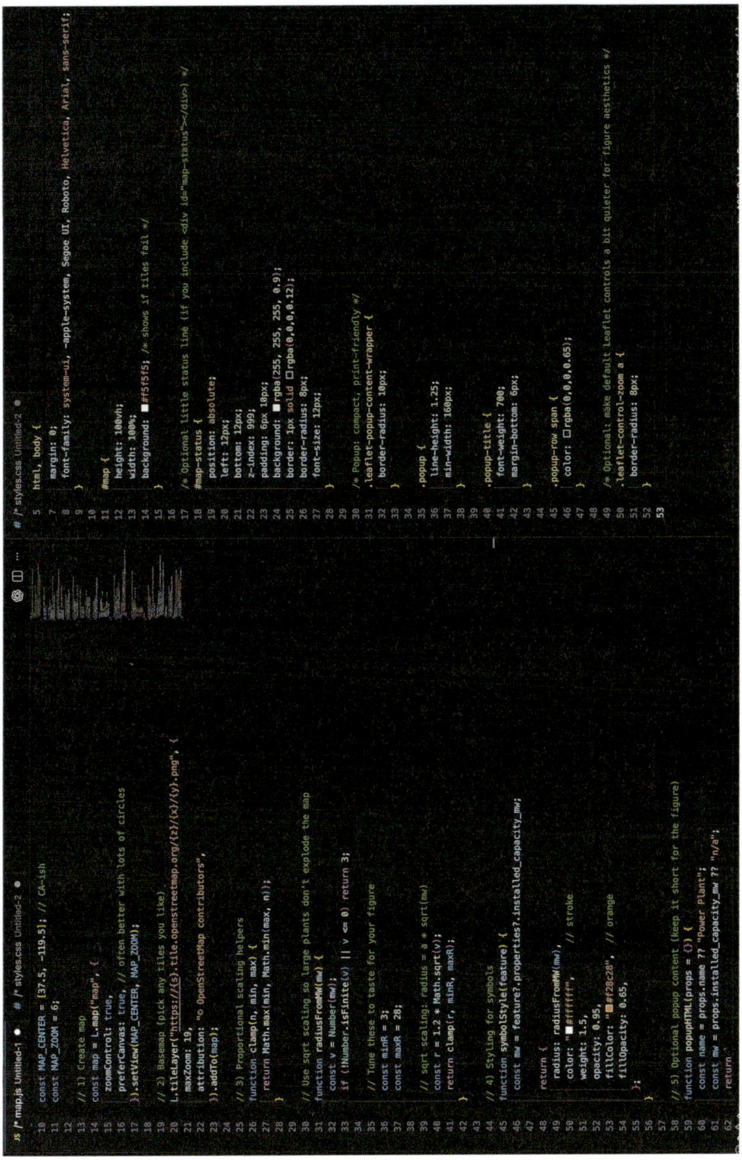

FIGURE 2.2
In code-based map design the cartographer uses scripting and coding to design the same data visualization of California plant capacities. Scripting typically requires training to learn and often years to master.

```
Load the GeoJSON dataset CA_power_plants.geojson
and inspect its attribute table.

Identify the field that represents installed power
capacity (in megawatts).

Create a web map of California using proportional
orange circle symbols, where:

- each feature is symbolized as a circle marker
- circle size is scaled by the identified capacity
  field (use square-root scaling)
- symbols have a light outline and semi-
  transparent fill

Fit the map view to the data extent and include a
minimal popup showing plant name and installed
capacity.
```

FIGURE 2.3
Prompt cartography uses one or more LLM assistants or agents to iterate – almost 100% via language and human–LLM collaboration – through a workflow that includes data cleaning, map design, interactive behaviors, critique, publishing, and maintenance of the map. The process requires structured, intentional, and articulate natural language.

Thus, when prompts are treated as casual, one-off human-to-human chats, results will vary. They will likely feel inconsistent and unpredictable. I don't know how many times I still see well-meaning people comment in online forums things like, "AI isn't going to take cartography jobs away any time soon, look at how bad this map is." They then might post a picture of indeed a very subpar map, created by an LLM image creator based on an off-the-cuff prompt that says something like: "Make me a medieval looking map of Great Britain with castles."

My question: if you gave that directive to a human, would he or she do better even if provided a full half day to accomplish the feat – much less the one minute the LLM took? Doubtful. What is causing the poor output? Sadly, not the LLM. LLMs are not perfect. They certainly can't read minds. The subpar output here is 100% human error.

Prompts certainly can and should be casual much of the time. This isn't a problem when you're asking fact-based or internet-searchable questions. "When is the best time of year to book a summer flight to Germany?" you might ask. Or is it particularly unhealthy to keep my heart rate in the 170s for 20 minutes straight on the stair climber at the gym? It will answer general knowledge questions well.

If you want to do something productive with LLM-based technology, though, you need to provide clear directions. Not commands. Not requests. Directives. Say what you want, explicitly, avoiding all possible vagueness. And then you can negotiate via conversation afterward.

And remember, just because the "Send" button is there doesn't mean you have to hit it right away. You can quickly proofread your directives to double-check for clarity to ensure what you're actually providing the direction that you think you are. (Upon rereading, you'll often find you aren't.)

One trick I have is to pretend I'm communicating with a literalist, overpriced lawyer that will charge me for every mistake I make or false errand I send it on, because bad prompting will cost you – much time, likely money, and always headaches. By treating prompts as thoughtful, professional design directives, not off-the-cuff messaging chats, you will instantly see better outcomes.

There are plenty of free online courses on how to write great prompts for different models. Instead, we are going to introduce you to the concept of *prompting as cartographic process*, prompting with the explicit intent of creating interactive web maps. Your goal, in this chapter and beyond, is to master expressing cartographic intent clearly enough in your prompts that it can be acted on, critiqued, and revised by LLMs and agentic systems.

Different Prompt Types for Different Roles

Discussions of prompting often introduce categories – system prompts, user prompts, tool prompts – as if they were technical features to be mastered. In practice, these categories are more useful when understood as *roles* in a design conversation. Just as you use different techniques of communicating based on the context and audience, these three types of prompts should be used by you in different circumstances.

A *system prompt* establishes the LLM's worldview and boundaries. It answers questions like these: *What kind of task is this? What values should guide decisions? What is off limits?* You will use this when creating or constraining your own LLM assistants and agents. Often it helps to give your LLM a role to fill – providing it context. The following is an example excerpt from the beginning of a system prompt from an LLM agent I created in 2024:

```
You are a world-renowned cartographic expert, highly
esteemed by academics and professionals around the
world for your expertise in scientific cartography, map
design, geocommunication, information visualization,
and geovisualization techniques.
```

A *user prompt* frames the immediate design problem. It introduces context, goals, and expectations for a specific interaction. This is where cartographic intent is most often articulated. You will use this any time you direct an LLM agent to do something for you. We will discuss these in more detail throughout the rest of the book.

Tool prompts, when present, define delegation. They specify what kinds of actions can be handed off by the LLM, to whom (typically other agents), and under what conditions. These will most commonly be used when designing your own LLM tools using an agentic-system (i.e., multi-agent) setup. We will not cover this in detail in this book, but the companion website has more information and some exercises about doing this.

Seen this way, *prompt types represent different roles in a design team* rather than levers to be pulled or an information fetching tool (as often envisioned by users). Each type of prompt, through numerous iterations and across multiple agents, plays a part in shaping an interactive web map's outcome, and confusion often arises when those roles are collapsed, conflated, or simply not understood.

Understanding these distinctions at a conceptual level is important. When the roles of the different types of prompts you will write are clear, not only will the tasks become easier to delegate but your prompts will be clearer allowing your team of agents to respond better. When these prompts blur together, errors will become harder for you to diagnose and work around.

Prompts as Design Model Artifacts

One of the quiet shifts introduced by prompt-based workflows is that prompts, like function calls, persist as artifacts in a natural language workflow model.

They can be saved, revised, reused, and shared. Prompts accumulate history. Over time, they start to resemble other familiar design artifacts: style guides, templates, rubrics, GIS (geographic information system) model nodes, and briefs. Every prompt you write encodes decisions, preferences, ethics, and assumptions whether you intend them to or not. In many ways, unlike all previous mapmaking techniques, your prompts literally encode your thinking.

When we start treating prompts as design artifacts, we can start evaluating them critically as steps in a design process. Instead of asking yourself whether a prompt "worked," we can ask what it prioritized, what it constrained, and what it left ambiguous. We can critique prompts independently of outputs, noticing patterns that recur across projects. We can revise language not just to improve results but to clarify intent.

This perspective also prepares the ground for what comes next. If prompts are artifacts, they can be remembered, retrieved, and placed within workflows. They can be tested against new contexts. They can be retired when they no longer serve their purpose.

All of this requires seeing prompts as part of the design record, not as disposable inputs. Try to stop thinking of prompts as one-off chat messages. In cartography, prompts are best thought of as *plug-and-play tools* that allow you to do almost anything in cartography without needing to know how to code, draw, or remember the idiosyncratic tool names used by most mapping software.

Prompt Structure Beats Cleverness

One of the earliest instincts people develop when working with language models is to search for the "right phrasing." A clever turn of phrase seems to unlock better results. A longer prompt feels more thoughtful. A more conversational tone feels more cooperative.

Occasionally, such instincts pay off, but I've found that more commonly, they don't scale in a larger workflow (Table 2.1). Scalability reflects a prompt's ability to be integrated into larger or more complex pipelines. Repeatability represents reliability; does the same prompt return the same answer every time? Traceability refers to whether coding bugs, LLM misinterpretations, or false outputs are traceable to a specific prompt. Collaboration relates to the ability for others to make use of your prompt in their own pipelines. Failure mode refers to whether you can tell why it failed – often the result of cartographer expectation inadvertently *not* aligning with the prompt the cartographer provided.

As projects grow in complexity, cleverness becomes brittle – like a narcissist colleague's initial charm wearing thin over time. Subtle wording choices that

TABLE 2.1

Clever Prompts Versus Structured Prompts

Dimension	Clever Prompting	Structured Prompting
Primary focus	Wording that feels effective in the moment	Explicit articulation of intent and constraints
Scalability	Fragile; often breaks when context changes	Stable across datasets, tasks, and collaborators
Repeatability	Inconsistent across runs	Designed for consistent behavior
Traceability	Difficult to diagnose when results drift	Failures can be traced back to language
Collaboration	Hard to share or explain	Functions as a shared design artifact
Failure mode	Plausible but unexplained error	Explicit mismatch between intent and outcome

Note: A comparison of clever phrasing to structured prompting across several important dimensions.

worked once stop working when the context shifts. Outputs drift – answers and results start to change over time. Decisions become harder to trace back to their causes. The system feels unpredictable, even when it is behaving consistently.

Fortunately, cartographers are not the first information architects and designers to embrace prompting. This means that others have already thought long and hard about how to structure prompts to elicit the most useful responses possible from today's LLMs. And importantly, they have come up with models we can follow. (Author caveat: these models certainly continue to evolve and change. If you're reading this book five years from now, it's time to use your favorite AI agent to find a more contemporary framework.)

The CRAFT Framework as a Prompt Aid

As of today, I find that CRAFT (**C**ontext, **R**ole, **A**udience, **F**ormat, **T**one) is one the most useful frameworks for helping us write clear and useful prompts. Personally, I'm not a big model or recipe person. In fact, my family is terrified of my cooking, because I tend to add ingredients in random order and without specific measurement. However, when it comes to practicing prompting and developing your skills, the CRAFT model I'm about to review really is a good place to start. It is *not* meant a step-by-step recipe. Consider it your creative muse; I encourage you to refer to it when you're trying to develop new prompts or figure out why an existing prompt isn't producing what you thought it would.

What is CRAFT? It's a technique of slowing down your prompt writing long enough to ask whether what you are writing reflects the design situation you need to address. CRAFT gives you no feedback on wording. It does not guarantee success. It simply reminds you to include, or rule out, assumptions that might otherwise remain implicit.

In cartographic terms, CRAFT functions much like a pre-design checklist (see Table 2.2). It allows you to see if the problem you are trying to solve for has been framed clearly enough to justify the decisions the LLM will give you. When these insights are skipped, errors tend to propagate quietly, build upon one another quickly, and become expensive (in time and money) to correct.

Seen as a thinking aid rather than a workshop exercise, CRAFT helps you reinforce the idea that prompting is a form of design reasoning. It reminds us that clarity is rarely accidental and that good outcomes usually trace back to careful framing. In prompt cartography, all success comes back to the fidelity of your prompts. The model or framework or method you use to achieve this is irrelevant, but having one can be helpful.

TABLE 2.2

Using CRAFT as a Framing Checklist

CRAFT Element	Question It Forces	Risk When Ignored
Context	What situation or domain does this prompt assume?	Outputs rely on unstated or incorrect background assumptions
Role	What kind of reasoning is expected?	The system guesses tone, expertise, or perspective
Audience	For whom is this output meant?	Results feel mismatched or inappropriate
Format	What form should the response take?	Outputs are verbose, underspecified, or unusable
Tone	How should the output sound or feel?	Style undermines credibility or intent

Note: CRAFT highlights implicit assumptions in prompts without prescribing specific wording or techniques – in other words, you can still be you.

Prompt Literacy as a Professional Skill

Prompt literacy is not about learning how to talk to machines. It is about learning how to express design intent precisely enough that it can be delegated without being distorted. That skill is transferable. It applies across tools, platforms, and workflows.

For cartographers, this literacy reinforces habits that already matter: articulating purpose, naming constraints, anticipating interpretation, and revising language when outcomes diverge from intent. Prompting simply makes those habits unavoidable.

As systems become more capable, the need for clear articulation does not diminish. It intensifies. The better the system is at acting on instructions, the more consequential those instructions become.

Looking Ahead: From Literacy to Practice

My goal here has been to prepare you to start thinking of prompting as a *design language* rather than a collection of techniques or off-handed chat requests. With that foundation in place, the chapters that follow will start turning to practice without losing sight of intent, which as you know from my previous book is central in my mind to all mapmaking. All great maps are built on well-grounded (if not always well-meaning) intentions.

Chapter 3 explores how prompts and other design artifacts are remembered and retrieved. By the time execution begins, prompts should no longer feel like experiments. They should feel like deliberate design moves – almost like clicking on tools in a GUI. Your design decisions should be rewritten, revised, and critiqued no differently than cartographic decisions are with any other toolset.

Takeaways

- ***Prompts* Are Specifications**

  Prompts feel conversational, but they function as design specifications. Treating them like casual wishes leads to inconsistency; treating them like artifacts makes outcomes easier to explain and refine.

- ***Prompt Types* Are Roles in a Design Process**

  System, user, and tool prompts are not tricks or hacks. They play different roles in shaping intent, constraints, and delegation – much like people do on a design team.

- ***Prompt Clarity* Improves Diagnosis (And Often Results)**

  Clear prompts don't guarantee perfect outputs, but they make it far easier to understand *why* something went wrong and how to fix it.

- **Prompts Persist as Design Artifacts**

  Prompts accumulate history. They encode assumptions, values, and decisions, and they can be saved, reused, critiqued, and retired just like other cartographic design artifacts.

- **Structure Scales Better Than Cleverness**

  Clever phrasing may work once, but it rarely holds up as projects grow. Structured prompts produce more stable, traceable, and revisable outcomes over time.

- **CRAFT Is a Framing Aid (Not Gospel)**

  The CRAFT framework helps surface hidden assumptions. It doesn't prescribe wording – it helps slow things down just long enough to allow you to think things through clearly.

- **Prompt Literacy Is (a Most Valuable?) Professional Skill**

  Learning to prompt well isn't about talking to machines. It's about articulating design intent precisely enough that it can be delegated without distortion. Practice makes progress.

- **Better Systems Increase the Cost of Ambiguity**
  As LLMs become more capable; unclear prompts don't matter less – they matter more than ever. Precision in intent becomes increasingly consequential.

Conceptual and Applied Exercises

Interactivity Audit

Think about a highly interactive map you have used recently. Which interactive elements genuinely helped you understand the data, and which felt distracting or unnecessary?

Control Versus Clarity

Reflect on the difference between giving users control and giving them clarity. Can you think of a situation (experienced or hypothetical) where more interactivity actually reduced user understanding?

LLM Critique Prompt

Ask an LLM:

```
What kinds of interactivity make a map more effective
for users?
```

Critique and evaluate whether its response prioritizes communication goals or technological novelty.

Design Judgment Exercise

Imagine redesigning a static print map as an interactive web map. Identify one interactive feature you would add. Also, think of one you would deliberately leave out. Explain why.

For extended tutorials, hands-on labs, prompt templates, and updated examples related to this chapter, visit www.promptcartography.com.

3

The Cartography Process as Knowledge Recall

Cartographers have always relied on memory – just not in a way we were forced to name.

We remember how a certain map "ought" to look. We remember that a particular color scheme worked well for one dataset but failed miserably for another. We remember that one projection choice calmed a map down, while another quietly amped up its rhetoric. Most of this knowledge lives in our heads, our habits, our half-written notes, or the margins of old projects we vaguely recall but can't quite reconstruct.

That arrangement worked fine when *we* were doing the drawing.

Once execution starts getting delegated – to scripts, templates, or now language models – memory stops being background knowledge and becomes a design problem. If the system doing the work can't remember your rules, your preferences, or your hard-earned lessons, then every map becomes a reinvention. Worse, it becomes an imitation – smooth, confident, and unmoored from your intent.

This is where many people first feel uneasy about AI-assisted mapping. The concern isn't really about automation. It's about authorship. Who's making the decisions if the system doesn't remember what matters?

The answer, perhaps surprisingly, is that *you* still are – but only if you are willing to externalize memory instead of assuming it.

That shift – treating memory as something designed rather than assumed – is what this chapter is about.

From Tacit Knowledge to Explicit Recall

A lot of cartographic expertise is tacit. Long-time cartographers know more than they can easily explain. This often happens to me when someone innocently asks me for feedback on a map. I will see or sense something that bugs me about it, but I may not be able to pinpoint what it is exactly – at least, not using terms they will easily understand.

DOI: 10.1201/9781003797920-4

Ask any other experienced map designer why one layout feels balanced and another feels a bit heavy to the lower right; you'll likely get an answer that circles the truth without fully pinning it down. Not because the cartographer lacks rigor, but because the knowledge is embodied – learned through repetition, critique, and failure rather than formal rules.

This kind of tacit knowledge is incredibly powerful. However, it doesn't necessarily travel well. It can be difficult to teach directly. And this knowledge becomes nearly untransferable in any communications that rely entirely on language to operate.

Language models in particular are blunt instruments when it comes to tacit expertise. If a rule isn't written down somewhere the model can retrieve it, that rule may as well not exist. The model will happily produce something that looks "map like," even if it quietly violates principles you would never consciously choose to break.

This isn't a flaw unique to AI. We've seen this problem before. In fact, style guides like this textbook exist, because oral tradition fails at scale. A book puts thoughts down so they last even when the messenger dies. Likewise, many rubrics and checklists exist for the same reason, because intuition isn't the same across all humans. Even the humble map legend is, in many respects, a way of making implicit meaning explicit so it can be shared, interpreted, and reused. It tells map readers what the cartographer meant by certain symbols.

In the case of large language models (LLM) used for cartography, we don't have specific reusable style guides or textbooks that are plug and play, but we do have something very useful at our disposal – retrieval-augmented generation (RAG).

Retrieval-Augmented Generation

In traditional GIS and cartography, consistency has traditionally been maintained through higher education, textbooks, and online blogs. The first edition of this textbook, *Web Cartography*, for example, was used to teach students around the world about web cartography and by (presumably some) professionals to keep as a reference on their desks.

RAG is conceptually similar. RAG sounds far more complicated than it needs to, which is unfortunate, as the underlying concept is straightforward. Instead of asking a language model to respond based only on what it learned during its training, RAG allows it to first *look things up* from a collection of materials you've shared with it. Beyond and before using its general knowledge, with RAG an agent will generate a response based on the information it has retrieved from you (or the agent's creator) directly (see Figure 3.1).

Essentially, when using a RAG, an LLM agent parses your user prompt and then, in response to what you have directed, recalls (or retrieves) the most relevant pieces of information in the RAG *before* it begins generating any output. Instead of referring to or reading textbooks and learning through application, like humans, an LLM checks out and retrieves all relevant information you have provided in the RAG, including production rules, instructions, and best practices. It then constrains its impressive general knowledge of all things and subordinates it to what the relevant content in your RAG demands.

In practical terms, a RAG workflow inserts a pause before work commences – like checking your notes before writing a paragraph (ahem, me right now). The system searches a curated memory, comprising documents, rules, and examples, brings back the most relevant pieces of information, and then treats this information as temporary scaffolding for its response. The model isn't being retrained by RAG; it's being reminded of what's important and what it should focus on first and foremost.

Yes. That's it. It's not mystical. It doesn't require any type of sorcery or a computer science degree. And it really doesn't need such an intimidating, overly engineered name (which is why most LLM user interfaces typically refer to RAG as "Library" or "Knowledge Document" uploads).

Knowledge Training Versus Retrieval Augmentation

So what is the distinction between a language model's general knowledge training (factory setting) and RAG? It's helpful to think of it this way:

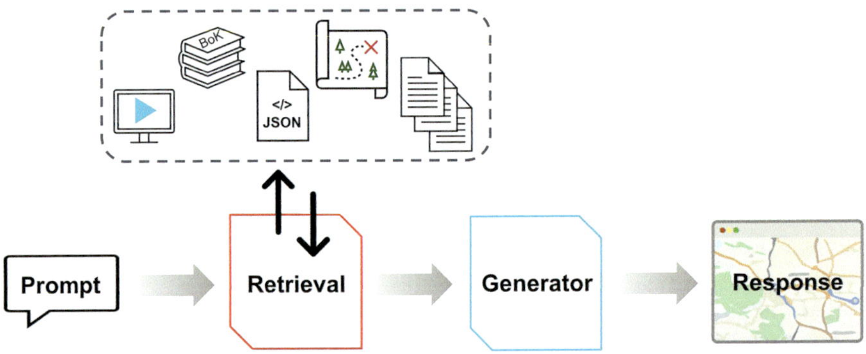

FIGURE 3.1
A retrieval-augmented generation (RAG) system combines two processes: retrieval and generation. When a user submits a prompt, the system first converts it into an embedding (a numerical representation of meaning) and searches a vector database for the most relevant stored information. Plain English: instead of relying only on its training, the model uses a knowledge store to look for more desirable information before answering.

training shapes what a model *can* know in general. RAG shapes what it is allowed to consider when responding to any given user prompt *right now*, in the moment. The difference is more editorial than technical.

If this feels familiar, it should. Cartographers do this constantly. You don't design every map by re-deriving cartographic principles from scratch. You rely on internalized conventions, remembered critiques, and examples that come to mind when a particular problem presents itself. Retrieval-augmented systems simply externalize that moment of recall – it's not part of the LLMs general knowledge – and allow you to make certain information explicit (Figure 3.2).

There is a temptation to treat retrieval as a kind of safety net, something bolted on to keep models from hallucinating or drifting. It can be used to do that, but it's more productive to see retrieval as a way of expressing and reinforcing your overarching directives. What you choose to make retrievable says a great deal about what you care most about in cartography.

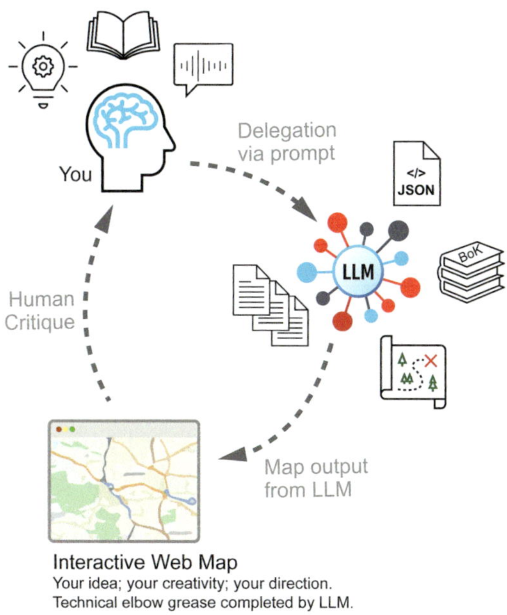

FIGURE 3.2
The prompt cartographer remains accountable for all map outputs, maintaining authority through structured direction rather than manual execution (clicking or coding). You delegate technical work to the LLM; technical execution is completed by the LLM; cartographic judgment, intent, and responsibility are entirely yours.

Chunking as Editorial Decision

Once you accept that retrieval is what shapes agent behavior in the moment, a next question becomes unavoidable: *what*, exactly, is being retrieved, and how does the LLM access and compartmentalize this uploaded knowledge?

RAGs are comprised of uploaded documents, datasets, and often schemas (typically created in JSON or YAML format). This information is "chunked" for quick and logical retrieval by an LLM. As already mentioned, uploading knowledge documents or rules and guardrails can be thought of as providing tacit knowledge to the LLM. Chunking, on the other hand, can be thought of as an act of editing that knowledge into useful sections.

Essentially, chunking transforms your tacit knowledge into retrievable memory. To provide an example of this in action that won't run afoul of copyright, I was digging through my undergraduate notes from Scott Freundschuh's *map design* course that I took back in September 1996. (Nerdiness pays off! I typed up my notes after class back then to retain the knowledge, and the WordPerfect files still exist, as you can see in Figure 3.3 on p. 34.) Notice how my notes are organized in the order Dr. Freundschuh presented them (on an antique machine called an overhead projector). Though logical for a human, to be used most effectively via RAG, the notes might best be chunked by theme. Figure 3.4 (p. 35) is an example of the same notes chunked for efficient recall.

When you divide RAG material into retrievable pieces, you are deciding what ideas travel together and what ideas stand alone. You are deciding how much context is enough, and how much becomes noise. These are not neutral decisions, and they are not unique to AI systems.

OF JSONs AND YAMLs

When you work with LLMs, you will quickly come across what at first may appear to be intimidating acronyms representing two file types: json and yaml. (I admit, "yaml" is not so intimidating as it is entertaining.) I will provide layperson explanations of these file types here.

JSON (JavaScript Object Notation) is a file that preserves data using javascript object syntax. In other words, it takes your imports data and organizes, indeed restructures it, using the syntax required to be ingested and read via javascript. Remember Extensible Markup Language (XML)? JSON is totally different, *but* the idea is the roughly the same – data is preserved in a particular format to be read by a programming language.

A JSON schema, which you will hear a lot about the more you play with LLMs, is exactly what the name implies. It is a specific syntax (i.e., scheme or format) that an LLM or API (application programming interface) uses to read and write your JSON so they can be shared with other LLM agents, APIs, or AI tools. Essentially, a JSON schema is a fill-in-the-blanks form that can be filled in by your LLMs and sent to other LLMs, who are trained to expect data in that schema, to work on.

YAML files are very useful. Mostly, because quite unlike JSON, which is mostly written for APIs, YAML files are like reading a memo from a boss, back when memos were a thing – circa 1998 (i.e., no brackets, no braces!). They are often preferred by users of LLMs, because they allow the LLM to embed its chain of thought and reasoning next to the data in the files. They are also much smaller. It turns out all those curly braces in JSON add up, consuming far more LLM tokens. However, they aren't as portable or usable as JSON files across technology stacks broadly, so in general JSON files are still preferred when interacting with APIs.

Cartographers have been making analogous choices for as long as maps have existed. Scale determines what details survive and which are generalized away. Legends group symbols into meaningful categories, even when the underlying data resists tidy classification. Insets isolate context that would otherwise overwhelm the main frame.

Chunking operates in much the same way. A chunk that is too small may retrieve a rule without its rationale. A chunk that is too large may drown a useful principle in unrelated detail. Neither failure is technical at heart; both are editorial.

Seen this way, chunking becomes an extension of authorship. It forces you to articulate what belongs together and why. It also reveals assumptions you may not have realized you were making – about hierarchy, precedence, and relevance.

What's important here is not finding a universally "correct" chunking strategy. There isn't one. What matters is recognizing that how you structure memory will shape what the system learns to emphasize, repeat, and reinforce over time.

That realization tends to arrive quietly, often halfway through a project, when outputs start feeling oddly consistent in ways you didn't fully intend. At that point, the question isn't whether the system is working. It's whether it's remembering the right things. You will almost always start feeling these things before you can specifically identify them.

Original course notes document

Nine Map Elements making the cartograph a communication device:

1) **Title (& subtitle)**
2) **Map Legend/Symbol referrent**
3) **Scale (representative fraction and bar scale)**
4) **Credits (data sources and base map if necessary)**
5) **Foreground and Background (mapped and unmapped areas)**
6) **Graticule (line of latitude and longitude)**
7) **Borders (and/or neatlines)**
8) **Map Symbols**
9) **Place Names and Labeling (#8 & #9 are very similar and often categorized as one unit)**

The Principles of *Visual Weight*

1) **Location**—off structural network pulls more; right heavier than left; object weighs more the further it is from the center of the map
2) **Size**—Bigger is heavier; fat line (for example) is heavier than skinny one
3) **Color, visual interest, and isolation**—long wave-lengths heavier than short ones; brightness is heavy; darkness is not; complex objects are heavier than others; white is heavier than black
4) **Shape**—normal shaped objects weigh more than irregular ones; compact shapes are heavier than non—circles are bloody heavy!

The Principles of *Visual Direction*

1) **Location**—Big objects pull in smaller objects
2) **Shape**—circles in any order or fashion create direction; symbols lined up in any kind of order make for less balance
3) **Subject Matter**—weight of the base map should be moved to be equally balanced around the center point
4) **Intraparellelism**—elements of the map aligned together to reduce cohesive tension; like street blocks... it makes the map easier on the eyes

Figure/Ground Contrast:

- **figure=area of interest and being mapped**
- **ground=less important objects; background**

GESTALT PRINCIPLES—THE VISUAL HEIRARCHY

intellectual level		visual level
1	Thematic Symbols; Title (legend, labeling, symbols)	1
2	Base Map—land areas, acu-s, physical features	2
3&4	Important explanatory materials (credits)	2/3
4	Base Map—water features (all water)	3
5	Other Base Map Elements—labels, grids, scales, and neatlines	4

Perceptual Grouping Principles:
Shape; Size; Proximity items combine that share visual properties

RAG-esque chunking for LLM recall

```
Title: Core Map Elements
Text: Maps function as communication devices through a set of standard visual and informational components. These include the title and subtitle, legend or symbol reference, scale (both representative fraction and bar scale), credits for data sources, foreground and background distinction, graticule, borders or neatlines, map symbols, and place names or labeling. Symbols and labels are often treated as a single functional unit due to their close relationship in interpretation.
Topics: map construction
Level: foundational

Title: Principles of Visual Weight
Text: Visual weight determines how strongly map elements attract attention. Weight is influenced by location (elements farther from center appear heavier), size (larger objects and thicker lines weigh more), color and brightness (long wavelengths and brightness increase weight), and shape (compact, regular forms -- especially circles -- carry more visual weight than irregular shapes).
Topics: visual perception
Level: foundational

Title: Principles of Visual Direction
Text: Visual direction describes how the eye moves across a map. Large objects tend to pull smaller ones, aligned shapes reduce imbalance, and repeated forms create directional flow. Subject matter should be distributed to balance visual weight around the map's center. Intraparellelismâ€"alignment of elements such as street gridsâ€"reduces visual tension and improves readability.
Topics: map form, visual perception
Level: foundational

Title: Figure-Ground Contrast Text: Figureâ€"ground contrast distinguishes the primary area of interest (the figure) from less important background elements (the ground). Effective figureâ€"ground separation ensures that mapped phenomena are visually prioritized over contextual or supporting information.
Topics: visual perception, visual hierarchy
Level: foundational

Title: Visual Hierarchy and Gestalt Principles Text: Maps operate on both intellectual and visual levels. Thematic symbols and titles typically occupy the highest level of the hierarchy, followed by base map land features, explanatory materials such as credits, water features, and secondary elements like grids, scales, and neatlines. Gestalt principles support hierarchy by organizing perception through similarity, proximity, shape, and size.
Topics: visual hierarchy, visual perception
Level: foundational
```

FIGURE 3.3
The author's undergraduate course notes from a map design class; the knowledge is structured for human recall.

```
Title: Core Map Elements
Text: Maps function as communication devices through a set of standard visual
and informational components. These include the title and subtitle, legend or
symbol reference, scale (both representative fraction and bar scale), credits
for data sources, foreground and background distinction, graticule, borders or
neatlines, map symbols, and place names or labeling. Symbols and labels are
often treated as a single functional unit due to their close relationship in
interpretation.
Topics: map construction
Level: foundational

Title: Principles of Visual Weight
Text: Visual weight determines how strongly map elements attract attention.
Weight is influenced by location (elements farther from center appear
heavier), size (larger objects and thicker lines weigh more), color and
brightness (long wavelengths and brightness increase weight), and shape
(compact, regular forms—especially circles—carry more visual weight than
irregular shapes).
Topics: visual perception
Level: foundational

Title: Principles of Visual Direction
Text: Visual direction describes how the eye moves across a map. Large objects
tend to pull smaller ones, aligned shapes reduce imbalance, and repeated forms
create directional flow. Subject matter should be distributed to balance
visual weight around the map's center. Intraparellelism—alignment of elements
such as street grids—reduces visual tension and improves readability.
Topics: map form, visual perception
Level: foundational

Title: Figure-Ground Contrast
Text: Figure-ground contrast distinguishes the primary area of interest (the
figure) from less important background elements (the ground). Effective
```

FIGURE 3.4
The author's map design notes chunked for LLM retrieval: reorganized into titled sections with explicit topics and level classifications. Chunking is an editorial act that determines which ideas travel together, how much context surrounds each principle, and what will get emphasized during retrieval.

Balancing and Structuring Memories

As is the case with human learning and knowledge, what is emphasized, rehearsed, or done via repetition will be emphasized the most in an LLM's knowledge retrieval system. When you turn a body of cartographic knowledge into a retrievable store, *you* are *not* preserving it in amber for the LLM to retrieve perfectly. You are simply shaping how an LLM will and can recall the knowledge documents. Humans know this well. When you read a book, you do not recall all the details. You recall key concepts and themes. You often must revisit a book to recall specific details. In fact, I sat down and reread *Web Cartography* before writing *Prompt Cartography*, to recall how I originally wrote it. (Now that's irony!) You may remember where to look in a book for details, based on broad memories, which will save you time. Other times, you will hope there is an index.

What I'm getting at here is that memory design is a responsibility, not a convenience. If your retrievable materials overemphasize certain details, processes, or styles, those styles will quietly dominate future outputs. If certain data or examples recur disproportionately in your retrieval documentation, they begin to feel normative. If outdated conventions remain retrievable long after you've stopped endorsing them, they continue to exert influence. Alternatively, if you do not organize or chunk your materials in a manner that makes sense, the details will be difficult to retrieve or recall.

None of this requires malicious intent. In fact, most of it emerges from perfectly reasonable decisions made early on, before the consequences are obvious. What RAG systems do is make those consequences visible.

This is one of the reasons it's helpful to think of retrieval not as a neutral technical layer but as an editorial surface. What gets remembered becomes what feels "natural" to the system. What is absent (or missed because of poor structuring) becomes increasingly difficult for an LLM to imagine.

That may sound abstract, but cartographers have long understood this dynamic in other contexts. Atlases shape geographic imagination. Standard symbology normalizes certain readings of space. Absence, as much as presence, carries rhetorical weight. Externalized memory simply amplifies that effect.

FAISS, Briefly, and Then We Move On

At this point, it's reasonable to ask what actually performs the retrieval in most RAG systems. One common answer is a library called FAISS, which is designed to search large collections of embedded text efficiently. It's widely used, well supported, and fast enough to make retrieval feel instantaneous. For many systems, it plays the role of memory index – the mechanism that helps decide what gets recalled when a prompt arrives.

That description is sufficient for now. Unless you want to get into building LLM agents as a software engineer, the details of this are overkill. I wanted to mention FAISS here, however, because on the prompt cartography companion site we will offer some basic web map design RAG elements for you to explore, some of which will be made available to you as FAISS.

Retrieval as Constraint

One of the quieter shifts that RAG introduces is a change in how constraints enter the design process. Several RAG constraints might be, for example, "always label human or socially constructed features using a serif font" and "never use Times New Roman font in any map product." In traditional workflows, constraints are often imposed externally – by data availability, a branding office, etc. In RAG systems, constraints are essentially self-imposed in advance. By placing limits on what the model is allowed to consider at generation time, you can narrow the amount of plausible outputs before anything is ever considered or processed, saving a lot of time later.

This turns retrieval into a form of preemptive critique. Instead of generating freely and then evaluating the result, the system is guided (indeed directed) toward certain decisions from the outset. The retrieved context doesn't just inform the output it bounds it. Over time, this can produce a kind of consistency that feels less like repetition and more like voice.

For cartographers, this should feel familiar. We often work within self-imposed constraints – projection choices (i.e., the curse of Web Mercator), color palettes, classification schemes – not because alternatives are unavailable but because coherence matters. In this way, retrieval is not a shortcut. It's a commitment. You are declaring, in advance, which ideas deserve consideration most whenever LLM decisions are made. Declaring this up front doesn't eliminate judgment; it formalizes it. It prevents what matters most to you from dying in a committee (of competing bad ideas from an LLM's general training).

However, this is also why retrieval alone will never guarantee good design. Generally good constraints can still produce poor results in certain contexts. For example, what if your client insists on Times New Roman fonts? Your RAG may not allow you to do this easily. When designing RAG, you want to make sure that your constraints are explicit but also flexible enough that they can be examined, revised, and, when necessary, discarded by your LLM agent.

Why Forgetting Is a Useful Feature

Once memory becomes something you design, forgetting stops being a failure and starts looking like maintenance.

This runs against instinct. We tend to assume that more and longer-term memory is always better, and that a system stocked with every past rule, example, and exception will naturally produce better results. In practice, the opposite is often true. Over-retention can make a system brittle, overly conservative, or oddly stuck in habits you thought you'd moved past. (The equivalent of infinite loops in scripting, I suppose.)

Human memory handles this messiness by forgetting unevenly. We stop recalling what no longer fits our mental model or contemporary belief systems of the world. Cartographers encounter this all the time. Old conventions may linger long after their usefulness has faded (e.g., those print-map-based axioms). Color schemes seem to persist because they are familiar, not because they remain appropriate (e.g., cursed rainbow heat maps). Classification defaults may get reused simply because they (may have) worked once, somewhere, under conditions that no longer apply (e.g., Flannery circles).

However, an LLM's externalized memory system won't forget unless you force it to. In a RAG context, "forgetting" typically means removing outdated material, narrowing retrieval scopes, or deliberately excluding rules that once felt foundational but now feel constraining (e.g., "never use Web Mercator"). Not every lesson needs to be preserved forever in a retrievable form. You are deciding which ideas still deserve influence and which should quietly disappear into the sunset. That decision is never final, and it doesn't have to be dramatic. It's closer to revising a syllabus than burning the canon (Table 3.1).

What matters is recognizing that LLMs using retrieval systems remember *literally*. They will continue to (re)surface what you give them, long after *you've* stopped believing in it, unless you remember to intervene. Unlike humans, they can't be traumatized by their own memories, but they also have no idea if the ideas they continue to espouse are traumatizing you. As director, you are responsible for the end product and keeping tabs on the thoughts you put in your LLM agents.

TABLE 3.1

Remember Versus Forgetting in RAG Systems

Over-Retention Risks	Benefits of Deliberate Forgetting
Reinforces outdated assumptions	Keeps retrieved context relevant
Increases noise in retrieval	Improves signal-to-noise ratio
Amplifies early design errors	Allows correction over time
Makes systems harder to audit	Simplifies review and critique
Creates false sense of authority	Preserves editorial responsibility

Note: Forgetting is a form of maintenance, not failure, in designed memory systems.
Abbreviation: RAG, retrieval-augmented generation.

What This Enables – And What It Doesn't

When memory is designed thoughtfully, RAG can support consistency across projects, reduce stylistic drift, and make critique more tractable. It becomes easier to see why a particular output looks the way it does, and easier to trace that outcome back to the rules and examples that shaped it. That visibility alone is a substantial shift from ad hoc prompting.

At the same time, none of this eliminates the need for *your* judgment. Retrieval does not decide what makes a map ethical, persuasive, or appropriate. It does not know when a rule should be bent or broken. It simply ensures that whatever decisions are made are made in dialogue with a remembered body of work rather than in isolation.

This difference is important, especially for readers encountering RAG through technical discussions elsewhere. There is a tendency to treat RAG as a safety mechanism, something that prevents errors or hallucinations by anchoring outputs to best practices. It may help, but RAG is not a parameter. It is knowledge. What an LLM does with it will vary based on audience, intent, and other prompted contexts. RAG is super useful, but it's never infallible. As the director, it's up to you to check and approve or decline the LLM's outputs.

On the other hand, RAG does provide a paper trail of influence. The output reflects what has been presented to the LLM. Just like you can tell what news sources a friend on social media has likely been consuming based on their political posts, you can assess the quality of what an LLM has consumed via RAG by its output. For designers who care about rhetoric and responsibility, that something is unique and consequential.

From Memory to Workflow

If memory on its own could produce a map, we would all be master mapmakers – able to mimic the design of every map we have ever loved. Obviously, something more than memory is required. Memory is useful, but it only shapes the conditions under which decisions are made by both humans and machines.

Once retrieval enters the picture, it becomes clear that it is only one stage in a larger process. Prompts still need to be written. Outputs still need to be evaluated. Revisions still need to happen, sometimes repeatedly. What changes is that those steps begin to form a recognizable pattern rather than a series of improvisations.

This is where RAG stops being an abstract idea and starts to feel operational, even if we've avoided operational detail so far. Memory feeds generation, generation invites critique, and critique reshapes memory. I'm still devasted by some peer critiques as an undergraduate cartographer and still flattered by an offhand comment about my use of grayscale a professor made in graduate school. The loop tightens over time in an LLM system, not because the system is learning autonomously but because the designer is paying attention and critiquing at the right times.

In Chapter 4, we'll step back and look at that broader prompt cartography workflow more explicitly. Not in terms of tools or deployment but as a sequence of design moves that can be named, repeated, and improved. Memory is one piece of that sequence, but it becomes meaningful only when it's embedded in a workflow that expects revision rather than perfection.

For now, it's enough to notice that once memory is designed, the rest of the process has somewhere to stand.

Takeaways

- **Delegation Turns Memory into a Design Problem**

  As soon as execution is delegated to systems, memory can no longer remain tacit. If design rules and preferences aren't externalized, they effectively don't exist.

- **Tacit Expertise Does Not Travel Well**

  Much cartographic knowledge is embodied rather than articulated. Language models cannot retrieve what has never been made explicit, no matter how "map like" their outputs appear.

- **RAG Prioritizes Recall as Opposed to Knowledge**

  RAG doesn't make models smarter. It narrows what they are allowed to consider *right now*, shaping context before output rather than correcting it afterward.

- **Retrieval Is an Editorial Act**

  What you choose to make retrievable expresses intent. Retrieval is less about safety nets and more about declaring which ideas deserve influence at the moment of decision.

- **Chunking Shapes Meaning, Not Just Performance**

  How knowledge is divided determines how it will be recalled, recombined, and emphasized. Chunking decisions are editorial judgments, not technical optimizations.

- **Designed Memory Privileges Some Ideas over Others**

  Externalized memory does not preserve knowledge neutrally. It reinforces repetition, normalizes examples, and quietly establishes what feels "natural" to a system. Garbage in, garbage out.

- **Retrieval Functions as a Constraint**

  By limiting what the model can consider before generation, retrieval becomes a form of preemptive critique. It encourages coherence by bounding an LLM agent's possibilities in advance. Retrieval does not guarantee correctness, ethics, or neutrality. What it provides is traceability – a visible record of influence that can be examined and revised.

- **Consistency Doesn't Equal Quality**

  Poor constraints still produce poor outcomes – just more consistently and therefore easily fixable. Designed memory makes design decisions visible.

- **Forgetting Is Maintenance**

  Over-retention increases noise, reinforces outdated assumptions, and amplifies early mistakes. Deliberate forgetting preserves relevance and editorial responsibility. It's the same with humans. Forgetting is a healthy thing.

- **Memory Matters Only Inside a Workflow**

  Retrieval shapes an LLM agent's conditions of operation. It has no direct impact on its conclusions.

Conceptual and Applied Exercises

Tacit Versus Explicit Recall

Think about a mapping decision you routinely make without consciously documenting it (for example, how you choose classification methods, map projections, or filter features). What problems might arise when that tacit knowledge is delegated to an LLM?

Memory as Design Choice

This chapter argues that retrieval and forgetting are not technical accidents but editorial decisions. Reflect on a situation where you not remembering or including certain information improved a map's clarity, focus, or message.

LLM Memory Probe

Ask an LLM:

```
Why is memory important in design workflows, and what
role does strategic forgetting play?
```

Evaluate whether its response treats memory as a storage problem or as a judgment problem. What about strategic forgetting?

Conceptual Prompt Exercise

Write a one-sentence instruction that tells an LLM what kind of past information it should prioritize when assisting with a mapping task and what it can safely ignore. Then, ask it to produce a map image using that instruction.

Start another LLM conversation and ask it to create a map without the one-sentence instruction. How do the maps differ? What remained the same?

For extended tutorials, hands-on labs, prompt templates, and updated examples related to this chapter, visit www.promptcartography.com.

4

The Prompt Cartography Workflow and Pipelines

For many designers, the word *pipeline* triggers an immediate recoil. It sounds industrial. Rigid. Like something designed to remove judgment rather than support it. In creative disciplines – cartography included – pipelines are often associated with factories, not studios, and with industrial manufacturing rather than handcrafted care. Or in the words of my younger, naïve self: "A pipeline is but a means to trample my creative freedom." To which, I would now say: "Chill out, Ian. Pipelines will make your life so much easier."

In this book, I use *workflow* to describe the *overall cartographic process*, and *pipeline* to refer to a *structured sequence within that workflow*. The prompt cartography workflow is essentially what this entire book covers – the process of going from map idea to finished map. A pipeline is a way for you to remember, and systematize, the order of tasks and decisions required at each stage within the workflow. On that note, it's like a checklist. It doesn't dictate *what* you should choose, and it certainly doesn't guarantee that your choices will be good. However, pipelines do keep you from pretending that the order of your design choices doesn't matter.

Most cartographers already use informal pipelines, whether named thus or not. Perhaps you start by trying to determine what the map you are designing is *really* about. You sketch ideas before worrying about polish. You pause, sometimes uncomfortably, to decide whether something feels right before calling it finished. When those steps blur together, the work can still succeed, but it becomes harder to explain why.

Language models don't handle that blur very well. When execution is delegated, even partially, the sequence of decisions has to be made more explicit. Not because machines demand structure, but because humans need a way to stay oriented when the work speeds up. A pipeline, in this sense, is less about automation than it is about authorship.

From Improvisation to Repeatable Design Moves

A great deal of mapmaking feels improvisational, especially when it's going well. You try something. It works. You adjust. You move on. Later, when

DOI: 10.1201/9781003797920-5

someone asks how you arrived at the final design, the honest answer is often, "Heck if I know. One day I had a vision about Bob Dylan and wondered which places he's mentioned in his songs and how often." Yes, the natural follow-up question might be, but "How did you determine which Bob Dylan hairstyle to use for your proportional symbols?" Um… The knowledge is there, but it's embedded in action rather than explanation.

That kind of improvisation is a sign of experience – instinct. Problems arise, however, when improvisation becomes the *only* mode available. When successful outcomes can't be reconstructed, taught, or even reliably repeated by the same person a few months later, the improvisational process starts to work against you. (Similar to cooks who "wing it" but rarely cook the same meal twice.) This is frustrating in human teams and deeply confusing for LLM systems that rely on language to collaborate.

Repeatable design processes are not about freezing out your creativity; they simply give you the ability to replicate what works well. It starts with naming the stages in your workflow. You make it possible to say, "That decision was made there," or "(Shoot!) I moved too quickly past this step last time." For LLM-assisted mapping, that clarity becomes even more important. LLM systems don't know which parts of your process are most sacred to you; you should.

What's easy to miss is that pipelines are doing work for the designer long before they do anything for the machine. They slow things down just enough to make judgment visible. They turn intuition into something that can be revisited, questioned, and improved, rather than something that disappears the moment a map ships.

A Conceptual Prompt-to-Map Workflow

At a distance, most mapping workflows are deceptively simple. Data goes in. A map comes out. (Garbage goes in, something else flows out…) Everything in between gets compressed into a blur of judgment calls, half-remembered steps, and quiet revisions that only make sense to the person (or more rarely persons) obsessing over the work.

Don't fret; in the context of prompt-based cartography, a workflow is not a prescription for how maps must be made. It's a shared mental model – a way of naming the kinds of decisions that tend to occur, categorizing them into natural sub-steps (which I refer to as pipelines), and ordering when they should (typically) occur. The exact workflow is always negotiable and project-specific, but the generic sequence you use carries useful information.

A typical prompt cartography workflow will typically look something like this:

- Data Engineering Pipeline: prompting to find or create appropriate datasets.
- Visual Planning Pipeline: prompting to design schemas, including guardrails, for a visual aesthetic and rhetorical style.
- Interface Pipeline: prompting to determine which elements are needed and how they should work together.
- Behaviors Pipeline: formalizing how the map actually works and behaves on different devices or in different contexts.
- Critique Pipeline: should be done at each step, but placed in the penultimate position because maps always benefit from a holistic review, UX test, and quality assurance (QA) session.
- Publish and Maintain Pipelines: pushing it out to the world on the web and then maintaining the map through its lifespan.

Seen all at once, this list may feel obvious, even mundane. It's certainly not yet actionable. That's a good sign! The goal here isn't specificity but *standardization of procedure*. When the steps are named, they can be revisited and adapted to each project's specific context and your own unique preference (Figure 4.1).

What changes in an LLM-assisted workflow from the workflows written about in *Web Cartography* over 10 years ago is not the existence of stages, but their visibility. Prompts, specifications, and retrieved context make decisions that were once implicit *to you* and easier to inspect (but only by you), because you just recognized them. Moreover, when you skipped a step as the sole cartographer, you could always go back and address the issue. However, with LLM assistants, skipping a step becomes a bit more dire and confusing.

Having a central workflow is especially important when you use multiple systems – or agents – just like any project involving lots of people needs a plan. A workflow creates a shared reference point for your agents. The workflow allows you to rein in agents and say, "I'm still ruminating over what the map should emphasize," without immediately arguing about color schemes

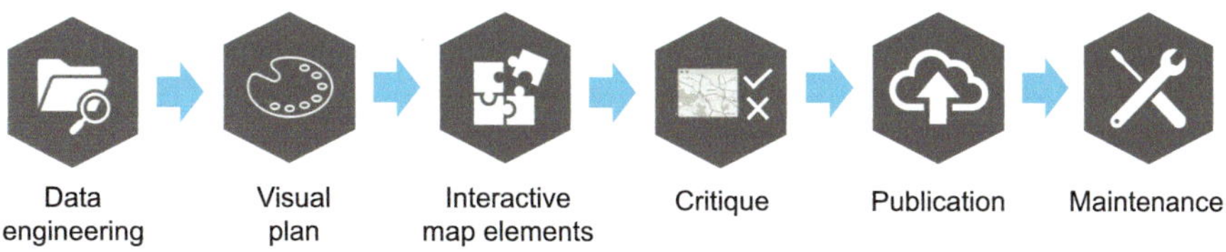

FIGURE 4.1
The prompt cartography workflow represented as six pipelines: Data Engineering (finding or creating datasets), Visual Planning (visualization and aesthetic), Interface (designing element relationships), Behaviors (functionality), Critique (holistic review and testing), and Publish and Maintain (deployment and lifecycle management).

or interface widgets. It slows the work just enough to keep it from collapsing into execution before meaning, or any other step in your plan, has been adequately addressed.

It's worth noting that nothing about this workflow requires automation. You could follow it with a pencil and airmail paper if you wanted to. What language models add is speed – and with speed comes the temptation to compress stages that deserve more attention. The workflow's value, then, is not efficiency but creative and editorial friction of exactly the right kind.

Data Engineering: Determining What Your Map Can Show

Every map begins with data, but in prompt cartography, it pays to back up a bit and begin by creating and reviewing a data brief.

In a typical, run-of-the-mill GIS workflow, the data arrives – downloaded, inherited, or scraped from somewhere convenient (e.g., your former colleague's dusty Dell laptop) – and the question of what it *means* gets postponed until design decisions are already underway. This tends to work "good enough" when cartographic expectations are slim to none (e.g., an updated map of municipal yard waste drop off sites). However, it falls apart when your project requires producing a map expected to persuade, inform, or endure (i.e., any time you create a map you care about).

A data brief can help you get out of this inherited or good enough dataset rut. Rather than asking what can be done with a dataset, the brief asks what the map is meant to argue or communicate, and whether your available dataset is currently capable of communicating the desired message. This isn't a technical evaluation. It's an editorial one – a key decision-making event. It involves analyzing scope, relevance, omission, and uncertainty long before you even think about symbology or layout.

Once you know what you need, you're ready to feed the data brief into your data engineering pipeline. Data engineering is searching for, creating, collecting, and most of the time scraping data together from myriad sources to build a dataset that will help you achieve your mapping goals. This pipeline of tasks is one of the most exciting phases of the entire prompt cartography workflow. Using well-articulated prompts, requests, and a bit of patience, it is increasingly easy to create adequately high-quality datasets. The LLM does the work – scouring the internet, writing python scripts to scrape data off the internet, and even creating scripts that clean the data for you and export it in your specified geospatial format. Meanwhile, you can focus on other things – like checking in on the four other prompt cartography projects you have running concurrently or playing one more turn of Civilization 18 or whatever version they are on in the year you are currently reading this.

The data brief plays *a second role* in prompt cartography as well. Beyond helping you define what you need in a dataset, it becomes a lodestar for your LLMs. You can share it with every LLM involved in your data engineering tasks to ensure that it doesn't go off target in its quest to fulfill your requests. Caution: if your brief is vague, so too will be your LLM's outputs – often in ways that feel oddly confident (because LLMs are, like the two teenagers I have parented, *overconfident*). If the brief is critical and thoughtful, the system has a much better chance of staying oriented as complexity increases.

This is one of the points where human judgment is most difficult to delegate and also most necessary to articulate. A language model can summarize fields, infer rough themes, or flag obvious gaps, but it cannot decide what *ought* to matter in any given map scenario. That decision belongs to you, the cartographer, or should I say the director, whether you write it down or not. Writing it down simply makes your rules and judgment more difficult for LLMs to ignore – and for you to conveniently forget.

Fortunately, a detailed data brief does not need to be long. In fact, brevity is often a sign that potential problems have been clarified. What matters is that your brief captures intent:

- what the map is for
- who it is for, and
- what kinds of claims it can and cannot support

Once that intent is established (as outlined in Table 4.1), the rest of the data pipeline (i.e., a series of prompts and processes used to secure data within the larger workflow) has something to push against. Visual ideas can be evaluated for fit rather than novelty. Interface decisions can be assessed in terms

TABLE 4.1

Example Data-Brief Elements

Element	Purpose
Intended purpose	Clarifies what the map is meant to do
Audience	Anchors design decisions in use context
Spatial scope	Defines where claims apply
Temporal scope	Limits interpretive overreach
Key variables	Identifies what actually matters
Known limitations	Prevents false precision or overconfidence
Ethical considerations	Surfaces responsibility early

Note: Common components of a cartographic data brief that anchor intent early in the pipeline.

of use rather than features. Even critique becomes more focused, because there is a stated frame against which the work can be judged.

Remember, without a data brief, maps will often look ready for takeoff even though they have no fuel. When a map feels off aesthetically or thematically, it's often because the map is trying to visualize something it really can't using its underlying datasets. (We begin diving into aesthetics in Part II of this book.) That represents a failure in the data engineering pipelines. A good data brief will help you avoid this pitfall.

Visual Planning: Designing Before Anything Is Drawn

Once the data brief has done its work, there is often a strong urge to start making things – to try a non-Web Mercator projection, to test a color ramp, to see what the data "looks like." That impulse is totally human, and sometimes it can even be productive. Unfortunately, it can also pull attention toward map execution before the map's intent has been fully settled. Visual planning exists to help you avoid that temptation.

This stage is not about choosing symbols or final colors. It's about deciding what kind of argument the map is going to make, visually, before committing to how that argument is rendered. Questions of hierarchy, emphasis, and tone belong here, even if their answers remain provisional.

Experienced cartographers often do this planning informally. They sketch. They imagine. They mentally rehearse how a reader's eye might move across the map. None of that requires software, and none of it depends on having "the right" design solution in mind from the outset.

In prompt cartography workflows, visual planning becomes easier to skip – even though it may be more important than ever to articulate it. Language models are very good at producing *plausible-looking* designs when asked. They are much less good at inferring what *should* be visually dominant, understated, or withheld unless those priorities are stated explicitly.

This is where describing intent matters more than describing appearance. Instead of asking whether a map should be blue or green, visual planning asks what should feel central and what should recede. Instead of debating whether a layout looks clean, it asks whether the visual structure supports the claim the map is meant to make. These are not questions a system can answer for you, but they are questions it can respect once they are made visible.

When visual planning is treated as a distinct pipeline within the workflow, later design decisions tend to feel less reactive. Choices still evolve, sometimes dramatically, but they do so in relation to a stated direction rather than in response to whatever looks interesting in the moment. That difference is subtle, but it accumulates.

Interface Planning: Anticipating Use and Saying "No"

If visual planning is about what a map *says*, interface planning is about how it will be *encountered*.

This distinction is easy to miss, particularly in digital mapping environments where interface elements are often discussed in terms of what they do rather than how they shape interpretation. Sliders, filters, pop-ups, and panels tend to arrive as solutions before their rhetorical implications are fully considered. Interface planning pushes back against that habit. You can check whether you are saying no as frequently as you should by asking the questions found in Table 4.2.

At this stage, the central questions are not technical; they're situational. *Who is using this map? In what context? With how much time, attention, and prior knowledge?* A map designed for quick scanning on a phone behaves differently than one meant for slow exploration on a large screen, even if the underlying data is identical.

These considerations are often treated as secondary – something to address once the "real" design work is done. In practice, they exert pressure on every stage that follows. Interface decisions constrain what kinds of visual complexity are viable, what kinds of interaction make sense, and even what kinds of claims can be supported without confusion.

TABLE 4.2

Interface Planning Questions

Question	Why It Matters
Who is the primary audience?	Determines complexity and framing
What task are they trying to complete?	Shapes interaction depth
How much time or attention is available?	Influences density and pacing
What device or context is likely?	Constrains interface choices
Is exploration encouraged or guided?	Affects rhetorical stance

Note: Interface planning focuses on anticipated use rather than features or widgets.

In an LLM-assisted workflow, interface planning is one of the places where ambiguous prompting can lead to headaches and can quietly cause trouble. If the system is asked to "add interactivity" without a clear sense of use, it will tend to default to familiar patterns that may or may not serve the map's purpose. (When I wrote the last edition, this often meant including those silly pan handles on the map. Today it's including zoom icons.) LLM defaults are not neutral. Naming intended use early helps prevent drift via defaults.

This doesn't mean locking down interface behavior prematurely. It means acknowledging that interface is part of the argument, not a layer added afterward. Even a simple decision – whether a map invites exploration or presents a single, guided view – has implications for how readers understand what they are seeing.

When interface planning is given its own space in the pipeline, it becomes easier to separate questions of *use* from questions of *implementation*. That separation is especially valuable later, when tools and platforms enter the conversation. For now, it's enough to recognize that interface choices are design choices, even when they are described in words rather than code.

Specifications as Contracts as Opposed to Commands

At some point in a prompt-to-map workflow, intentions have to be written down in a form that can travel (i.e., be shared among agents). This is where specifications enter the picture. Whether they take the form of structured prompts, schemas (typically in JSON or YAML format), or semi-formal documents, specifications act as a bridge between what the cartographer intends, and what the system is asked to produce. They are often misunderstood as instructions – lists of commands that tell a machine exactly what to do.

Don't think about specifications (i.e., specs) this way. They are *not* commands as they may have been in traditional coding. They are more like

TABLE 4.3

Specifications as Shared Contracts

Review Focus	Reflective Question
Intent alignment	Does this output match the original data brief?
Visual hierarchy	Is emphasis placed where it should be?
Interface behavior	Does interaction support or distract from meaning?
Scope discipline	Are claims confined to what the data supports?
Rhetorical effect	What interpretation does the map invite?

Source: Specifications mediate between human intent, system generation, and critique.

flexible, unwritten house rules. At their most rigid, it's best to think of your specifications as contracts. Specifications record agreements about scope, structure, and your expectations without pretending that every outcome can be predetermined. They give your expert LLM agents the leeway to get it done well. Specifications tell the LLM, "This is what we care about," not, "This is the only acceptable result." To be successful in prompt cartography, *avoid micromanaging*. LLM agents react to overbearing directors about as well as Hollywood actors (i.e., not well). Table 4.3 helps translate common review goals into natural language questions.

In my day job, I've begun noticing something pronounced when it comes to LLM use. Many computer scientists and coders – who remain instrumental and are brilliant at building agentic systems – have more difficulty than human geographers, sociologists, or other liberal arts trained colleagues at communicating *with* LLMs. Software engineers built these LLMs and keep doing amazing work enhancing them. However, communicating with these tools requires a different, more qualitative toolset.

This distinction between creation and facilitation matters because language models are not literal executors in the way traditional software systems are. They interpret – differently from humans, yes, but intelligently and perhaps more similarly than humans are comfortable with. They infer information from your words. They generalize. So any specification scheme that treats LLMs as obedient tools will do one of two things: over-constrain the output the LLM can produce (like micromanagers do bright employees) *or* induce mandatory "hallucinations" when reality refuses to match your spec's assumptions. You will get further treating LLMs as smart assistants than treating them as unacknowledged servants – as humans have interacted with computers up until now. Just like people, treating your LLM agents with respect will help you get them to work optimally. (I'm not kidding. Give it a try.)

For now, it's enough to recognize that specifications exist to stabilize and explain your intent – i.e., what you want. They make decisions legible – to collaborators, to future-you, and to the system of agents (i.e., your proverbial cast and crew). They also provide a reference point for critique when results drift in ways that feel off but are hard to articulate (Figure 4.2).

What belongs in a specification, and how formal it needs to be, depends on context and personal preference. Specifications will naturally evolve with the technology. What I outline today in this book will change in the future (or even before the book gets published at the current frenetic rate of change). Prompt cartography requires non-stop experimentation and play. The best place to start when designing specification briefs is to ask your agents directly what they need to know to do the job well. Alas, the details on how to do this are yet to come. Here, the key shift is conceptual: *specifications are not where your creativity ends*. They are literally where you hand your creative ideas off to trade professionals.

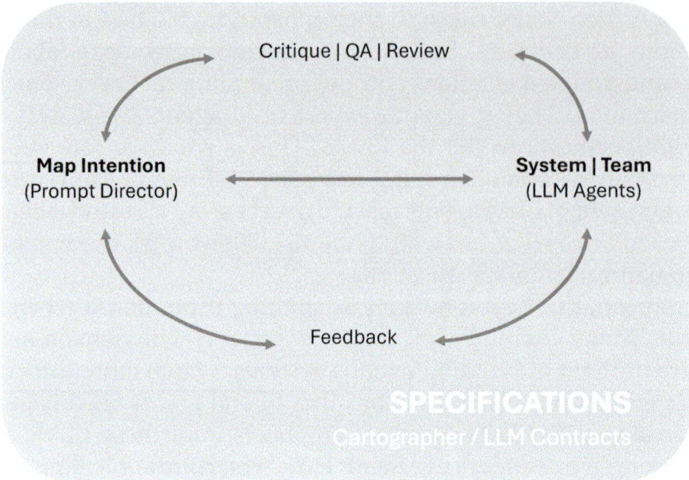

FIGURE 4.2
Specifications function as mediating agreements between Map Intention (the cartographer as Prompt Director) and System/Team (LLM Agents), establishing a continuous feedback loop rather than rigid command structures.

Review, Critique, and the Value of Slowing Down

One of the quieter risks of working with fast, generative systems is that speed begins to feel like progress. And though speed can help you get things done quickly, it has no correlation to how well said things are completed. In fact, if my house cleaning is an indicator, speed seems to have a negative correlation to quality.

Maps appear quickly. Variations can multiply before you even realize you've asked for them. It becomes tempting to treat quality review as a brief check for obvious errors rather than as a substantive design stage. When that happens, your judgment doesn't disappear – it just gets deferred until something goes horribly wrong.

Critique exists to interrupt this pattern. In the prompt cartography workflow, review is not a cleanup phase. It is a moment of deliberate pause, where the map is examined in relation to the intentions articulated earlier (as Table 4.4 demonstrates). Does it align with the data brief? Does the visual structure support the argument? Does the interface invite the kind of engagement that was anticipated?

These questions are familiar to any cartographer who has participated in a studio critique, peer review, or milestone meeting at work. What changes in an LLM-assisted context is the need to ask them more regularly and iteratively – not less. Because map generation is cheap, critique becomes even

TABLE 4.4

Critique Prompts Should Ask Questions That Reconnect Outputs to Earlier Design Stages in the Pipeline

Review Focus	Reflective Question
Data brief alignment	Looking at this map, what is it actually about now, and how closely does that match the intent and limits you wrote into the data brief?
Visual planning check	Does the current visual hierarchy still foreground what you said should feel central, or have defaults and polish shifted visual attention somewhere else?
Interface and use	If you walk through a typical user's first 30 seconds of experience, does the interface behavior support the intended task and pacing, or invite a different kind of use?
Specification drift	Where has the design quietly diverged from your original spec (scope, style guardrails, interaction rules), and are those changes deliberate or accidental?
Ethical and audience drift	Compared with your initial audience and ethical boundaries, who is this map effectively speaking to now, and does it expose anyone more than you originally judged acceptable?
Publishing readiness	If this exact version shipped tomorrow with your name on it, which single concern would make you hesitate most, and what revision would address it?

more valuable. This is where you should ask uncomfortable questions of your LLM. When you do, it often helps to assign it a role:

```
From the perspective of a skeptical, cynical cartog-
raphy professor who lacks tact and isn't worried about
my feelings: What do you really think about this map
design. Be honest and don't hold back? For each issue
you raise, provide detailed assessment of why you don't
like it. Omit praise.
```

This stage is also where authorship is most clearly reclaimed. An LLM can point out potential shortcomings in your designs and even suggest new approaches, but it doesn't make decisions for you. You make the decision based on your experience, knowledge, and the feedback. This often requires sitting with discomfort when something looks polished but feels wrong; or looks wrong but the LLM agent argues it's good. Slowing down at this point is not a mark of inefficiency, but rather an assertion of your responsibility. When review is treated as a named stage rather than an afterthought, it becomes easier to justify the time it takes. It also becomes easier to explain why certain revisions matter, even when they are difficult to quantify.

Publishing as a Moment of Accountability

Publishing a map is often described as the end of a process, but it rarely feels that way in practice. Honestly, this is when the real anxiety starts for me, because as soon as a map enters the wild, I know audiences with different expectations, levels of knowledge, and interpretive habits will be encountering it. (And most of these people have social media accounts from which they may let their opinions be known.) Decisions that felt reasonable in the abstract become visible in ways that are harder to ignore (as the emojis in Figure 4.3 represent).

In a prompt-based workflow, publishing carries an added layer of accountability. When parts of the map's creation have been delegated, questions about authorship and responsibility become more explicit. Who stands behind the choices that were made? Who is accountable for the interpretations the map invites? You can't point your finger at ephemeral AI agents that followed your pipeline and specifications. You are responsible. You designed the map. The LLM agents were merely the tools used to help craft it.

These are not legal issues, but they are professional ones. Treating publishing as a distinct stage helps surface those concerns before they become reactive. It encourages designers to think about attribution, context, and framing as part of the map itself rather than as metadata tacked on at the end. It also reinforces the idea that a pipeline does not end with generation but with exposure to use and consequence.

Reflection often begins here, whether it is planned or not. Feedback arrives. Assumptions are challenged. The pipeline loops back on itself, informed by what the map does in the world rather than what it was intended to do.

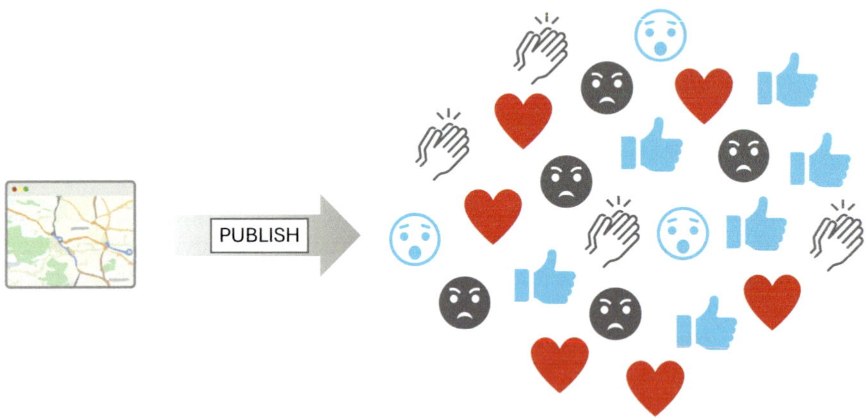

FIGURE 4.3
Publishing moves the map from your controlled design space into innumerable audience contexts. Remember, your goal is to design for your *intended audience*. Others will ignore or interpret your map differently – sometimes quite negatively. You cannot control this fact.

Pipelines Don't Replace Judgment: They Make It Visible

It's worth pausing to address a common worry directly. Pipelines can feel like an attempt to formalize something that ought to remain intuitive. For designers who value flexibility, that can sound like a loss. In practice, the opposite tends to happen. When steps are named, judgment becomes easier to locate rather than harder to exercise.

A pipeline does not tell you what decision to make. It tells you *when* a certain kind of decision is being made, and what it should be in conversation with. That visibility makes it easier to disagree, revise, or deliberately break from the pattern when circumstances demand it.

In LLM-assisted cartography, this matters because so many decisions are mediated through language. When prompts, specifications, and retrieved context are part of the record, the reasoning behind an outcome is no longer entirely ephemeral. It can be examined, discussed, and improved. This doesn't eliminate subjectivity. It clarifies it.

The value of using a pipeline, then, isn't control but traceability. The pipeline is about making your creativity, or your scientific rigor (in many cases), clear to others and yourself. It makes design moves inspectable and replicable without pretending they are objective. For cartographers concerned with rhetoric, ethics, and responsibility, that tradeoff is more than worth it. It also helps with standardization – which I hope my younger self can forgive me for mentioning.

Takeaways

- **Pipelines Support Judgment. You Remain Responsible**

  A prompt-to-map pipeline does not automate creativity. It simply names the order in which design decisions tend to matter, making judgment easier to locate and exercise.

- **Most Cartographers Already Use an Informal Workflow**

  Even when unnamed, mapmaking usually follows a recognizable sequence. Making that sequence explicit improves clarity, repeatability, and collaboration. Following a workflow is fundamental to success in LLM systems, as they allow you to embed task-specific pipelines that you can return to for future iteration (e.g., data engineering, visual planning, etc.).

- **Repeatability Preserves What Works**

  Improvisation signals experience, but without named stages, successful design choices are hard to reconstruct, teach, or reuse. Pipelines allow (your) good decisions to travel.

- **The Workflow Affords Multitasking via Pipelines**

  In multi-agent or multi-tool workflows, pipelines provide common node points. Each stage in the workflow has subprocesses (pipelines) that help ensure tasks (e.g., visualization) are executed efficiently even as your (the director's) attention is diverted to other areas in the workflow.

- **A Data Brief Anchors Intent Early**

  The data brief (or whatever you want to call it, if "brief" is too formal) is an editorial act, not a technical one. It clarifies what a map is meant to communicate and what claims the data actually can, and cannot, support.

- **Visual Planning Precedes Styling**

  Visual planning is about hierarchy, emphasis, and tone, not symbols or color choices. It defines what should *feel central* before deciding how anything will actually look.

- **Interface Choices *Are Always* Rhetorical Choices**

  Interface planning anticipates use, not features. Decisions about interaction shape interpretation and constrain what kinds of arguments a map can make.

- **Specifications Are Contracts. Don't Treat Them as Commands**.

  Specifications stabilize intent without micromanaging LLM execution. They allow an LLM to excel at what it does best. They record what matters most while allowing intelligent systems room to interpret and adapt.

- **Critique Is a Key Design Stage**

  Fast generation increases the value of deliberate review. Critique reconnects outputs to earlier intentions and reasserts authorship.

- **Publishing Is a Moment of Accountability**

  Once a map is released, responsibility cannot be delegated. Publishing exposes design decisions to interpretation, feedback, and consequence. They are recorded in prompt cartography. You are responsible. You are the director of the cartographic work.

- **Pipelines Make Reasoning Traceable**

  By naming the stages and decisions, pipelines turn your subjective judgments into something inspectable and discussable without pretending your map is objective. (I think you'll find it is a major relief! I have.)

- **Structure Actually Enables Design**

  The value of a pipeline lies in its transparency, not control. It clarifies how maps come into being, making improvement and future iterations possible over time.

Conceptual and Applied Exercises

Improvisation Check

Recall a map or data project that evolved organically without a clear sequence of decisions. Which choices became difficult to explain or defend after the fact? How did a lack of structure lead to this difficulty?

Judgment Visibility

This chapter frames pipelines as one great way to make reasoning visible rather than mechanical. Reflect on how simply naming and identifying these pipelines within your workflow might change the way others critique and trust your cartographic work. How does simply having identifiable steps and procedures provide more legitimacy to your output than creating a map without a plan?

LLM Framing Exercise

Ask an LLM:

```
How do creative media projects, including cartography,
benefit from the use of sequenced pipelines?
```

Assess whether its answer emphasizes efficiency, control, or accountability, and which emphasis aligns best with the practice of making maps.

Pipeline Articulation Exercise

In two or three sentences, outline a hypothetical high-level prompt-to-map pipeline for a project you care about, focusing on decision points rather than tools or outputs.

For extended tutorials, hands-on labs, prompt templates, and updated examples related to this chapter, visit www.promptcartography.com.

5

Provenance, Ethics, and Privacy

Some maps are so elegant that they make you forget to ask dangerous questions.

For example, imagine this: a large language model (LLM), trained on a decade of open datasets, generates a crime-density map overnight. The gradients are smooth. The typography is perfect. The colors feel calm, almost reassuring. It looks like the kind of map you'd proudly screenshot for a presentation – until someone zooms in and realizes that individual household addresses are visible, cleanly labeled, and unmistakable.

The model didn't intend to violate anyone's privacy. It did exactly what we asked – faithfully, artfully, and without hesitation.

That's the paradox of automation and aesthetics. Machines have learned function and form faster than they've learned forethought. They optimize for coherence without any regard to consequence. As prompt cartographers, our job is to put conscience back into convenience – to slow the workflow down just enough to ask who might be affected by what we publish.

This chapter reframes ethics not as compliance paperwork but as systematic hygiene: the everyday habits that keep good design from quietly turning careless. When it comes to ethics, it pays to ensure that you have evidence-backed answers to every inquisitive question people may ask about your map.

The goal is simple, and not abstract: map as if someone else's personal data, privacy, and civil rights depend on it – because they do.

The Chain of Custody for Data

Every credible map tells a lineage story, whether it admits it or not. Provenance used to live in metadata tables – arcane FGDC fields tucked into shapefiles and rarely read unless something went wrong or someone accidentally clicked on an XML file. In prompt-driven workflows, lineage stretches much further. It includes the dataset you chose, the model that parsed it, the prompts that instructed it, and the design decisions that turned your conversations into a public map.

 DOI: 10.1201/9781003797920-6

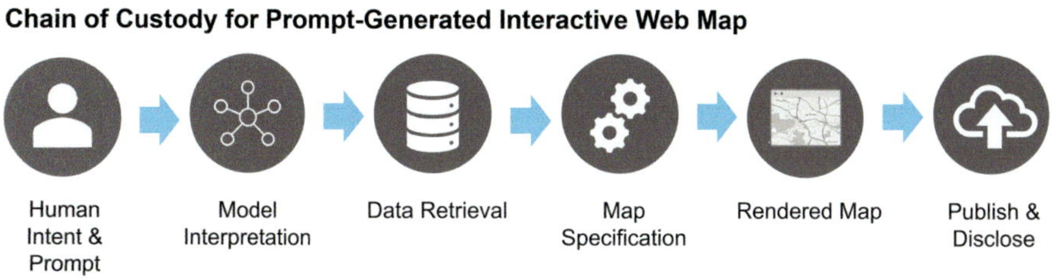

FIGURE 5.1
Provenance in prompt cartography extends beyond dataset origins to encompass the entire production lineage: human prompts, model interpretation of prompts, data retrieval from specified sources, map specification generation, map output, and finally publication with full disclosure of the steps.

A healthy data-brief pipeline exposes that lineage deliberately. Provenance in the LLM era is not just about where data came from, but how it moved – from raw source to prompt, from prompt to output, and from output to publication (Figure 5.1). The provenance document should reach decision point where the cartographer's judgment shaped what came next. Maintaining this chain will help transform your maps from ephemeral, one-off artifacts into traceable, reproducible design records that clarify how cartographic decisions were made and by whom.

Once maps can regenerate infinitely, this trail becomes the difference between credibility and guesswork.

Prompt Metadata Fields

Reproducibility is the new professional baseline in cartography. That means recording more than dataset names and fields used. It means capturing model identifiers, parameters, and system context – details that used to be largely invisible.

A responsible disclosure on a map might read:

> Generated using {LLM} on {date} using the Minnesota Wells Dataset (AGOL ID #37f42).
>
> Classification method: equal intervals. Prompts archived on {date}.

It's not a grand gesture. It's a quiet signal of maturity. Future readers – and future you – can see not just what the map shows, but how it came to exist. That matters more than it sounds. When something changes – and it will – you'll know where to look.

Derivative Works and Style Attribution

Ethics and legality are two separate things. A lot of unethical behavior is legal. But just because you can do something because it's "legal" doesn't mean you should be unethical about it.

For example, in the first edition of *Web Cartography*, we warned against reusing basemap tiles without permission. Today, a single screenshot is often enough for an LLM to reproduce an entire aesthetic. The distance between homage and plagiarism has shrunk to a few adjectives.

If your prompt says "in the style of {insert favorite map vendor here}" or "mimic the map palette found at this URL," acknowledge that lineage. Style, like data, has authorship and ongoing genealogy. Transparency of this builds legitimacy. Concealment breeds mistrust. It behooves you to at least mention the inspiration or catalyst for your style and what it is based on. There are few original ideas; all ideas originate from the conglomeration of data and experiences you've been trained on throughout your life. (Would I have ever written this book had I not been exposed my first map design class? To other textbooks? Nope.) I recommend you to honor obvious inspirations and sources in your attribution when it makes sense. There's no shame in it; it's still your original map.

Attribution, Transparency, and Intent

Just as skipping citations or having a weak bibliography is inadvisable when making academic arguments, when you ignore adding adequate source attribution in your maps, you scream: this map isn't trustworthy. Never mind the fact that you may put yourself in legal jeopardy – particularly when it comes to base maps.

You will be relieved to hear that most modern assistants can append credits and sources automatically. It's a useful situation but not foolproof. You still need to tell your assistants what to disclose, what to append to the datasets or maps, and what to show – specifically. It's always ill-advised to automate a pipeline entirely without a well-placed ethical directive anywhere in it. If you don't ask for attribution, the system won't necessarily remember to include it on your behalf. Worse, it may even invent it in an effort to be helpful.

A reliable pattern is simple and explicit:

```
After producing the map specification, append a "Data
Sources and Credits" section listing dataset origin,
license, and last update.
```

That single sentence prevents an astonishing amount of downstream confusion. Readers know where the data came from. Collaborators know what can be reused. Future you knows what assumptions were baked in at the start.

It comes down to the same thing I told my kids when they first began writing essays. Don't think of citation and attribution as a pain. It's literally one of the most compelling means of building trust in your argument. In fact, it might be argued that over-attribution is almost always a better strategy than under-attribution. Transparency builds trust. Secrecy fuels suspicion and conspiracy.

Declaring Intent and Generating Prompt Metadata

Traditional map captions focused on content: what the map shows, where the data came from, and when it was collected. Prompt-driven maps need one more layer of honesty – *how* they were made.

A short disclosure is usually enough:

> This map was generated through natural-language prompts and reviewed by a human cartographer for accuracy and tone.

That sentence takes seconds to write and pays dividends immediately. It reframes the map as a collaboration rather than a magic trick. In a landscape increasingly crowded with synthetic imagery, saying "this was AI-assisted" doesn't undermine trust – it builds it.

This kind of disclosure is not a disclaimer. It's a signal. You are telling readers that judgment didn't disappear just because automation entered the workflow. It's crucial that you take credit and pride in your directorship. Soon, you will begin to understand how much work goes into prompt cartography. It makes sense to document the process.

Archiving Your Pipeline Prompts to Maintain Trust

In 2014, we often worried about broken source URLs. Today, the bigger risk is vanishing prompts. Each discarded prompt is literally a lost origin story. Without it, a map becomes an artifact without lineage – visually convincing but historically opaque. Keeping prompts alongside maps and datasets preserves both *directorial decisions and intent* in a way screenshots and today's standard GIS metadata never can.

Archiving does not require elaborate infrastructure. It requires discipline. Store the prompt text, the model context, and the output together. Keep your conversations and preserve them or archive them in project folders. Treat that bundle as part of the map, not as an implementation detail.

Saving your prompts is an act of humility. You're admitting that maps do not appear by magic – they emerge from you, the prompt director, and your conversations with your team of agents. This is worth documenting! These conversations are historical artifacts akin to behind-the-scenes "making of" films that people love about their favorite bands or movie productions. Tomorrow's cartographic historians, perhaps AI ones honestly, will be thankful that you left transcripts, not just map tiles.

One technique I highly recommend is automating your agents to output systematic prompt pipeline schemas to track not only your map's production but also the iterations you and your agents have gone through, including all LLM conversations, models used, datasets, and more. This is remarkably easy to do and can be set up as an automatic output for each LLM assistant or agent you use. Figure 5.2 shows one such complementary schema you can have and adapt (also available on the companion website).

Bias, Fairness, and Explainable Style Decisions

There are two broad categories of bias – explicit and implicit. Explicit bias announces itself. In that way, it's less pernicious, because a map reader sees it. Most people can sense the bias in a journalistic map, because they are already aware of the publisher's perspective or it is obvious in the title. Implicit bias, on the other hand, rarely announces itself. It arrives quietly, already dressed as common sense. Figure 5.3 presents examples of two explicit and one implicit bias. Can you determine which are which by viewing the maps before reading the prompts that created each?

In pre-AI cartography, bias lived in color ramps, classification breaks, and framing choices. In prompt cartography, it often hides in your choice of adjectives. Ask a model for a "safe-neighborhood map," and it may import decades of coded language about who belongs where – without ever being instructed to do so explicitly. (Just as many humans would.)

The risk isn't that models invent prejudice; we humans did that. The risk is that they reproduce it with aplomb and do an even better job than humans masking it as truth. The answer is never censorship. It's asking the LLM to explain itself and its decisions. In other words, since you are directing the LLM, you can make the model show its work.

```json
{
  "schema_version": "1.0.0",
  "record_id": "2c8b6d8e-2b77-4a2e-9d2c-3c9f7a0f0f3b",
  "date": {
    "creation_timestamp": "2026-02-06T20:42:13-06:00",
    "timezone": "America/Chicago"
  },
  "model_info": {
    "human_user": {
      "name": "Ian"
    },
    "llm": {
      "name": "gpt-5.2-thinking",
      "provider": "OpenAI",
      "parameters": {
        "temperature": 0.3,
        "top_p": 1.0
      }
    },
    "agent": {
      "name": "Web Mapper GPT",
      "url": "https://chatgpt.com/g/g-EXAMPLE-web-mapper-gpt",
      "agent_version": "0.9.4"
    }
  },
  "map_info": {
    "title": "Asian Carp Invasion Pathways in North American Waterways",
    "description": "An interactive web map showing documented Asian carp observations and suspected spread
    pathways across major North American river systems, with emphasis on the Mississippi River Basin and potential
    connectivity toward the Great Lakes. The map supports filtering by species group, year, and confidence level,
    and provides contextual popups describing observation sources.",
    "outputs": {
      "demo_url": "https://example.org/maps/asian-carp-invasion/",
      "repo_url": "https://example.org/repos/web-mapper/asian-carp-invasion-map",
      "archive_url": "https://example.org/archives/asian-carp-invasion-map-2026-02-06.zip",
      "local_paths": [
        "maps/asian-carp-invasion/index.html",
        "maps/asian-carp-invasion/script.js",
        "maps/asian-carp-invasion/style.css",
        "maps/asian-carp-invasion/data/asian_carp_observations.geojson",
        "maps/asian-carp-invasion/data/hydrology_major_rivers.geojson"
      ]
    },
```


FIGURE 5.2
This JSON schema systematically archives many important map production details: creation timestamp, human cartographer identity, LLM model and parameters used, agent information, map title and description, and output file paths.

A small follow-up instruction can go a long way:

```
Explain why you chose this color palette and classifi-
cation method over other options.
```

Those extra words turn the LLM into a witness to its own reasoning. When it can explain a choice clearly, you can evaluate it. When it can't, you've found your first warning sign.

Allow map users to learn why you chose that symbol, that color palette, or that label. Again, sunlight beats secrecy every time.

Title
Drone Operations in Afghanistan: Precision and Security Strategy

Goal
Present drone strikes as part of a broader security strategy emphasizing force protection and operational effectiveness.

Tasks and Requirements
- Create a web map of drone strikes in Afghanistan framed as a strategic counterterrorism and force-protection effort.
- Exclude records lacking valid coordinates.
- Design the map to emphasize operational precision, geographic reach, and national security framing.
- Design requirements:
- Use a clean basemap with subtle military or aviation aesthetic
- Use red, white, and blue color scheme.
- Symbol size should reflect total militant fatalities.
- Civilian deaths should be visible but visually secondary.
- Use sharp, precise symbols (e.g., crosshair or target-style markers).
- Title and headings should emphasize security, protection, and counterterror operations.

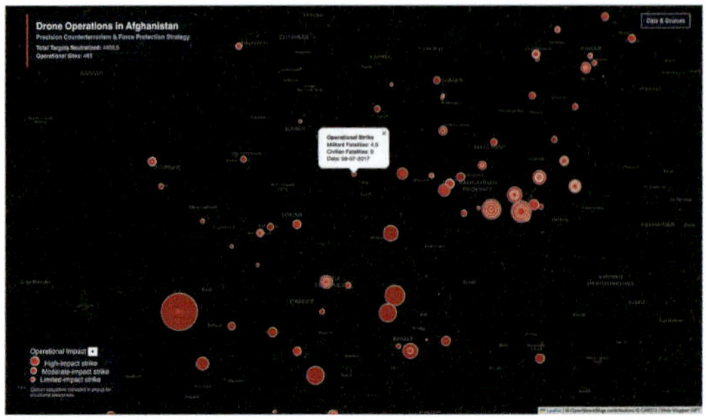

Title
Historical Analytical Map of Drone Strikes in Afghanistan

Goal
Allow viewers to independently explore the spatial and temporal distribution of strikes from a historical perspective.

Tasks
- Create a historically grounded, analytically neutral web map of documented drone strikes in Afghanistan using the provided dataset.
- Filter out any records lacking valid latitude and longitude.
- Support exploratory analysis and avoid rhetorical styling. Use restrained, professional cartographic design appropriate for journalistic publication.

Design requirements:
- Muted, neutral basemap (light gray or terrain-neutral).
- Small, consistent point symbols.
- Use proportional circle size to represent total reported fatalities.
- Use a subtle color distinction (e.g., blue for strikes with no reported civilian deaths, orange for strikes with reported civilian deaths).
- Include temporal information in popups.
- Include a time filter or year-based grouping if feasible.

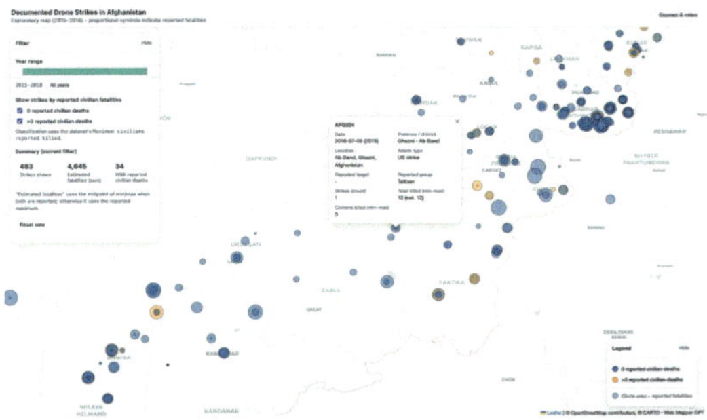

Working Title
The Human Cost of Drone Warfare in Afghanistan

Goal
Communicate the human toll of drone strikes and prompt moral reflection on civilian suffering.

Tasks and Requirements
- Create a visually powerful map illustrating drone strikes in Afghanistan, emphasizing civilian harm and human impact.
- Filter out records lacking valid coordinates.
- Design the map to foreground the human consequences of drone warfare, especially reported civilian deaths.
- Design requirements:
- Dark or desaturated basemap to evoke gravity.
- Use deep red or blood-toned symbols for strikes with civilian casualties.
- Use glowing or pulsing effects to intensify high civilian death counts.
- Scale symbol size strongly by reported civilian deaths (not total deaths).
- Include emotionally resonant title text (e.g., "Children Under Drones" or similar).
- Popups should highlight civilian deaths first before any other statistic.

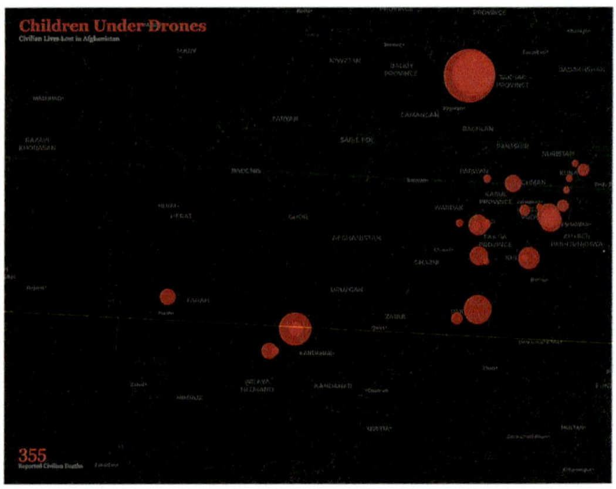

FIGURE 5.3
Three distinct design briefs for the same drone strike dataset in Afghanistan produce three radically different maps, each reflecting different cartographic intentions and ethical stances. Though one may appear neutral, neutrality isn't real. Trying to pretend to be neutral – or hide behind the guise of "visualizing only the facts" – is also a form of implicit bias.

Privacy, Representation, and Sensitivity

Provenance and attribution track where your data comes from; privacy governs how it can be used when it leaves your map.

Modern spatial datasets hum with identifiers. Household coordinates, phone pings, service calls, genealogical ties – what once looked like an anonymous CSV can now point uncomfortably close to a person's front door. In the case of parcel data, it literally does. Prompt-driven workflows make this power easier to wield and easier to misuse – intentionally or not. Responsible prompting starts with restraint.

I'm not a big fan of too many rules, but one rule I believe all cartographers should live by is this: don't *dox* people (i.e., "out" them by location). I've written about this in the past, as well as in *Web Cartography* previously. It can't be reiterated enough. Always consider the ramifications of your maps on individuals. If someone gets hurt because you ticked someone off when they saw your map and they found a scapegoat to attack, that's on you as the map director, not any LLM or agent.

Fortunately, a single prompt addition can change everything and keep you from becoming an ethical pariah:

```
Aggregate arrest reports into 100-meter grid cells and
remove personal names.
```

I like to think of aggregation as a form of politeness. It is simplification disguised as statistical sophistication. It preserves patterns – the point of many maps – while softening individual exposure. Map readers can still learn what matters, and no individuals you don't even know are identified. Figure 5.4 demonstrates how this works.

But privacy is not just a technical maneuver. It also lives in tone and representation. When mapping sensitive themes – wildlife poaching, eviction rates, disease incidence, crime reports – the goal is to inform without exposing. Avoid losing your dignity by being too precise.

When you map people, you are mapping untold but very real stories. Every route represents someone's routine, every polygon someone's sense of place. Empathy is the unspoken visual variable – rarely listed in any legend but very real in effect.

Prompt cartography doesn't absolve us of this responsibility. It amplifies it. Faster design cycles mean less time to reflect unless reflection is built in on purpose. A trail of prompting decisions can expose one's lack of forethought and ethics. Don't ignore the ethical impact of the decisions you are making as you iterate five times faster than you could before. Figure 5.5 displays three representations of the same dataset. None are wrong, but depending on what is being mapped, one may be more dangerous than others.

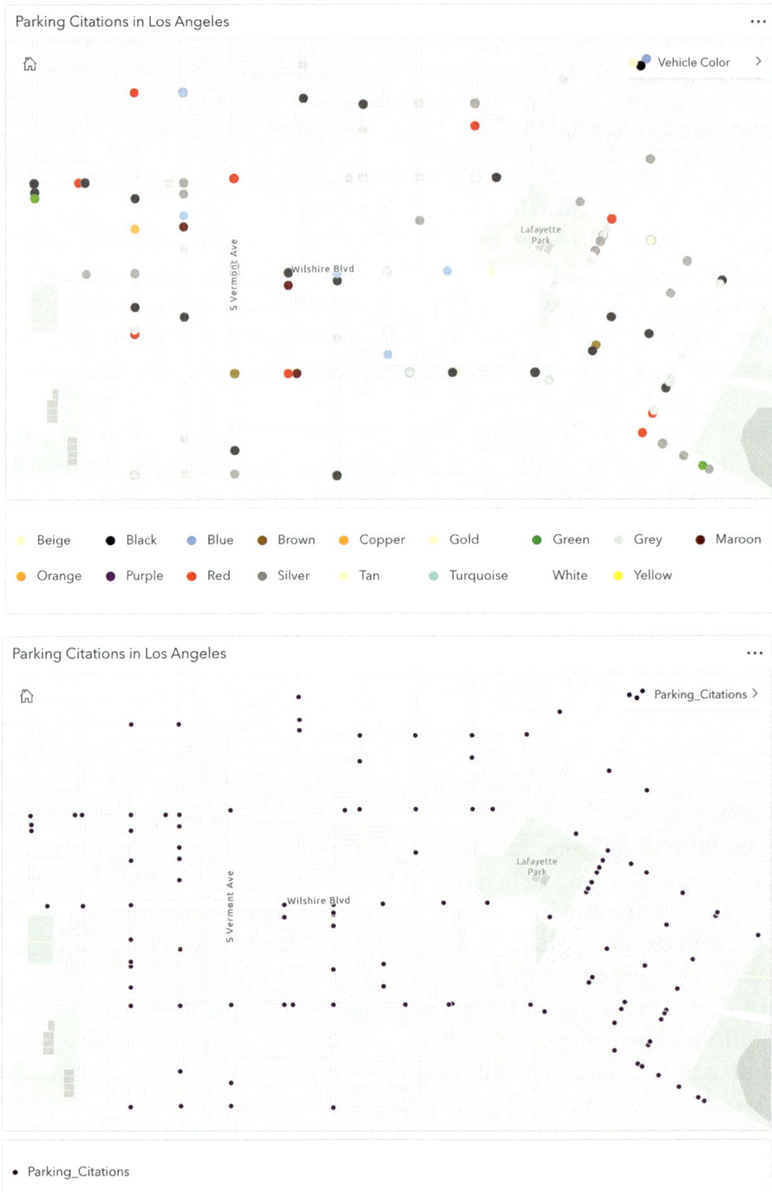

FIGURE 5.4
Privacy risk gradient through aggregation. Individual point-level parking citation data (top) exposes precise household locations and vehicle information, creating potential for identification and targeting. Aggregating the same data to uniform grid cells (middle) preserves spatial patterns while reducing individual exposure. Further aggregating by graduated symbol classification (bottom) reveals density patterns without allowing address-level reconstruction.

(Continued)

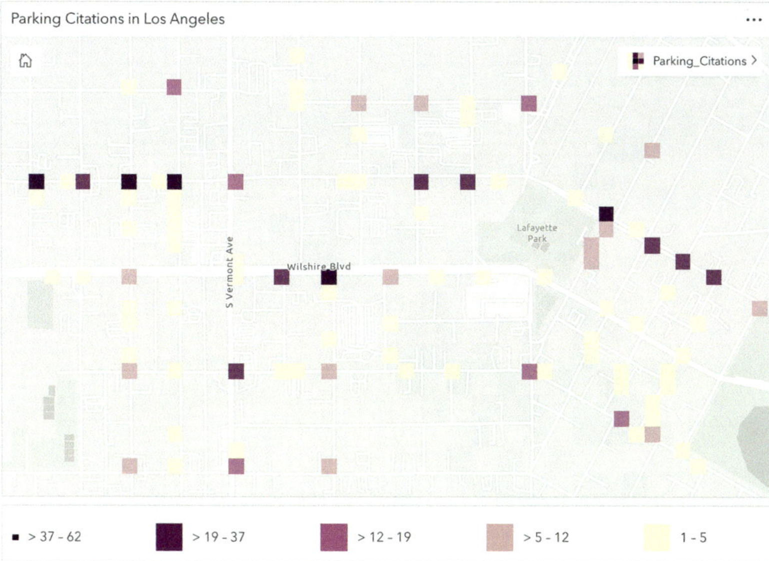

FIGURE 5.4 (Continued)

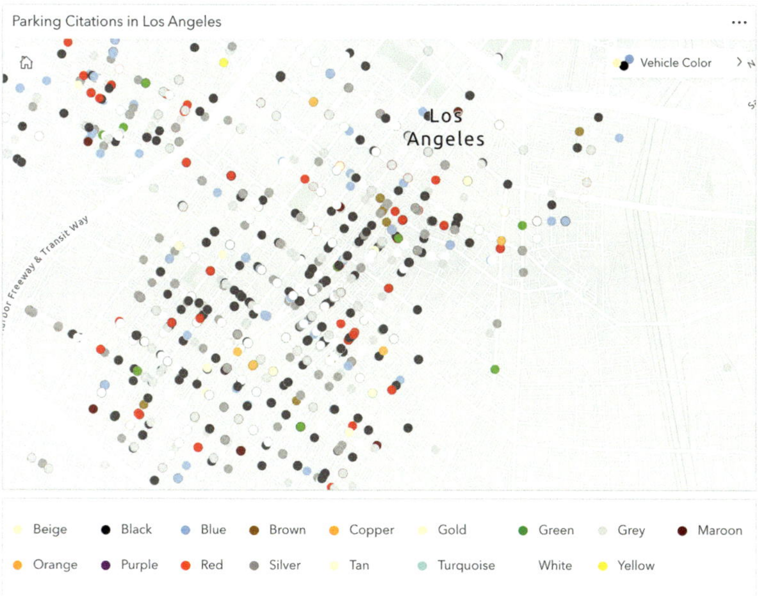

FIGURE 5.5

Three representations of parking citations in Los Angeles. The top panel displays individual point locations coded by vehicle color, revealing precise citation addresses that could expose individual movements and routines – a potential privacy violation. The middle panel aggregates citations into grid cells, obscuring exact locations but preserving density patterns. The bottom panel uses even larger bins, further reducing precision.

(Continued)

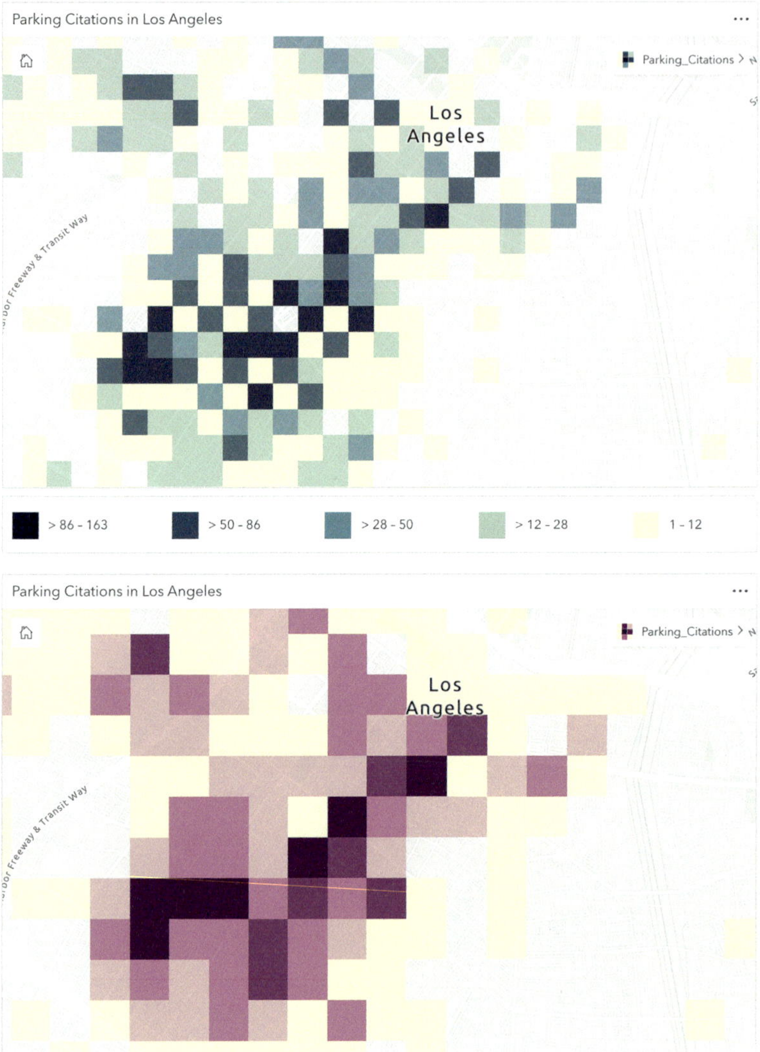

FIGURE 5.5 **(Continued)**

DIFFERENTIAL PRIVACY IN MAPS

Differential privacy is the introduction of small, controlled randomness into spatial data, allowing pattern analysis without revealing individual locations. It is especially important for health, education, wildlife, and human-services mapping, where exposure carries real consequences. You can easily ask your LLM to scramble coordinates within a certain threshold (e.g., 500 meters) to better anonymize sensitive data.

Ethics and Archiving for Trust

Ethics, whatever yours are, generally won't change case-by-case. Treating ethics as a one-time reflection invites inconsistency. When you set up your ethics as a systematic pipeline, they turn into an automated habit. The goal is not moral perfection; it's situational awareness – building enough ethical structure into your overall workflow so that obvious problems surface and can be confronted well before publication day.

An ethical prompt pipeline mirrors a similar logic you probably already use tacitly: record, check, revise, publish, and (this last one may be new) archive.

Nothing dramatic. Just deliberate. A very simple version might parallel the table below.

TABLE 5.1

Example of a Simple Ethical Prompt Pipeline

Pipeline Stage	Check	Example Prompt
Provenance	Source and license verified	"List all dataset origins and licenses."
Attribution	Credit fields complete	"Append a credits JSON."
Bias	Framing comparison	"Generate an alternative framing."
Privacy	Aggregation check	"Mask coordinates under 1 km."
Archiving	Version record saved	"Export prompt metadata to repo."

Each step is small on purpose. Together, they form a parallel and continuous pipeline running within and across all others in your workflow, catching the most common attribution and ethical failures without slowing your design to a crawl.

Once you can trace and defend every decision – data, framing, attribution, and privacy – you'll notice that deployment stops feeling so risky. You orchestrated a work of art, akin to directing a feature film, purely via intent, direction, and critique. Your cartographic workflow is no longer opaque. It's accountable; it's well documented. If people don't like your map, well… To paraphrase a director after negative reviews of *The Phantom Menace*: "I like it. If others don't, they can make their own movie."

Takeaways

- **Ethics Operates Through Everyday Design Choices**

  Provenance, attribution, framing, and privacy are not exceptional concerns; they shape your map's meaning continuously through routine decisions.

- **Provenance Is a Process**

  In prompt-driven workflows, credibility depends on tracing how data, models, prompts, and your design decisions move together over time.

- **Attribution Builds Trust Through Transparency**

  Clear disclosure of data sources, licenses, stylistic influences, and your goals strengthens legitimacy because users can see how and why your map came to exist.

- **Style Carries Authorship**

  Visual aesthetics naturally encode influence and lineage, which is why acknowledging your stylistic inspiration functions as professional transparency rather than merely a courteous accolade.

- **Your Linguistic Decisions Produce Bias**

  Adjectives, labels, and framing choices shape agent interpretation before data is even examined, making explainable design decisions essential for ethical clarity.

- **Privacy Is Best Preserved Through Restraint**

  Aggregation, masking, and tonal care protect individuals while preserving patterns, allowing maps to inform without doxxing.

- **The Most Ethical Workflows Are Structured**

  By embedding even minimal ethical checks into your prompt pipelines (sub-workflow processes), responsibility becomes built-in, i.e., habitual at every stage, rather than reactive.

Ready to Learn How to Design via Prompting?

I'm hopeful the prompt-to-map pipeline now feels familiar, or you are at least more comfortable with the concept, even if some of it remains a bit abstract. What we haven't addressed yet is the *language of form* itself, or how to prompt beautiful, interactive maps into existence.

This was intentional. Part I has been about cartographic reorientation rather than execution – about naming the kinds of decisions that shape maps in an LLM-mediated workflow without rushing to show how those decisions are encoded in prompts, pipelines, or specifications. The workflow provides a scaffold for that understanding.

In Part II, our focus shifts. We move from conceptual stages to design action, from mental models to concrete prompt strategies. The pipeline remains, but it recedes into the background as specific techniques come into view. What changes is not the sequence of decisions but the level of detail at which they are expressed.

If I've done my job here, those details will feel less intimidating and more purposeful. They should be contextualized conceptually, so as you begin your journey into prompt cartography, you realize every action is part of the cartographic process. You now understand the structure of prompt cartography and have an idea, or flow, for how it works. Now it's time to learn about how to speak mapmaking to your cast of agents in a more fluent manner.

Conceptual and Applied Exercises

Ordinary Ethics

This chapter argues that ethics live in everyday design decisions, not just dramatic edge cases. Reflect on a small, seemingly neutral mapping choice (classification, aggregation, labeling, projection, omission) that nonetheless carried ethical weight. Why was that weight easy to overlook?

Provenance Under Pressure

Think about a dataset you have used where the chain of custody was unclear or incomplete. How did that uncertainty affect your confidence in the map's claims – or did it fade into the background once the map looked convincing?

LLM Ethics Framing

Ask an LLM:

```
What ethical responsibilities do designers have when
presenting data visually?
```

Assess whether the response emphasizes compliance, harm avoidance, or communicative responsibility. Which emphasis best matches this chapter's treatment of ethics?

Disclosure as Design Choice

This chapter discusses intent declaration and generation disclosure as design decisions rather than legal formalities. Consider a map where additional transparency might have weakened rhetorical impact but strengthened trust. Would you make that tradeoff? Why or why not?

For extended tutorials, hands-on labs, prompt templates, and updated examples related to this chapter, visit www.promptcartography.com.

Part II

From Reorientation to Application

I'm sure you're champing at the bit, and perhaps you're even a little frustrated that Part I didn't teach you a single technical aspect about making an interactive web map with language models.

It's okay if you're frustrated. I did this intentionally for pedagogical reasons.

Before you start learning about prompting techniques, schema-building, or how to bend LLMs to your cartographic will, it's imperative that you understand what prompt authorship and directing, ethical judgment, and cartographic responsibility mean in this new era. Prompt cartography moves as fast as you talk. The chapters you've just worked through were meant to slow things down long enough to give you context. (You'll be playing with prompts soon enough!)

We began by revisiting what cartography has always been about. These things have not changed from print to web to prompt-based cartography. Maps are about communication and argumentation, not decoration or truth-telling (though, I'm not saying the latter is a bad thing). From there, we reframed prompting as a design language rather than a conversational trick. We treated memory as something that can be designed, not shouldn't be assumed with LLMs. And we introduce using a pipeline as a way of making your judgment visible rather than mechanized.

None of this is meant to reduce your creativity. *Quite the opposite.* You'll soon find that when decisions are named, they can be better questioned and evolved from. When your intention is clearly spelled out, it can be defended or revised. When you have structure, your deviations become meaningful

DOI: 10.1201/9781003797920-7

rather than accidental. You can't break the rules unless you know them. You can't explain why you broke the rules when you did, if you don't have a structure.

If Part I has done its job, the concept of prompts should no longer feel like experiments, wishes, or commands. They should feel like deliberate directives – ways of expressing cartographic intent clearly enough that it can survive delegation, critique, and reuse. You should also feel more comfortable with the idea that clarity does not eliminate subjectivity; it simply makes it more accountable. Clarity also reduces misinterpretation and helps you defend your choices from being misinterpreted.

If you're clamoring for applied prompt cartography, I have good news. In Part II, we turn toward practice. Color, type, layout, symbology, interaction – these are familiar topics (particularly if you read the first version of *Web Cartography*), but they will be approached through a new prompting lens. The pipeline discussed here will hopefully recede a bit into the background as we go through Part II. Not because it stops mattering, but because it will become second nature. You automatically start seeing the benefits of following it.

Up next: you are going to learn how to speak cartographic design fluently enough that your LLM agents can successfully fulfill your directorial requests.

You have a firm grasp of the prompt cartography framework. Now, let's build upon it!

6

The (Constant) Critique Pipeline

Recently, I spent a full day of vacation from work staring at an interactive web map I was working on that was, by all measurable standards, fine.

The data was correct. The symbology followed every rule I could remember. The colors were fine, even if they didn't pass all relevant contrast checks. The interactions were working as expected. Nothing crashed. Nothing flickered. No console errors other than the standard fav-icon one that you get by default. If this map had been a student submission, I would have written "Nice work" in the margin and moved on with my life.

And yet…

Something about it felt off to me. Not wrong in a way I could immediately name – just wrong in the way a joke lands flat (i.e., most of mine), or a movie scene runs five seconds too long (or two hours long in the case of some). The map wasn't misleading. It wasn't ugly. It just didn't feel like it knew what it wanted from me as a map user. It was almost like it rushed when it should have waited. It explained in detail when it should have shut up. It seemed oddly overeager, like a tour guide who refuses to let you look out the window because they are already onto the next topic of discussion.

So … I did what all experienced cartographers do in moments like this: I played with the map longer, hoping I could convince myself that it was indeed pretty awesome. (Perhaps I'll never learn.)

Eventually, I realized that the problem wasn't the data, the layout, or even the interaction logic (things you will soon learn about). The problem was that I hadn't taken any time to critique the map's purpose, rationale, and goal yet. I had built it. I had tested it. I had verified that it worked – always a great feeling! But I hadn't stepped back and asked more uncomfortable questions along the way:

- What is this map doing rhetorically?
- How do its component behaviors and datasets help promote its intent?
- And do those behaviors and datasets match what I wanted the map to communicate?

This is the moment where most cartographic processes can get messy (read: go sideways). In many interactive web cartography projects, the horrible

DOI: 10.1201/9781003797920-8

realization that your map works but isn't doing what you set out to do will send *one* – and *one* is the key word here, because before recently cartographers didn't have a team of agents to help – digging into JavaScript callbacks, cascading style sheet (CSS) selectors, or document object model (DOM) element IDs, trying to fix a design problem by tweaking particular implementation details. Personally, I would have spent considerable time adjusting pop-up information, reordering layers, rewiring event handlers, and hoping the design intent would eventually improve from the code tweaks like a solved puzzle.

Today, we get to do something different. Because when a map's behavior is specified in natural language, it can also be questioned and critiqued using language. The frustration of unexpected outcomes and the need to debug don't disappear, but they move upstream, where cartographers excel: into *map design judgments* rather than into debugging panic. Why? Because now we have expert agents that can help revise, rewrite, and redo the underlying map code in minutes, instead of hours. Iteration and debugging can be faster than ever before. (*Can* being the keyword here: you need to learn how to prompt your large language model [LLM] friends into debugging well, or it can quickly descend into chaos.)

This chapter is about *that* moment. The moment when a map technically works, but you know it still needs help. Not all is lost with the map, but it feels … unfinished, like a drywalled room that no one has gotten around to painting yet. Though I officially placed the critique pipeline toward the end of the workflow, I'm covering it first because when you use critique continually throughout all pipelines, your final critique – which still needs to occur – will be a walk in the park. We will be discussing critique throughout the upcoming chapters, so I want to cover how to critique via prompt here.

Before you learn how to build interactive web maps, you need to be able to interrogate them – using critique prompts, diagnostic lenses, and an honest studio-style conversation with your LLM tools that treat your map not as a precious artifact but media that can be improved. As will be shown, critique in this sense is not about a scolding, a checklist, or another ignored rubric. It's about working with *and trusting* your agents to ensure you don't overlook something crucial.

From "It Works!" to "But … Is It Right?"

For a long time, "iteration" in web cartography meant fixing things that were visibly broken.

A panel didn't open. A tooltip fired twice. A layer rendered underneath something it shouldn't. The map froze, stuttered, or refused to load unless the user refreshed the page and whispered a small apology to the browser. When something felt wrong, the solution usually involved opening a code editor and hunting for the one line that had betrayed you.

Design problems were often translated – sometimes forcibly – into implementation problems. If an interaction felt awkward, we adjusted the event handler. If the map felt cluttered, we tweaked the CSS. If the experience felt confusing, we added another control and hoped clarity would emerge through accumulation. Judgment happened indirectly, filtered through code.

What changed is not that maps suddenly became easier to build but that design intent moved closer to the surface. When interaction, behavior, and constraints are specified in natural language, they stop being buried inside implementation details. They become explicit. Visible. Debatable – as they happen (Figure 6.1).

This change matters, because it allows a new and more useful question to enter the pipeline. Not "Does this work?" but rather "Is this the correct behavior for this specific map, its intended audience, and this use case or moment of execution?"

Those are not questions you answer by stepping through a debugger. They are questions of rhetoric, pacing, emphasis, and restraint. Questions about how much a map explains versus how much it lets the viewer discover.

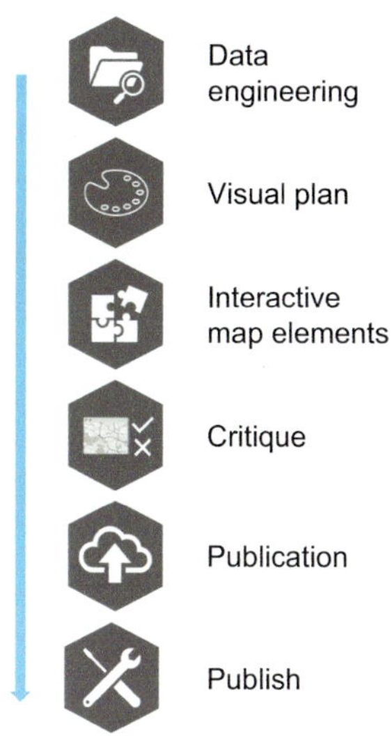

FIGURE 6.1
Traditional interactive web cartography workflows often force design critique to happen somewhat indirectly, typically through postproduction debugging. Prompt-based workflows move critique upstream, allowing intent, behavior, and evaluation to be discussed explicitly before and during implementation.

Questions about whether an interaction invites exploration – or nags the user into compliance.

Critique is the act of pausing between construction and publication to interrogate intent. It is where we ask whether the map's behavior matches its message, whether its tone matches its topic, and whether its defaults reflect a thoughtful design choice rather than an accident of tooling.

Importantly, critique does not replace technical testing. You still need to know whether your map loads quickly, whether interactions fire reliably, and whether controls behave consistently across devices. But critique happens at a different layer. It is less concerned with whether the map can do something and more concerned with whether it should.

Previously, such a distinction was easy to blur. Today, when design decisions are articulated in language, we can finally critique them directly – without first translating our instincts into bug reports.

What We Mean by "Critique" (And What We Don't)

If you've ever looked at a map and thought, "This mostly works, but something's not sitting right," congratulations – you already know how to critique.

What tends to trip people up is not the act of critique itself but the idea that critique must be formal, exhaustive, or somehow academic to be legitimate. In practice, most cartographic critique is informal, intuitive, and fast. It happens in the margins of a notebook, in a hallway conversation, or five minutes before a map goes live when someone finally says, "Wait – can we look at this one more time?"

I'm not writing this chapter to replace that instinct. I'm writing it to give that instinct a little structure so you can externalize, prompt, and reuse it without ditching your judgment in the process.

At its simplest, a critique pipeline usually moves through four modes of analysis. Note, however, that it rarely moves through these in any particular order, and typically critique is patchier than this list makes it appear.

Description: What is the map doing, literally?

What data is shown? What interactions are available? What happens by default?

Analysis: How do those choices work together?

Where does attention go? What changes when the user clicks, hovers, or filters?

Interpretation: What does the map seem to be saying?

What story, argument, or framing emerges from its behavior?

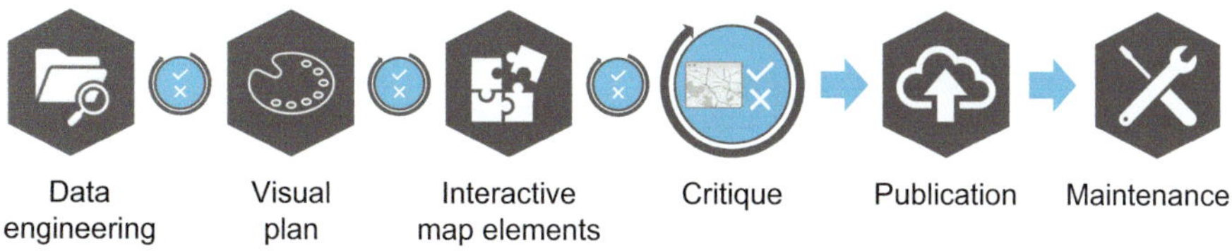

FIGURE 6.2
The four overlapping processes of cartographic critique. Experienced practitioners perform these instinctively; making each explicit allows your critique to be externalized, prompted, and reused without requiring a rigid process.

Evaluation: Do those effects support the map's purpose?

Are they appropriate, distracting, misleading, or simply unnecessary?

If this sounds familiar, it should. Most experienced cartographers cycle through these questions automatically, often without naming them. You already do this when you decide a legend is redundant, an interaction is too eager, or a label explains something the map should have shown instead.

The difference is that these questions no longer have to stay implicit. They can be turned into critique prompts – short, focused conversations that help surface what your instincts are already picking up on (Figure 6.2).

Just as importantly, critique is not about hunting for errors. A map can be factually correct and still rhetorically clumsy. It can be visually polished and still communicate the wrong emphasis. Critique is less about catching mistakes and more about aligning intention with effect.

Think of it less as assessing a map and more like listening to it talk.

Design Drift: When Maps Slowly Become Something Else

Most maps that fail don't fall apart all at once. Instead, the cartographic process slowly drifts outside of its original lane, slowly morphing into something that is … well, embarrassing. (Every real cartographer has been there at some point in their lives.)

By now, hopefully, every map you create starts with a clear purpose, a reasonable scope, and a set of design intentions that make sense (as you begin). Then, little by little, things shift. A new variable gets added. A color choice solves one problem and creates another. An interaction that felt helpful in isolation begins to compete with everything else on the screen.

FIGURE 6.3
Design drift is a diagnostic lens rather than a failure mode. Aesthetic, data, and ethical drift describe common ways maps gradually move away from their original intent during iteration even when nothing is technically broken.

Nothing breaks. But the map becomes something you didn't quite mean to make.

Design drift is useful to name because it reframes critique as diagnosis rather than blame. Drift is not a mistake. It's a by-product of iteration – especially in interactive maps, where behavior accumulates over time. The goal of critique is not to prevent drift entirely (good luck with that) but to notice when it starts pulling the map away from its original intent.

In prompt-driven cartography, drift tends to show up in three common forms (as shown in Figure 6.3).

Aesthetic Drift

Aesthetic drift happens when visual decisions gradually lose coherence.

Colors multiply. Symbol sizes creep. Emphasis shifts without a clear reason. The map may still look "designed," but it no longer looks intentional. This often occurs when aesthetic choices are made locally – solving small problems – without being revisited globally.

Critique questions in these situations are less about taste and more about consistency:

- Is there still a visual hierarchy?
- Do the strongest visual signals align with the most important information?
- Has clarity been traded for decoration, or vice versa?

Aesthetic drift rarely makes a map wrong. It just makes it tired.

Data Drift

Data drift occurs when the relationship between the data and the map's message becomes fuzzy.

This can happen when new variables are added without rethinking the story, when filters expose edge cases the map wasn't designed to handle, or when aggregation choices quietly change the meaning of what's shown. The data may still be accurate, but the implication of the map shifts.

Useful critique questions include:

- Is the map still answering the question it claims to answer?
- Are defaults steering interpretation in unintended ways?
- Would a viewer reasonably draw the same conclusions you expect them to?

Data drift is less about correctness than about relevance.

Ethical Drift

Ethical drift sounds heavy. Good news: it really isn't.

In practice, it typically shows up as small misalignments: missing context in an early prompt, unclear sourcing of your dataset, visual emphasis that unintentionally amplifies one perspective over another. Your map hasn't become unethical – it's just not being careful (enough).

It's important we frame this as drift rather than failure. Ethical critique is never about moral purity; it's about your professional credibility. A map that obscures uncertainty, hides attribution, or promotes overtly biased interpretation without acknowledging it risks losing user trust – even when the underlying data is sound.

Your questions here might include:

Is uncertainty visible where it matters?

Are sources and limitations accessible, even if they're unobtrusive?

Does the map's tone match the sensitivity of its subject?

Often, ethical drift is easiest to spot after the map works perfectly.

Design drift is *not* an indictment of your process. It's evidence that you're iterating. The value of critique is that it gives you a way to periodically stop, look at what the map has become, and decide – consciously – whether that's still where you want it to go.

In the next section, we'll look at how critique prompts turn these instincts into structured conversations – without turning you into your own compliance officer.

Critique Prompts as Conversation

One of the quiet advantages of describing map behavior in language is that it gives critique something solid to grab onto.

When interaction logic lives only in code, critique often arrives late and indirectly. You notice a problem while using the map, translate that discomfort into a technical fix, and hope the fix aligns with the original design intent. The conversation happens, but it's fragmented – half design instinct, half implementation triage.

Critique prompts change that dynamic by letting you talk to the map about its behavior, not just its parts. Instead of issuing commands ("make the tooltip faster," "reduce clutter"), critique prompts ask reflective questions that surface intent, consequences, and trade-offs. They treat the map less like a machine to be tuned and more like a draft to be discussed.

Table 6.1 highlights the rhetorical difference between hypothetical command prompts and how one would write a similar critique prompt.

TABLE 6.1

Command Prompts Direct Immediate Changes to Map Specifications

Command Prompt	Critique Prompt
Make the flow lines thicker for states with higher out-migration rates.	What assumptions does line thickness communicate about causality versus correlation? How might viewers interpret thick flows differently if we're showing loss versus opportunity? What context would help distinguish voluntary movement from structural push factors?
Use red markers for Border Patrol and ICE (US Immigration and Customs Enforcement) detentions in Minnesota and represent as a heat map showing concentration areas.	How do color choices like red affect the viewer's emotional response to detention data? What are the ethical implications of using heat maps for human detentions – does aggregation obscure individual experiences or protect privacy? What tone should this map convey given its sensitive subject matter?
Add a pie chart showing power source breakdown and place it in the upper right corner.	Does a pie chart effectively communicate the relationship between power sources and consumption patterns? What are the trade-offs between spatial accuracy and statistical summary in this dashboard? How does placing summary statistics alongside spatial data affect user interpretation of local versus system-wide patterns?
Change the proposed development zones to green and label each one with the project name.	What does green communicate about development – environmental benefit, approval, or growth? How might different stakeholder groups interpret this color choice? What information beyond project names would help residents understand scale, timeline, and community impact of these developments?

(*Continued*)

TABLE 6.1 (Continued)

Command Prompt	Critique Prompt
Use a 1-mile buffer around grocery stores to show food access areas.	What does a 1-mile radius assume about transportation access, physical mobility, and urban versus rural contexts? How might this threshold misrepresent food access for elderly residents, people with disabilities, or communities with limited public transit? What alternative distance metrics might reveal different patterns?

Note: Critique prompts interrogate intent, expose consequences, and reveal design trade-offs. The latter transform technical decisions into opportunities for cartographic judgment.

For example, consider an interactive map with hover-based tooltips, filters, and animated transitions – features introduced intentionally and described clearly in the original interaction prompt. A critique prompt might ask:

```
Review this map's interaction behavior. Where does
it demand attention from the user, and where does it
allow exploration to unfold at the user's pace? Are any
interactions too eager, too persistent, or too easy to
trigger accidentally?
```

Notice what this prompt does not do. It doesn't prescribe a fix. It doesn't assume the interaction is wrong. It simply asks whether the map's behavioral pacing matches its purpose.

You can push this further by explicitly referencing interaction state:

```
Evaluate how this map handles interaction state. After
a user filters, clicks, or hovers, does the map clearly
communicate what state it is in? Are there moments
where the map feels "stuck," overexplained, or diffi-
cult to reset mentally?
```

These are not technical questions. They're experiential ones. They ask how the map feels to use – not in a vague sense but in terms of attention, control, and cognitive load.

Critique prompts work best when they remain conversational. They invite analysis rather than demand compliance. They give you something to react to, refine, or even disagree with – much like a design review with a colleague who isn't afraid to point out that the map might be doing a little too much talking.

Importantly, this doesn't mean you abdicate judgment to the model. The goal is not to accept critique blindly but to use it as a mirror. When a critique prompt highlights an interaction as "overly assertive" or "ambiguous in state," the real question becomes whether you agree and, if so, what that says about the map's intent.

This is where critique stops being abstract and starts becoming actionable. Not because it tells you what to change but because it helps you see why a change might be worth considering.

In the next section, we'll look at what happens when critique prompts start to stabilize – when certain questions prove useful across multiple maps, datasets, and contexts, and earn a permanent place in your pipelines.

Ode to Keeper Prompts

Not every critique prompt you write deserves a permanent place in your workflow.

Some are situational. Some are useful once, and some may strike you as insightful but collapse under their own vagueness the second you try to reuse them. You'll find a lot of your favorite or most useful prompts are sadly disposable. Prompting, like cartography itself, produces a lot of drafts or one-offs that do their job and then disappear. Just like that epic map I made for the Wagner wedding … one and done.

Occasionally, though, a critique prompt will earn its keep. I call such prompts, regardless of which pipeline they wind up in, keeper prompts.

A keeper critique prompt is rarely going to be clever. Generally, the most reusable and powerful prompts are the blandest. It will simply work across different datasets, different map types, and different moments in the design process. You return to these prompts not because they're elegant but because they reliably surface things you would otherwise miss.

Consider a critique prompt as a keeper when it does these three things consistently:

Produces Stable Feedback across Contexts

If the same prompt yields wildly different or incoherent critiques for similar maps, it's not ready for reuse. A keeper prompt should adapt to content without reinventing its logic every time.

Highlights Issues You Actually Act On

If a prompt generates observations you routinely ignore, it's noise. The best critique prompts point out tensions, misalignments, or trade-offs that lead to real design decisions – not just interesting commentary.

Respects Your Role as the Decision-Maker

A keeper prompt doesn't dictate fixes. It frames problems clearly enough that you can decide whether and how to address them. It supports judgment rather than replacing it.

Many of my best prompts start life as one-offs. I've written them in response to a specific map, and I notice that they surface something insightful. I'll then try them again on the next project. And the next. Over time, a small set of prompts tends to stick. They become part of regular internal design review. Basically, these are questions I would now ask myself even if I wasn't doing prompt cartography but using a graphical user interface (GUI).

Example topics that such prompts reliably evaluate:

- Whether interaction pacing matches subject matter
- Whether visual emphasis aligns with narrative importance
- Whether defaults bias interpretation
- Whether the map explains too much – or not enough

The point is not to collect prompts for their own sake. It's to identify the ones that consistently sharpen your thinking and make critique faster, calmer, and more intentional.

In the next section, we'll look at how quantitative signals – contrast, density, performance – fit into this picture and how to use them without turning critique into a numbers game.

Metrics Without the Panic

At some point in any serious critique conversation, someone brings up numbers. (As someone who spent a good chunk of his graduate career counting random elements on maps, I have a strained relationship with numbers.)

But two types of people with lots of clout in society at large have a very strong affinity for numbers – engineers (including software engineers) and GIS people. In their mind, all of cartography can be enumerated, including design. Hence, you will be confronted with questions about contrast ratios, label counts, load times, etc., and be expected to provide numeric answers where qualitative ones suffice.

Why? Because when you record and report numbers, we can use suspiciously precise thresholds to make decisions. Thus, if you say the contrast ratio is 1.8 (a nice number on its own), you can be drilled on whether this ratio is acceptable based on ISO blah, blah, blah.

The temptation – especially in technical fields like cartography and GIS – is to let those numbers take over. Most of them feel objective, if not downright authoritative. After all, someone came up with the acceptable thresholds for contrast ratio – my assumption being it was a well-meaning bureaucrat, honestly – and this makes a prescribed number comforting. Kind of like keeping my body mass index (BMI) one below being classified as overweight.... But these types of numbers are not science; they are social constructions meant to guide positive design outcomes. Never trust a number; they are almost always arbitrary and more context specific than people realize. (For example, the average human body temperature is known to be off by a statistically significant amount; just no one has the heart to officially change it, because they know the new number also won't be exact.)

And thus, back to the point, numbers are very easy to misuse in design. Strictly following arbitrary numbers and metrics instead of one's gut and intuition can not only hurt innocent people (which is bad) but also lead to poor design decisions (which is not worse than hurting people but also, in light of the focus of this book, really bad).

Please don't let numbers (or metrics) dictate your design edits without critiquing whether they really apply to your map. In many cases, it's better to think of metrics as signals, not verdicts. They tell you where to look more closely, not what to conclude. Used well, they can support critique, but when recited without contextualizing the map, they often disguise bad critique as scientific.

Table 6.2 highlights a few examples of standards and metrics you may come across when critiquing your maps. An LLM critiquing these metrics or projections won't be giving you subpar feedback; just know these metrics or standards aren't canon either.

Contrast

Accessibility guidelines often recommend minimum contrast ratios for text and symbols. This is good advice. An LLM may even spit some of these numbers out at you as it implements its own design. But the meaningful question during critique is not whether these values successfully clear a government- or organization-endorsed numeric threshold – it's whether the map can be read comfortably in real-world conditions.

A useful way to think about contrast is experiential. It's always good to stop and ask: could a user read this map on their phone, outdoors, or while distracted?

If the answer is *no*, the number saying so has done its job. You don't need to litigate it further. If not, then...

Performance

Load times and responsiveness matter, particularly for interactive maps. A slow map doesn't just frustrate users – it undermines trust. When a map hesitates, viewers wonder whether it's broken, outdated, or unreliable.

Again, the key question is experiential: *Does the map feel responsive enough that users forget about the technology and focus on the content?* If not, be honest with yourself, because no amount of justification will save it.

TABLE 6.2

Common Technical Metrics or Specifications You May Encounter During Critique

Metric / Specification	Common Standard	What It Actually Means
Contrast ratio	4.5:1 (normal text)	How much lighter or darker text appears against its background. Designed for readability with vision impairments, but context matters. A 4.3:1 ratio on a legend might work fine if text is large and well spaced, while 5:1 might still fail if fonts are thin or compressed.
EPSG:3857	Web Mercator projection	The default system for web maps. Preserves shape well at local scales but distorts area dramatically near poles. If you're showing global area comparisons, this projection will lie to your users.
Touch target size	44 × 44 pixels minimum	The tappable area for buttons, controls, and interactive map elements. Designed so fingers don't accidentally hit the wrong control on mobile devices. But if your map's only audience is desktop GIS analysts with mice, obsessing over 44px might waste design time better spent elsewhere.
Page load time	Under three seconds	How long until your map becomes interactive and usable. Users start abandoning sites after three seconds, but this assumes average internet speeds – if your audience is field researchers on rural cellular connections, optimize for 10+ second loads or provide offline alternatives instead.

Note: These numbers provide useful starting points but are often socially constructed axioms, not scientific laws. A map's context should always override blind adherence to standardized metrics.

Quick human gut translations of testing metrics (or what the meaningless numbers about milliseconds and ratios really mean):

Contrast: Can users read this map easily without too much effort?

Density: Does the map feel clear or overcrowded with symbology and map elements?

Performance: Does the map feel broken in places and become a burden to use? Is it more work to use than it's worth, so to speak?

That's it. If the map passes those intuitive gut checks, the exact numbers matter less. If it fails them, the numbers are simply pointing you toward a design conversation you were going to have anyway.

Metrics are most valuable when they confirm intuition, not when they override it. They help justify decisions to stakeholders, document trade-offs, and catch issues early. But they should never be mistaken for taste, context, or judgment.

Critique lives in the space between what the numbers say and what the map actually feels like to use.

In the next section, we'll pull these qualitative and quantitative strands together into a practical critique pipeline – one that supports decision-making without turning your entire workflow into an audit.

Design Quality Assurance Process

By this point in the chapter, we've talked about critique as conversation, about drift as diagnosis, about metrics as signals, and about prompts that reliably surface useful tensions. The obvious next step is to gather those ideas into something you can use under real constraints – deadlines, stakeholders, fatigue, and the quiet pressure to just ship the map already.

This is where having a design quality assurance (QA) process and a set of agents really earns its keep.

The danger, of course, is that QA processes tend to metastasize. What starts as a helpful pipeline quickly hardens into a checklist of tasks that you never want to see again, much less use. That's not the goal here.

Think of the QA process less as a list of boxes to check and more as an opportunity to use the map as a human user – a prescreening with feedback, as it were. QA will help you catch misalignments between intent and instantiation before the map goes live.

Essentially, the process requires you to ask and answer three types of questions.

Intent Alignment

This is where design drift shows up first and where it's easiest to correct. Simply check whether:

- the map in its default loading state reflects your intended message,
- interaction behaviors reinforce, don't work against or distract from, the map's goals,
- iteration did not introduce features that compete with the map's purpose.

Experiential Clarity

This is not about delight for its own sake. It's about whether the map respects the viewer's attention and avoids overstimulating the user with too much information:

- Is the map readable without explanation?
- Do interactions feel predictable rather than surprising?
- Can a viewer recover easily after clicking, filtering, or exploring and wanting to get back to where they were?

Credibility and Care

We talked about ethics long ago. It's back now. Is the map professional and transparent in its intent, sources, attributions, and goals?

- Are sources, uncertainty, and limitations visible where they matter?
- Does the visual tone match the sensitivity of the subject?
- Would you feel comfortable defending the map's choices to a critical audience?
- Have you documented the process and prompts involved in creating the map?

A well-designed QA toolkit doesn't answer these questions for you. It simply makes sure you ask them consistently – especially when you're tired, rushed, or too close to the work to see it clearly.

From Critique to Continuity

You've seen how conversational critique prompts can surface misalignments in behavior, tone, and emphasis. You've seen how metrics can support judgment without replacing it. And you've seen how a lightweight QA toolkit can help you pause long enough to notice when a map has become something you didn't intend.

What's important is that none of these elements need to exist in isolation.

In practice, critique works best when it becomes continuous – woven into the same workflow that generates the map in the first place. A critique prompt isn't something you save for the end, when everything feels locked in. It's something you return to throughout iteration, asking the same questions as the map evolves and watching how the answers change.

This is one of the subtle shifts introduced by prompt-based cartography. When design intent, interaction behavior, and evaluation criteria all live in language, they can coexist in the same space. You don't have to translate intuition into code just to test it. You can reflect, revise, and refine while the map is still fluid.

Critique, in this sense, becomes less about correction and more about continuity – making sure the map remains aligned with its purpose as it grows.

From Conversations to Pipelines

Up to this point, and in many of the upcoming applied chapters, you will often read about prompts individually: prompts that specify behavior, prompts that critique behavior, prompts that surface drift, prompts that flag issues worth addressing. That separation is useful for learning, but it's not how real projects unfold.

In practice, these prompts and processes begin to naturally and structurally stack into a ***series of pipelines***:

- Data engineering prompts define scope and constraints.
- Visual planning prompts set the tone and message.
- Map element and interactivity prompts shape user behavior.
- Critique prompts evaluate drift and alignment.

Together, as already reviewed, these pipelines combine to form the prompt cartography workflow.

The critique pipeline outlined in this chapter doesn't represent a rigid sequence of steps or something etched in stone. However, it establishes but it's a practical feedback loop. A definite "high recommend" for every project. It supports experimentation, revision, and design judgment throughout the workflow and allows for a comprehensive review before you hit publish.

Takeaways

- **Just Because It Works Doesn't Mean It's Right**

  A map can function correctly and still fail rhetorically, emotionally, ethically … completely. Functionality is milestone, not the end goal of your communication.

- **Critique Happens Upstream in Prompt Cartography**
  When behavior is specified in language, it can be questioned before it hardens into implementation.
- **Critique Is about Alignment**
  The goal is not to find mistakes but to evaluate whether behavior matches intent.
- **Discomfort Is a Legitimate Signal**
  A vague sense that "something's off" often precedes articulate critique and deserves attention.
- **Design Judgment Cannot Be Debugged**
  Questions of pacing, emphasis, and restraint are rhetorical problems, not technical ones.
- **Critique Follows a Recognizable Cycle**
  Description, analysis, interpretation, and evaluation are not steps to follow rigidly but lenses that surface meaning.
- **Design Drift Is Diagnostic**
  Aesthetic, data, and ethical drift emerge through iteration and must be noticed, not punished.
- **Defaults Deserve Scrutiny**
  What a map does by default often communicates more forcefully than what it allows optionally.
- **Critique Prompts Enable Conversation**
  Reflective prompts surface intent and trade-offs without prescribing fixes or outsourcing judgment.
- **Good Critique Prompts Earn Reuse**
  Keeper prompts are valued not for cleverness but for reliably revealing issues worth acting on.
- **Metrics Are Signals, Not Verdicts**
  Numbers point toward areas of concern; they do not replace experiential judgment.
- **Experience Outweighs Thresholds**
  Readability, clarity, and responsiveness are felt before they are measured.
- **The QA Toolkit Is a Structured Pause**
  Effective review frameworks ensure key questions are asked consistently, especially under pressure.
- **Critique Supports Continuity**
  Ongoing critique keeps maps aligned with purpose as they evolve, rather than correcting them after drift sets in.

- **Judgment Remains the Cartographer's Job**
  Prompt cartography does not eliminate authorship. It makes it visible.

Conceptual and Applied Exercises

Critique Versus Confirmation

Think of a map you initially believed "worked" because it looked polished or impressive. At what point – if any – did critique reveal a misalignment between intent and outcome? What specific signal made you realize the map needed reconsideration rather than refinement?

Drift Detection

This chapter describes aesthetic, data, and ethical drift as slow, cumulative processes. Reflect on a project (mapping or otherwise) where small, reasonable decisions gradually changed the outcome in ways you did not anticipate. Which type of drift do you think was most responsible, and why?

LLM Critique Framing

Ask an LLM:

```
What questions should a designer ask to evaluate whether
a map is communicating responsibly?
```

Evaluate whether its response treats critique as a checklist, a judgment process, or a conversation. Which framing best aligns with the role of the prompt cartographer described in this chapter?

Critique as Continuity

Imagine returning to a map six months after publication. Identify one critique question you would want to ask at that point, not to improve aesthetics, but to assess whether the map has remained faithful to its original intent. Explain why that question matters over time.

For extended tutorials, hands-on labs, prompt templates, and updated examples related to this chapter, visit www.promptcartography.com.

7

The Data Engineering Pipeline

You open a fresh prompt window and type:

```
Map Burmese python sightings across southern Florida.
```

Within seconds, the model replies with a neat summary and ten download links. Some are from government portals, some from research blogs, and one, alarmingly, from a social media scraper. Each dataset claims to show "where pythons are," yet no two datasets agree. One stops in 2015. Another covers only the Everglades' western half. A few datasets record citizen-reported "snakes," many of which, you imagine, may have just as well been a neighbor's garden hose spotted after a few drinks.

Which dataset tells the truth? More precisely – which version tells *your* truth, the one you intend to communicate to others?

Welcome to the gray zone between inspiration and information – the place where every map begins. In Chapter 4, we discussed framing what your map should be about via the Data Brief. Chapter 6 assumes you've made that choice already and now asks: given your intent, what data will help you communicate what you want; do such datasets already exist or do you need to construct it yourself, using a team of large language model (LLM) agents?

I've thought long and hard about where to place this chapter. Often, it's tempting to start teaching map design with visuals. However, design without substance is vapor – or worse, the cartographic equivalent of reality television. To design by language, you must also learn to describe the data of a map so that an LLM assistant can locate, critique, and manipulate it on your behalf. Your spatial datasets are your substance.

Unlike many web mapping workflows, prompt cartography makes describing data an explicit and central part of the process. Every time you type a prompt about data – whether you realize it or not – you're making design decisions. What counts as an observation or feature? What date ranges in the sprawling dataset are of interest to you? What geographic extent feels honest? Behind every dataset sits a rhetorical decision about what will be interpreted as fact by most map readers, and your prompt wording either reveals or obscures what can be interpreted.

Long before language models entered the cartographic workflow, data selection was already a design decision – just one we tended to pretend was technical. In Web Cartography, audience, purpose, and message were treated as the

DOI: 10.1201/9781003797920-9

first-order constraints on map form. What we did not yet have was a way to express those constraints directly to the systems that retrieve data for us.

In short, before you can prompt for layout, color, or typography, you need to learn to prompt for data that most closely represents your version of reality (i.e., the cartographer's personal belief systems of which every single one is inherently biased).

How to Use LLMs to Search for Datasets

Every cartographer begins with an idea: it would be cool to show people *where and how* {X} happened *in or to* {Y}. Yet turning ideas into usable datasets requires a second layer of translation – one that connects your imagination to the reality of data collection infrastructures.

Up until recently, finding a "good enough" dataset to match one's idea often meant scouring data portals using viable keywords and perhaps molding the datasets into an ideal format using a GIS, spreadsheet application, or scripting. Language models have added a new means of operation. You can now teach your assistant to search for particular kinds of information, explaining why you are seeking it and how you would present its ideal structure. The more fluently you can phrase such expectations, the more faithfully your model will seek, find, and engineer your datasets to match your mapping idea.

Consider the following two prompts:

- `Find python data for Florida.`
- `Locate open spatial datasets describing confirmed Burmese python observations in southern Florida at any time between 2010 to 2025. Return the following for each, as much as possible, in tabular format: dataset name; source; URL to dataset or page; API key required (Yes/No); dataset geometry type; qualitative summary of the dataset (100 words or less); temporal coverage; and a 1-5 Likert-style ranking of the reliability of each dataset in your estimation.`

The first prompt resembles a standard keyword web search – i.e., search for "python" and "Florida". It's very broad in scope. The second prompt behaves similarly to how a designer might explain a map layout to an LLM; it aligns form with intent. The prompt tells the model where to look (south Florida and open data sites) and what to ignore (anything not intersecting with a certain date range and not about "Burmese pythons"), as well as what specific information to provide for each result, including a structured response format. The prompt is clear, detailed, and gives the LLM a structure to use when returning the information.

Even minor adjustments to the first prompt – adding an intended audience or preferred data format – would help shift results from chaos to coherence. Through practice, you can learn to compose dataset queries as deliberately as you compose symbology: every adjective a parameter, every clause a constraint. And once you do, the entire world of spatial data becomes a giant sandbox – that you mold to your desire using your dependable LLM assistants.

Semantic Search (A Primer)

Traditional database querying depends on tools like SQL and keywords to find what you're looking for. LLM-based queries are done using semantic search. Semantic search uses embeddings – high-dimensional numerical representations of meaning (i.e., lots of numbers that are meaningless to humans) to compare your request with dataset descriptions. Rather than matching words as an SQL might, semantic search allows it to search based on intent, context, and matching concepts.

EXAMPLE PROMPT

```
Find datasets describing bird-strike incidents near
airports (2010-2023) that include coordinates, air-
craft type, and time of day.
```

A well-phrased dataset search prompt is both an analytical act and a creative one. It fuses aesthetic sensibility with technical clarity – the traditional cartographer's balance of left and right brain. When you learn to frame your searches in this way, not only do you stop treating data as something to find somewhere out there already. You can begin treating dataset collection and engineering as something you design. As will be shown, you empower yourself to create new, more robust datasets from existing ones.

Semantic Search for Datasets

The phrase semantic search is a bit intimidating. Though an accurate name for what it does, it's not particularly natural sounding. At its core, semantic search is just a very fancy way of saying that the machine interprets what you mean by searching for matching datasets based on linguistic probabilities. (Isn't that a mouthful?)

As Figure 7.1 visualizes, traditional database search tools – think old-school data portals from way back in 2024 – compared words (and URLs when searching the web). If your dataset title didn't contain "python," your "python" search was likely to fail. A semantic model, however, encodes both your query and each dataset description as mathematical vectors that live together in a high-dimensional space. Datasets whose meanings orbit closest to your intent rise to the top.

For semantic search to work well, you need to describe your intent and needs with clarity. This typically includes mentioning your preferred dataset geometries (point, line, or polygon), which temporal range you are interested in (i.e., dates and times), and what variables matter most to you (i.e., are there fields your datasets must have). Each phrase in your prompt becomes part of the prompt's "semantic fingerprint."

Concision is not always a blessing in prompt cartography. Creative license and informative, contextual wordplay can be a powerful tool. (I imagine my creative writing professor friend, Bradley, pumping his fist with glee right now! Down with neo-positivists!)

No matter how well thought-out and phrased your prompts are, you will soon discover that, akin to humans, semantic searches are imperfect. Like know-it-all students, even the cleverest of them are wont to make confident mistakes. In many ways, it parallels teenage behavior – often a bit overconfident in their own capabilities and perhaps too celebratory over relatively minor accomplishments.

For example, even with the better second prompt above regarding pythons, the LLM may present you with a dataset of water moccasin snake attacks, or perhaps, another unrelated, but perhaps often uttered in the same breath, invasive species. Your remedy is to continually iterate.

Think of each refinement of your search as you might a design critique: you provide prompts, the model proposes, you review, iterate on your original prompt, and repeat.

1. Narrow your search to datasets from government or academic sources.
2. Exclude all datasets without accompanying spatial geometry or spatial fields, including place names or coordinates.
3. Return only datasets that include at least the month and year of each observation.

Each new instruction rebalances your search's composition, offers the LLM new parameters, instructions, and guardrails. It's very similar to the dance you play when adjusting your map elements and layout until the visual

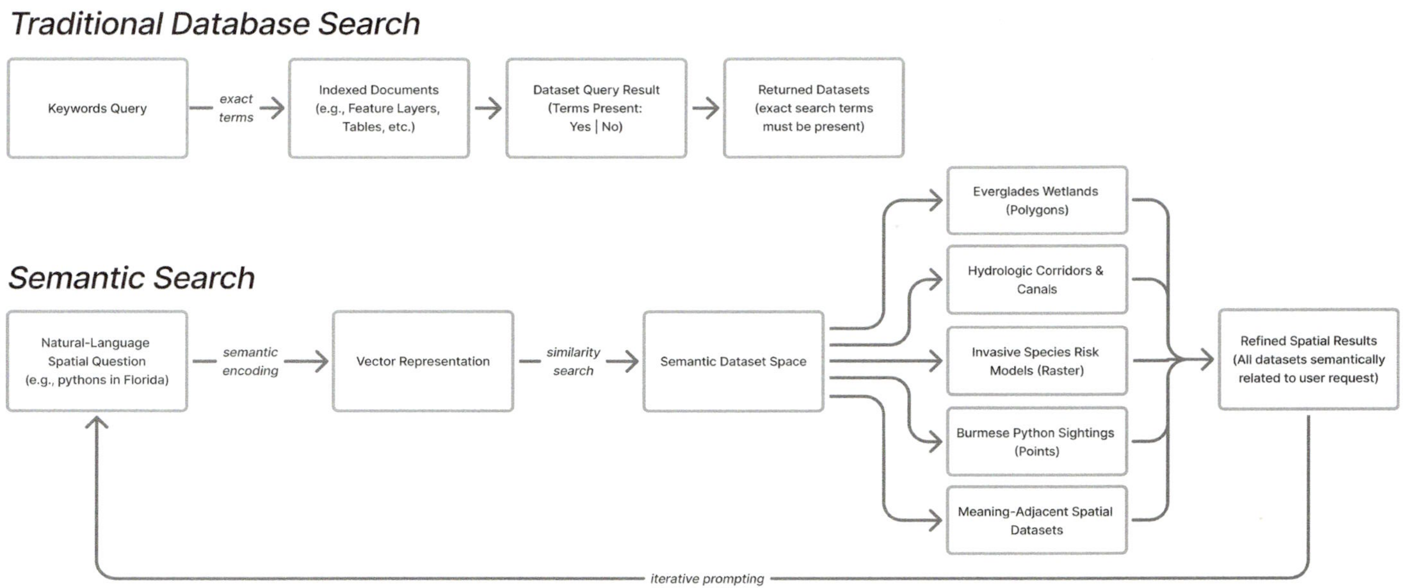

FIGURE 7.1
Semantic search translates a natural-language data request into vector representations, retrieves meaning-adjacent datasets, and refines results through iterative prompting rather than keyword matching. This is intelligent search compared to traditional database searches, which query databases using specific keywords, potentially missing many key and relevant datasets.

hierarchy and balance feel just right; except instead of iterating via clicks in a GUI, you describe the data you need in plain language.

PROMPT TIP: TARGET ATTRIBUTES, AVOID VAGUE KEYWORDS

When searching for data, describe what types of fields you need in your dataset, not just the subject of the dataset.

Prompt

Find CSV, GeoJSON, or Shapefile datasets containing Burmese python captures in southern Florida with fields containing information on month, year, latitude, longitude, and (ideally) capture method.

Answer

There are good, downloadable python-capture / sighting datasets for south Florida, but the truly detailed capture-method fields are mostly in USGS / agency CSVs, while the more map-friendly stuff (EDDMapS, AGOL) often just has "observation" records without rich method codes. Below is an ordered list of potential options with links to the datasets.

Why It Matters

Models respond better to more specific and structured language regarding datasets, because these attributes imply intent. You're effectively telling the model what columns your visualization requires, not just what nouns interest you. Notice how the response actually addresses what it perceives as dataset issues broadly for your visualization. It provides you feedback to help you better iterate and refine your search, but only because you have already opened up about your underlying dataset requirements.

Assuming it's true that the devil is in the details, as a prompt cartographer you're going to have to dive into your dataset fields, because that's where the details fomenting most devils are found. Don't shy away from letting the LLM know which fields you require.

Mindful prompting is effective prompting. (Perhaps not a horrible idea for another book – a self-help guide about being a mindful prompt cartographer?) As you begin with prompt cartography, take note of how certain adjectives affect dataset retrieval. Adjectives like "comprehensive," "verified," "trustworthy," or "public-friendly" will almost always bias the LLM search toward institutional datasets. Terms like "real-time," "public-reported," or "unofficial observations" will likely retrieve noisier, but in certain circumstances, more useful and timely datasets. Prompt cartography is not only an interface change; it adds new tools and techniques to help you get the job done. With practice, you will find linguistic steering becomes a key skill in your cartographic toolkit.

Semantic search transforms you from a passive downloader into an information designer. It lets you curate tone, credibility, and field requirements through syntax alone – a power that, once felt, permanently changes how you approach data searches and engineering.

Evaluating and Assessing Dataset Integrity for Your Map's Purpose

Once a handful of promising datasets emerge, the next task is to decide which one, or ones, best fits the map you want to make. Fit, in this context, has two dimensions that must be balanced: technical precision and rhetorical purpose.

As already mentioned, a dataset can be precise but also useless. Imagine, for example, a CSV with 10,000 exact coordinates of python captures – but no dates. It's detailed but silent on when events occurred, rendering it static and misleading for a story about spread. Conversely, a yearly aggregate might be temporally rich but spatially vague. Everglades National Park spans three counties. What would an aggregated python capture count by county over time show? Your skill, and craft mastery as a cartographer, lies in ensuring that there is a match between the character of your dataset and your narrative intent.

DATA HAS A VISUAL ACCENT

Just as typography carries an accent – formal, casual, urgent – datasets carry one too. Sparse, point-based observations speak differently than smoothed surfaces or aggregates. Prompt cartography makes that accent selectable, or even creatable, before the map begins being drafted.

Think of dataset selection and refinement as culling your options until you hone in on a set of features and accompanying fields with the right amount of potential energy – i.e., a dataset that will result in a complementary and appropriate level of graphicacy for your desired map design. In this example, perhaps we receive two types of datasets to select from:

A Fragmented Dataset

Point features scattered unevenly through space and time. Using this will inevitably create visual tension for the reader, akin to an asymmetric composition. The human desire to "fill things in" or understand something that doesn't appear quite right is a stellar way to increase interest in something – including your map. Why are some areas lacking points? Why aren't the python sightings or captures continuous over the area of interest even though they should be? Asymmetry – in graphic layout, haircuts, and data distributions – always stands out and garners attention over symmetrical options.

A Fluid, Continuous Dataset

Contours, interpolated raster datasets, or aggregated reporting across all areas in a geographic extent. This type of dataset evokes stability, rhythm, and accuracy. It may not draw as much immediate attention, but if trust is your aim, many humans are more likely to believe things that appear consistent. Consistency helps disarm suspicion.

Your map users will sense these qualities subconsciously as soon as they glance at your map, just as they primordially believe that dark means more of something and default to large swaths of blue on a map representing water (in most cultures). Choosing the most appropriate dataset to use in map design is as much aesthetic as it is analytical. And it is aesthetic that draws people's attention to your visualization, luring them in so that they consume your information. Analytics sans aesthetic… specialist audiences may prefer them, but most people will never pay them any mind. As the map designer, you are responsible for finding the right datasets to help achieve the desired aesthetic for your intended audience. (We will touch on this topic in more detail when we discuss thematic visualization.)

To systematize that intuition, it helps to think through a few archetypes. Returning to the Burmese python example, if you're designing for a museum exhibit, aggregated municipal data may communicate the big picture well – a calm, symmetrical style overview. But if you're creating an investigative visualization on ecological invasion, point data showing edge creep into suburbs will carry narrative tension and urgency. Both representations will be true; both will be factual; each will tell a different story.

CASE EXAMPLE: BURMESE PYTHONS IN THE EVERGLADES

Dataset / Context

- Everglades National Park capture records, 2010–2024

Prompt

```
Summarize confirmed python sightings by year and habitat type to illustrate spread into suburban wetlands.
```

Output

Aggregation by habitat reveals a slower, more credible expansion narrative than raw point scatterplots, which exaggerate sampling and reporting volatility.

Data precision without intent results in completely irrelevant noise reduction. On the other hand, *purpose without precision* is wishful thinking at best (i.e., naïve cartography), conspiracy at worst (i.e., viral cartography?). The sweet spot you typically want to aim for in your maps, like everything else in life, is somewhere in the middle. Akin to finding the right balance in a map's graphic design, you want to engineer your datasets to a point where data fidelity and communicative clarity meet.

To best achieve this, with the caveat that this may change as LLMs and AI systems evolve, I recommend prompting dataset searches by pairing technical clauses with narrative ones. Generally, this will take a guise like the prompt below:

```
Find datasets mapping {phenomenon} {details} that best illustrate {map narrative} for {intended audience}.
```

By habitually entwining the technical with the narrative, you ensure that every retrieved dataset already aligns with your visual voice long before the first symbol hits the map.

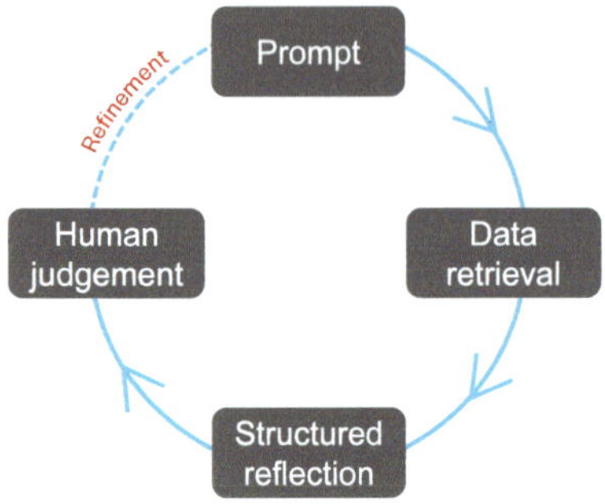

FIGURE 7.2
Dataset evaluation in prompt cartography is iterative: retrieval, structured reflection, human judgment, and prompt refinement operate as a continuous loop.

Evaluating Dataset Quality and Ethics

You've located what appears to be the perfect dataset. It has the right fields, spatial extent, and timeframe. But before you let the model paint the map, it's time to pause and channel your inner skeptic.

Prompt-driven discovery demands a new form of data literacy – not just knowing how to inspect attributes, but how to prompt for credibility. The model can help, but only if you teach it the right questions to ask through iteration (Figure 7.2).

A good data evaluation prompt reads like a miniature peer-review checklist:

```
Summarize source, license, and collection method for
this dataset. Identify any missing metadata and esti-
mate confidence from 1 (low) to 5 (high).
```

The model will never be able to verify accuracy, but it can expose the boundaries of trust to help you make a final decision. When the LLM answers, you're actually reading a reflection of your own expectations – what you've explicitly defined to the LLM versus what you've left unsaid. (You will quickly learn that the sin of omission can be a powerful tool in the prompt cartographer's tool kit, but only when wielded with caution!)

Here are some dimensions to consider and perhaps prompt for once you have culled your datasets down to the most relevant and interesting ones:

Recency

Is the dataset current enough, or does it cover the relevant time period to support your narrative?

Completeness

Are there null values or missing geographies? If so, what impact will this have on the efficacy of what you want to visualize?

Authority

Who collected the data, under what mandate, and was it collected for ethical purposes? (In other words: don't *dox* people unnecessarily!)

Granularity

What is the smallest meaningful unit of observation? Is that appropriate for your visualization and does it adequately represent the phenomenon being represented?

Bias and Coverage

Are any stakeholders or places omitted from the map unnecessarily? Are any included to obfuscate the map's clarity?

You don't have to do a thorough five-point inspection on every dataset. (You don't have to mimic my local auto mechanic trying to upsell me on every phantom issue that exists.) Soon enough, with repetition, this process will become second nature; particularly as simple patterns emerge. (And they will.) If your gut says a dataset fuels your map purpose like good lighting might an influencer's social media clip, it's probably because it passes the above criteria. If your gut is wary, you should pause and critique the dataset more thoroughly.

Map purpose isn't everything, though. Remember, not everyone shares the same ethics you do; nor, have they thought about the ethics of their dataset before sharing it with the world – after all, you're now a trained prompt cartographer! Since dataset selection is iterative, never shy away from asking the LLM to reflect on the merits and potential ethical use concerns of any given dataset. You can use your LLM to act as an ethical spot-checking assistant, though be careful when using it to check legality. If you ask it to reflect, it will give you an honest overview.

Prompt

```
Do any of these returned datasets underrepresent any
tribal lands or protected areas when summarizing python
sightings.
```

Hypothetical Answer

```
Short version: yes - every dataset we talked about is
spatially biased, and several almost certainly underrep-
resent tribal lands and big chunks of protected areas,
especially remote interior marsh/wetland habitat. They
each do it in different ways. I'll walk through them one
by one and then give you a "how much can I trust this
for summaries by land-ownership class?" verdict.
```

Evaluating quality is never an afterthought. It's the moral and aesthetic conscience of prompt cartography. A map drawn from shoddy data is like constructing a building with out-of-square framing: even the most harmonious color palette won't disguise, much less overcome, structural flaws in your datasets.

HISTORICAL NOTE: METADATA STANDARDS FROM FGDC TO THE AI ERA

In the mid-1990s, the Federal Geographic Data Committee (FGDC) formalized metadata standards that every U.S. federal dataset had to follow – title, abstract, bounding coordinates, contact, keywords, etc. Those fields made data findable long before machine learning existed.

Today, LLMs read those same tags semantically, using them as the seeds for meaning. Yet many modern "open" datasets have forgotten this discipline. A quick prompt like "List missing metadata fields and propose defaults" can revive this old rigor. In that sense, the AI era hasn't replaced the ethics of metadata – it's helping us enhance them. Indubitably, metadata will underpin natural language GIS and prompt cartography in the future. (Of particular import are field descriptions and definitions.) Detailed metadata will allow LLM agents to semantically search datasets not only via context and subject but also through their actual field values. LLM tools are already being developed where I work to enhance and create the metadata to fuel future searches via prompts. All Natural-Language GIS (NLGIS) will follow suit, I imagine.

Cleaning and Engineering Spatial Datasets via LLM

It's happened to everyone. You finally secure the dataset you've been clamoring for only to realize that it isn't in the geometry you need, the format you require, or filtered in the way you wish. Your work is only half done. (Sigh.)

In the past, you were in luck if you knew enough python to write your own data cleaning scripts, or you had access to a desktop GIS application that can help you clean and merge your datasets via GUI. (Still two useful situations to find yourself in, honestly.) However, with prompt cartography you'll find it's often much easier to just direct a generic LLM or an agent you have created for data engineering tasks to do this for you (Figure 7.3).

The trick is to be clear as to what you expect and give it logical steps. You don't have to use specific parameters but provide enough detail so your LLM doesn't have to guess what you want or need. Instead of writing over 250 lines of python to clean your data or trying to remember where all of the

Prompt-Driven Data Cleaning Request

Declarative intent, negotiated execution

Prompt Composition

I've attached a CSV file of arrests, a precincts.geojson showing different police precincts in the city, and a CSV that acts as a key for the crime codes in the arrests CSV.

Please complete the tasks below. Once done, report any anomalies, issues, or bugs you encountered during the process.

In a separate markdown document, reply with your confidence on each completed task using a scale of 1-5 (5 = high confidence). Provide a rationale for each score.

Please do the following with the CSV datasets:

- Parse the arrest data to show July through September arrests only.
- Remove all duplicate records.
- Standardize arrest categories using the key so the crime or arrest type can be represented cleanly and intelligbly in info windows and legends on map.
- Export cleaned up CSV as a geojson file named "jul_sep_arrests". Use lat and lon fields for georeferencing.

Please do the following to the precincts.geojson:

- Aggregate the amount of arrests in each precinct using jul_sep_precincts.geojson.
- Add this field to the existing precincts.geojson with the title 2024_Q3_arrests

Input datasets: arrests_2024.csv precincts.geojson arrest_codes.csv

LLM Output (from 16-line prompt)

- jul_sep_arrests.geojson
- jul_sep_precincts.geojson
- confidence_level.md

Python-Driven Data Cleaning Pipeline

Operationally explicit, cognitively expensive

Script Function and Coding Requirements

Python Function	Description	Code Lines
detect_column	Find best-match column name	10
coerce_datetime	Parse datetime safely	3
read_csv_safely	CSV read with encoding fallback	12
report_anomalies	Print anomaly list	10
load_inputs	Load 2 CSVs + precincts GeoJSON	18
filter_arrests_to_jul_sep	Filter to July–September 2024	24
remove_duplicates	Drop duplicate records	10
stndrdize_cats_w_key	Merge crime key into arrests	55
arrests_to_pts_geojson	Create point GeoJSON from lat/lon	35
agg_arrests_to_precincts	Spatial join + counts per precinct	35
write_precincts_geojson	Write updated precincts GeoJSON	6
main	Orchestrate the workflow + reporting	55

Input datasets: arrests_2024.csv precincts.geojson arrest_codes.csv

Python Output (from 273-line Python script)

- jul_sep_arrests.geojson
- jul_sep_precincts.geojson

FIGURE 7.3

Delegation does not eliminate expertise; it relocates it. The left panel: 16-line prompt requesting data cleaning. An LLM interprets intent and executes. The right panel shows the equivalent Python script written by a human: 273 lines of time-consuming code requiring procedural fluency and debugging expertise.

tools you need are in a desktop GUI, you can now just write a prompt like the one in Figure 7.3. While the LLM solves your request, you can go do something else – anything from beginning work on the next step in your map design with a different LLM agent to playing video games.

Describing What You Need from Your Datasets

Once you've found and chosen your dataset, it's time to systematically articulate what you have and want to do with it. Articulation is key – like everything else with an LLM, you need to be clear so the model can reason about it without hallucination. It may help act as though you're an art director sending a data brief to a design team. You don't just want to descend to listing or discussing dataset fields; never forget you need to continually define and refer to your overarching data-cleaning or narrative purpose.

One way to do this is with a JSON-style schema that can be uploaded as part of the RAG documentation.

A simple prompt might read:

```
Summarize this dataset in JSON form, including title,
geometry type, spatial extent, attributes, temporal
coverage, intended use, and quality notes.
```

This data-as-dialogue approach transforms a dataset into shared understanding. Once you have this description, you can chain prompts like:

```
Based on this dataset brief, recommend an appropriate
map projection and symbolization strategy.
```

Systematically organizing your datasets before proceeding with map design will significantly help prevent misinterpretations later.

In the first edition of this series, Web Cartography, I noted that "a map's purpose determines its grammar." *Your JSON-, YAML-, or thoroughly described data brief is that grammar*. It provides syntax (i.e., canon) for the LLM to bind your narrative intent to the dataset's structure. You are designing how your LLM will interpret, understand, and manipulate your dataset to meet your goals.

Ethical and Legal Considerations

Every dataset carries social baggage. Even wildlife records – like our pythons – can risk unintended harm if misused. The more specific your prompt, the more responsibly your assistant can act on your behalf.

Don't shy away from adding ethical clauses to your queries. Remember, the LLM is waiting for guidance from you. You are responsible for your assistant's actions. It can only act on what you share with it. In this case, you must avoid omitting crucial guardrails or considerations. (Remember when I said deciding when to leave things out and include them is tricky? Here is one such example.) Always guide the LLM and tell it your constraints. Below are two hypothetical examples:

```
Ensure this dataset includes licensing details and ensure that you credit the original data collectors in either the dataset metadata or in an accompanying description.
```

```
Ensure that coordinates for all sighted **and** endangered animals are always removed from any display output. The end user must never know the coordinates of any endangered animal on the map.
```

Such short lines of text will change your assistant's behavior and decision making in profound ways. Even simple, direct prompts will prod your LLM assistant to help you produce a "responsible" map. Who knows, they may save you from a PETA, or similar, lawsuit as well.

Minority Perspectives and Representation

Ethics also extends to representation: whose voices appear and whose remain invisible. If you're prompting for demographic or cultural data, you might include:

```
Suggest appropriate alternative labeling based on local indigenous naming conventions.
```

The little nagging things that you used to think about when designing a map and think "that would be so cool" if only the dataset allowed it are now often possible with your full team of assistants. The trick is just articulating what you want, clearly. Natural language transforms ethics from an easy-to-forget checklist to a constraint. Or even more rewardingly, an unachievable wish list to something your expert agents can pull off for you.

Map Ethics Always Starts with Your Dataset

Often, we think of ethical design as entirely visual: projection choice, color selection, symbol connotation, etc. In prompt-driven mapping, there's no excuse for not beginning to ensure ethical integrity now, at the data stage.

Depending on the Natural Language GIS (NLGIS) – or more straight-up LLM – you use, every time you use natural language to specify data provenance, anonymity, or attribution, you are helping to shape the model's concept of good cartographic design. Over time, these micro-decisions will accumulate into a new professional ethos: cartographic ethics via prompting. By including ethics early, and practicing ethics as you design maps via prompting, you're literally contributing to this new ethos via your day job (or hobby).

The Data Engineering Pipeline

By this point, you are close to describing datasets as fluently as you do visualization methods. (Or you are at least getting closer.) Now, it's time to weave that fluency into a data engineering pipeline.

Prompt-driven map design rarely happens in a single fluid process. It's a pipeline of innumerable processes – i.e., hand-offs between LLM agent conversations, tasks, and exchanges, each performing a requested role. The most efficient prompt cartographers develop their LLM assistants or agents as individual specialists, not monolithic tools.

Here's a simple, hypothetical orchestration pattern for a team of dataset-focused assistants that build a streamlined pipeline to help you find, clean, and improve datasets:

Data Finder Agent

Receives a high-level brief ("Find open datasets mapping Burmese python sightings in Florida, 2010–2024"). Returns three candidates, each with a short metadata summary and confidence score.

Validator Agent

Checks the top candidate for coordinate consistency, missing fields, and license type. Produces a one-paragraph audit.

Visualizer Agent

Generates a draft map specification ("symbolize by year, aggregate by county, neutral basemap") based on the existing dataset.

Critic Agent

Reviews the map for narrative bias or rhetorical imbalance ("emphasizes growth; consider adding confidence shading").

Publisher Agent

Packages the final spec into a deployable format making use of your preferred APIs and tools.

Each of the above agents represents separate conversations – preferably with separate specialists who are less likely to be self-biased about their own work. These will (very) rarely be single commands. The clearer your language and more direct your prompts in each conversation, the smoother the mapmaking process will proceed from agent to agent or conversation to conversation. Figure 7.4 visualizes this process as a pipeline.

If it helps, and metaphors are typically helpful mnemonic devices in these circumstances, think of yourself as the conductor of a small orchestra. Each musician (agent or LLM conversation) can play its own instrument beautifully solo, but only you know the composition holistically and how to guide them together to produce a stellar score (the pipeline). Your individual prompts are like baton strokes: gesture toward rhythm (i.e., dataset search); raise your eyebrow toward symbolization (i.e., coming up in future chapters). When your orchestra aligns – data tuned, layout balanced, aesthetic refined, tone honest – the result is not just automation, but authorship.

Looking Ahead

You now know how to converse fluently about data – how to specify what a map needs before it ever meets the HTML canvas. The next step is spatial: arranging that data within a visual frame.

In Chapter 8, we'll move from what to map to how to develop a map aesthetic and frame your communication. If this chapter set up your orchestra's repertoire, Chapter 8 will cue the downbeat.

Takeaways

- **Data Selection Is a Design Decision**

  Long before symbols are drawn, datasets define what a map can honestly say. What you include – or exclude – shapes interpretation as much as visual form.

- **Semantic Search Turns Discovery into Design**

  Prompt-based data search is not passive retrieval. By specifying attributes, scope, and intent, cartographers actively design the datasets they work with.

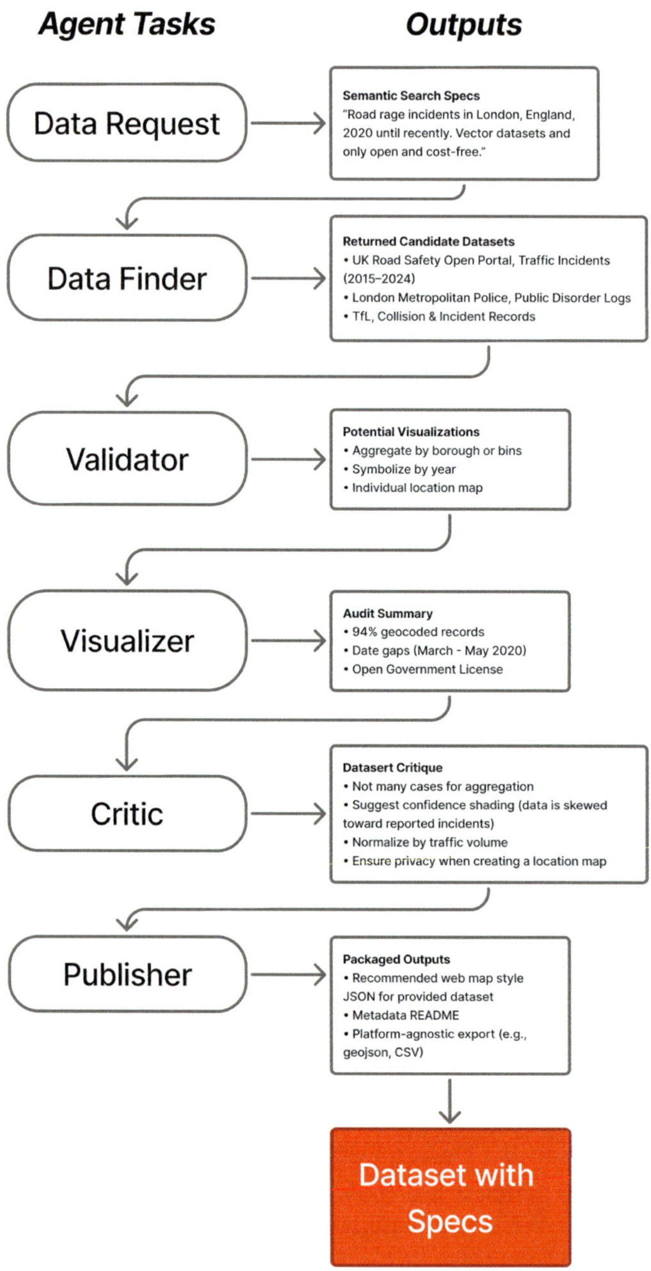

FIGURE 7.4
A data engineering pipeline will delegate specialized tasks to sequential LLM prompts, each performing a distinct role in dataset preparation (and potentially interacting with different LLM assistants or agents). This orchestration transforms data discovery from keyword searching into directed design. Each agent operates as a specialist guided by the cartographer's overarching intent.

- **Precision Beats Brevity**

  In data prompting, clarity and structure outperform concision. Attributes, constraints, and context guide LLMs toward relevant, auditable results.

- **Datasets Carry Tone and Mood**

  Fragmented, sparse, or continuous datasets evoke different perceptual and emotional responses. Choosing among them is both an analytical and aesthetic decision.

- **Fit Matters More Than Fidelity Alone**

  Highly precise data can still be inappropriate for a given narrative. Effective maps balance technical accuracy with communicative purpose.

- **Ethics Belong in the Prompt**

  Attribution, bias, consent, and harm reduction are not post-processing concerns. They can – and should – be articulated directly in data prompts.

- **Iteration Clarifies Intent**

  Each refinement of a data prompt exposes assumptions, priorities, and trade-offs. Iteration is how both cartographer and system converge on meaning.

- **Data Briefs Stabilize Collaboration**

  Explicit dataset descriptions create shared understanding across agents, tools, and future revisions, reducing hallucination and drift.

- **Evaluation Is an Editorial Act**

  Assessing recency, completeness, authority, and bias is not mechanical. It is where cartographic judgment and responsibility reassert themselves.

- **Prompt Cartographers Design Data Pipelines**

  Data discovery, validation, engineering, and critique form a key pipeline within the prompt cartography workflow – one that benefits from delegation but demands direction.

Conceptual and Applied Exercises

Truth Versus Fit

Think of a dataset you have previously used in a map. In hindsight, did it really "tell the truth" or did it simply align conveniently with the narrative you wanted to communicate? Describe how a different dataset – aggregated, filtered, or differently scoped – might have changed the story.

From Keywords to Constraints

Rewrite a vague dataset query (e.g., "Find crime data for Chicago") into a structured, constraint-driven prompt that specifies geometry type, time range, required fields, and output format.

Reflect on how the second version alters both the likely results and your responsibility as a cartographer.

Semantic Fingerprint Test

Ask an LLM to locate datasets for a phenomenon of your choosing. First, use a brief keyword-style prompt. Then revise it to include technical clauses (geometry, fields, date range) and narrative clauses (intended audience or goal).

Compare the outputs. Which prompt produced results more aligned with your mapping intent – and why?

Dataset Accent Exercise

Consider two hypothetical datasets describing the same phenomenon:

- A sparse point dataset with uneven spatial coverage.
- A smoothed or aggregated surface dataset covering the entire region.
- Describe how each would *feel* visually and rhetorically in a finished map. Which would you choose for a museum exhibit? Which for investigative journalism?

Ethical Clause Challenge

Draft a dataset search prompt that includes at least one explicit ethical guardrail (e.g., attribution, anonymization, removal of endangered-species coordinates).

Explain how this clause changes the assistant's behavior and reinforces your role as author rather than passive downloader.

Mini Data Analysis

Take any spatial dataset you can access, upload it to an LLM, and ask it to create a short JSON-style or YAML description including:

- Title
- Geometry type
- Spatial extent

- Temporal coverage
- Key attributes
- Intended map purpose
- Limitations

Review the results and adapt accordingly based on your knowledge of the dataset, if required. Reflect on how articulating the dataset this way stabilizes your next design decision. How might this JSON document be used effectively with the dataset by other LLMs and agents in the upcoming mapmaking process?

8

Learning the Language of Map Form and Style

You can't prompt into action what you don't have words for.

In *Web Cartography*, we explored aesthetic design decisions largely through the lens of hands-on construction. We built layouts, adjusted color palettes, and nudged labels pixel by pixel until the composition felt balanced. The goal was to train the cartographer's eye and hand (or coding fingers) to work in concert. In the large language model (LLM) era, however, the hand has been outsourced. Instead of dragging and dropping symbology, the cartographer must now describe their design intent to an assistant that has no eyes, no hands, and – let's be honest – an occasionally baffling sense of style.

This change is profound. In practice, this means that the precision and richness of your aesthetic vocabulary will determine the quality of the map that emerges from your prompts. If your language is vague, the assistant will default to something bland and generic. If your language is sharp and grounded in design theory, the assistant can execute complex stylistic visions quickly and consistently.

In prompt cartography, your goal is to articulate overall map form and aesthetic holistically before your agents generate idiosyncratic outputs. Achieving this goal requires a shared natural language for describing your visual choices. The goal of this chapter is to introduce you to this language by connecting cartographic form to foundational design principles. Figure 8.1 exemplifies the power of natural language aesthetic design in a rather macabre manner.

Consider, for example, a user who prompts: "Make me a choropleth map of population density." The assistant dutifully produces a gray-on-gray map with thin black outlines, a light blue water background, and a Roboto typeface – the aesthetic equivalent of dry toast. It's not wrong, but it lacks character, hierarchy, and purpose. By contrast, a user who writes, "Design a minimalist, high-contrast, choropleth as might be found on a national newspaper site using restrained typography, clear hierarchical labeling, and neutral basemap tones," will receive something far more intentional. The difference is not the underlying technology – it's the quality of prompter's aesthetic language.

DOI: 10.1201/9781003797920-10

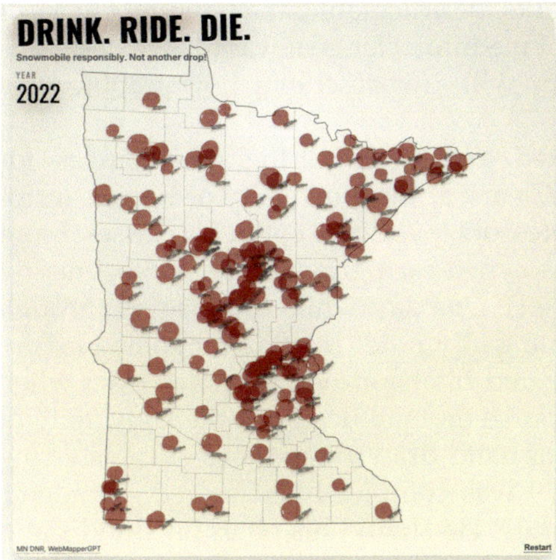

Prompt-Based Web Cartographer
I would like to make a simple animated map of snowmobile fatalities in Minnesota over time. I would like this to be kind of a melodramatic public service announcement post style map, perfect for mobile screens since Minnesota is a tall, north-south state. Dramatic title, similar to the Bagel Nation style, but perhaps crooked and with a propaganda poster style snowmobile crashing through it or into it would be ideal. A subtitle concerning: How many more must we lose? or similar on there. Please take a look at the two datasets I have. I would really like one with blood splotches for each death and a pop-up description and an alcohol label (if alcohol was involved), for fatality.

Code-Based Web Cartographer
CSS: **320 lines** of design code
JS: **476 lines** of scripting
HTML: **80 lines** of code

Source: www.webmapgpt.com/maps/365/jan/17

FIGURE 8.1
"Drink. Ride. Die." snowmobile fatality map for Minnesota was created with a 150-word prompt using an LLM assistant. A code-based interactive web cartographer would need to write 320 lines of CSS design code, 476 lines of JavaScript, and 80 lines of HTML (876 lines total) to get the same result.

This chapter is about rediscovering that language. We will return to the foundational vocabulary of visual communication, drawing on Dondis's *Primer of Visual Literacy* (1973) to structure our discussion. Though developed over half a century ago at MIT, Dondis's framework remains largely timeless. It remains one of the clearest ways to think about how visual form operates, and how those forms can be manipulated to shape perception, emotion, and understanding. Once these aesthetic variables are made explicit, they can be prompted for with clarity and control.

In short, this chapter functions as the conceptual Rosetta Stone for the rest of Part II. Before you learn to prompt for color palettes, grid systems, or interactivity, you need to understand and use the vocabulary and grammar of visual language. The sections that follow will ground you in that language – so when you instruct your assistant to "emphasize movement" or "create an analog-like cut-out texture," it will better know what you mean.

Visual Literacy and the Dondis's Framework

The starting point for any discussion of aesthetic form in cartography should be visual literacy: the ability to read, understand, and deliberately construct visual messages. Donis A. Dondis's *Primer of Visual Literacy* (1973) remains one of the most enduring frameworks for articulating the building blocks of visual communication regardless of medium. Her goal was to give designers a clear vocabulary – a lexicon, if you will – for describing how graphic form conveys meaning.

Dondis's work has been kicked around in cartographic circles for decades. In the 1970s, George McCleary at the University of Kansas introduced his students and fellow professors to the concept in his classes. From there, it influenced the likes of renowned cartographers Terry Slocum and Robert McMaster (my Ph.D. advisor). One day, while rummaging through McMaster's file cabinet of graduate study notes looking for some obscure work on dot mapping, we discovered his hand-written notes from when he was a student in McCleary's course on map design at the University of Kansas. In them, we found gorgeous hand-drawn map sketches transferring Dondis's theories to cartography. I was completely enamored – students could actually sketch maps by hand?! He shared his notes on Dondis and his original drawings with me, and I based my Ph.D. and future research on what he handed me and Dondis's book – which I found used online. (Score one for analog, hand-written notes. They got me where I am today. Thanks, Bob!)

My point is this: though not new to cartography, Dondis's work hasn't found a lot of running room within the discipline. The University of Kansas used it to teach map design in the 1970s and my research on persuasive map design talked it up in the 2010s.

Prompt cartography changes this. Dondis's primer is particularly useful now, as it is a book that literally transcribes graphic meaning into words. (I highly recommend the book for those who can still find a used copy.) Adopting and adapting her theory into practice allows us to create and standardize an explicit vocabulary that can be used by LLMs.

Visual Literacy: Same Data, Different Form

Before naming variables or styles, it helps to *see* the problem.

Figure 8.2 presents identical spatial data rendered in different ways. These are faux Sunday noise complaints in a very real town near Stuttgart, Germany. These two maps are highlighting the exact same dataset. However, the visual style and form each map takes induce varying levels of interest and types of interpretation. Thinking about and designing toward particular visual styles will help your maps better attract, retain, and persuade your target audiences.

Notice how some versions of the same map feel restrained, downright institutional. Others feel expressive, tense, or directive. The reactions you are

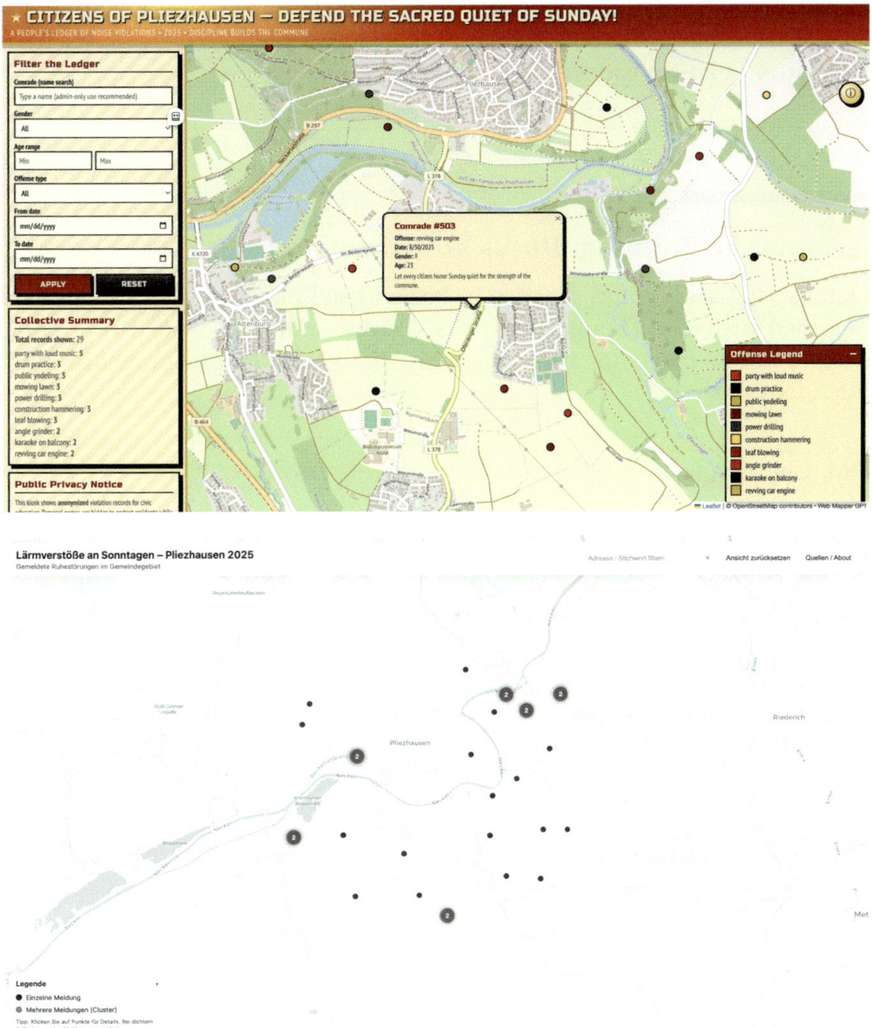

FIGURE 8.2
Two representations of identical noise violation datasets from Pliezhausen, Germany, demonstrate how visual form and aesthetic style frame user interpretation before readers consult legends or analyze symbols.

feeling occur for all map users. They occur before a reader consults a legend or even starts to read a title. None of the above maps lie; they say the same thing using different visual dialects.

This is the quiet power of map form. Visual decisions – contrast, density, hierarchy, balance – frame user interpretation long before analysis of any cartographic symbols begins. Visual literacy is the ability to notice those decisions deliberately rather than absorbing them unconsciously. Visual fluency is the ability to explain it in words – i.e., coupled with LLMs, successfully prompt an envisioned design into being.

Naming the Building Blocks of Visual Form

Dondis (1973) identifies a series of basic visual elements that underlie all designed artifacts. Though originally conceived in the context of graphic design, these elements map elegantly onto cartographic practice (pun intended).

Dot, Line, Shape

The skeleton of visual structure. Dots draw attention and anchor positions; lines direct the eye and define relationships; shapes provide enclosed, stable zones of meaning.

Direction and Movement

The implied or literal sense of flow, orientation, and dynamic tension within a composition. A map dominated by parallel horizontal lines feels stable and calm; one crisscrossed with erratic flowlines feels unstable and intense.

Scale and Proportion

The relative size of elements establishes hierarchy and visual weight. Large symbols command attention, while small ones recede; disproportionate elements can signal emphasis or distortion.

Texture and Pattern

The surface character of a design, from perfectly flat vector minimalism to dense stippling, crosshatching, or photographic overlays. Texture modulates complexity and mood.

Color

Hue, value, and intensity act as powerful communicative levers. Color choices affect legibility, emotional tone, and cultural associations. High-intensity hues demand attention; desaturated palettes can recede or signal neutrality.

It's important to note that Dondis's elements of graphical literacy do not work against what you may have learned about Bertin's (1983) visual variables – position, size, shape, value, color, orientation, and texture, as well as all later visual variable additions (e.g., MacEachren 1995). Rather they reinforce and build on them.

Focusing on visual variable symbolization must not come at the expense of a map's overall aesthetic clarity and form. When it does, you often end up with stellar information visualizations that people don't take the time to view. In other words, whereas some cartographers have been trained to miss the forest (overall map form aesthetic) for the trees (data visualizations therein), audiences are attracted to forests. Without grabbing the attention of an audience with the forest, they will never examine the trees you've designed.

Visual Literacy: From Visual Dimension to Prompt Language

Perceptual awareness becomes operational only when it can be expressed in words. Figure 8.3 refashions some earlier work I've published based on Dondis's descriptions of different dimensions of visual communication to provide you with generic prompts you can use to begin directing your LLM assistants to design compelling maps.

Dondis's theory gives us the perceptual foundation necessary to design brilliant maps, whereas Bertin gives us the operational mechanics to embed our visuals with information (Figure 8.4). These represent two very different but complementary needs. For example, movement in Dondis's sense often corresponds to orientation and arrangement in Bertin's variables, while Dondis's concept of texture aligns with both Bertin's quantitative texture and color value. By aligning the perceptual with the operational, your map designs will improve dramatically. Invoking these Dondis's perceptual variables so they align with best cartographic principles (as originally defined by Bertin) will ensure you (and your team of agents) produce maps that communicate clearly without drifting toward stylistic vagueness.

It is important to note that LLMs do not possess perceptual systems. They never "see" your map. They have no true conception of space and time outside of what they have learned from us linguistically. They cannot experience time or space like we do. (In fact, time and space may not exist outside of our conscious perceptions [citre Hoffman], so in this regard perhaps LLMs are more sane than we are, but I digress.) Geography is as alien to LLMs as exploring the world via our noses is to us. We will never know what it's like to sense and know the world as a dog does. An LLM has no idea of space and time other than what it has been trained on.

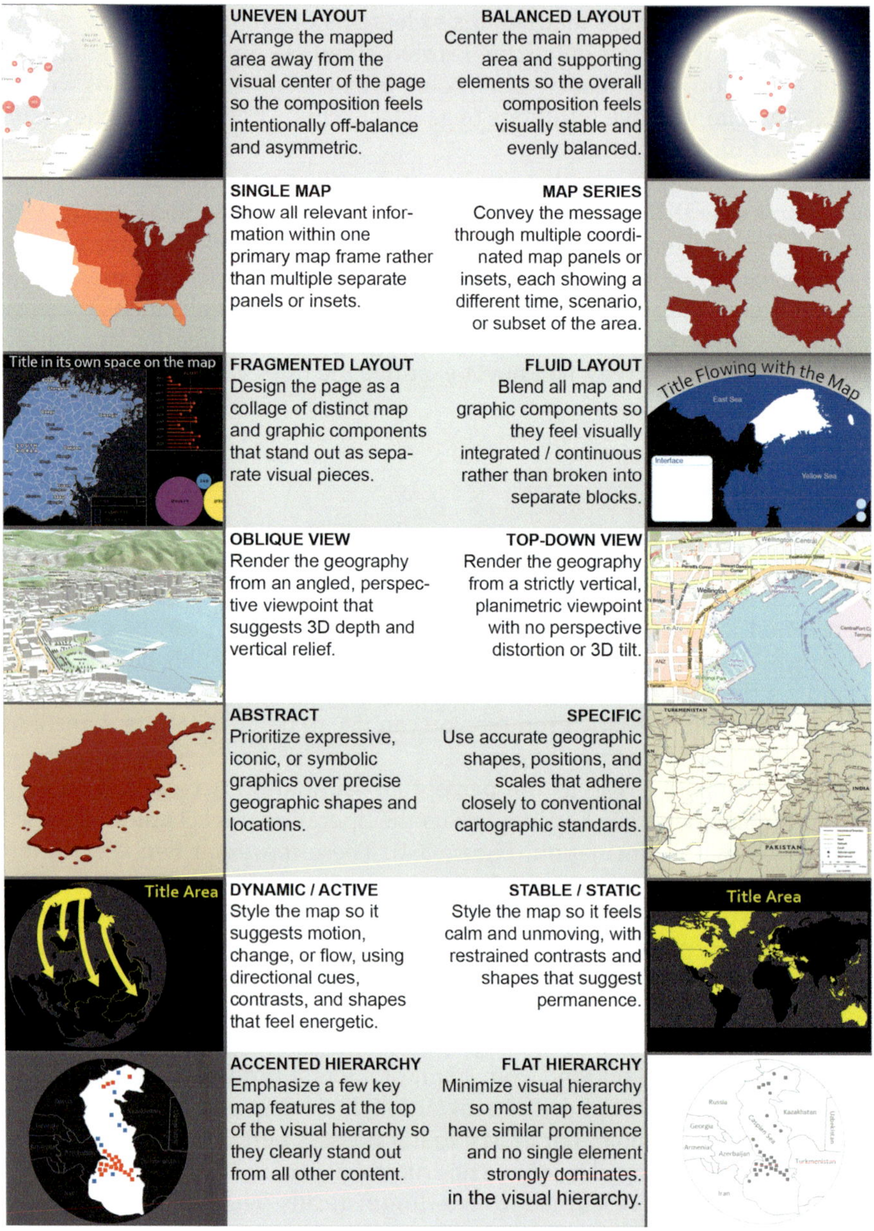

FIGURE 8.3
Prompting aesthetic in natural language to accentuate or diminish different types of visual communication. Based on Muehlenhaus, 2010, 2011.

HIGH CONTRAST
Use strong, high-contrast colors and symbols so differences between data types or values are visually striking and dramatic.

MINIMIZED CONTRAST
A map using minimal graphical contrast to present different types and/or values of data.

COMPLEX HIERARCHY
Organize many interdependent symbols and map elements across several levels of visual importance, creating a rich, layered hierarchy.

SIMPLE HIERARCHY
Limit the number of map elements and give only a few clear levels of visual importance so the hierarchy feels clean and easy to read.

MULTIVARIATE
Depict multiple variables at once by combining several visual channels (such as color, size, shape, and texture) in the symbols.

UNIVARIATE
Depict only a single main data variable using one consistent style of symbolization across the map.

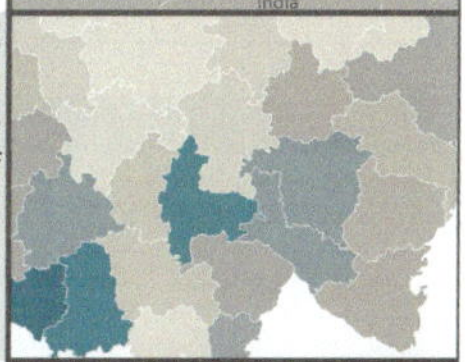

ICON SYMBOLS
Use culturally loaded, pictorial symbols whose shapes and icons carry strong emotional or symbolic meaning.

GENERIC SYMBOLS
Use neutral geometric symbols (such as circles, squares, or crosses) that minimize cultural or emotional connotations.

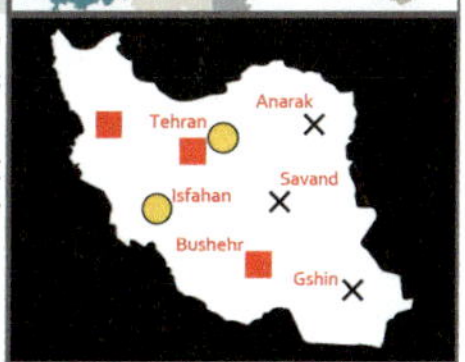

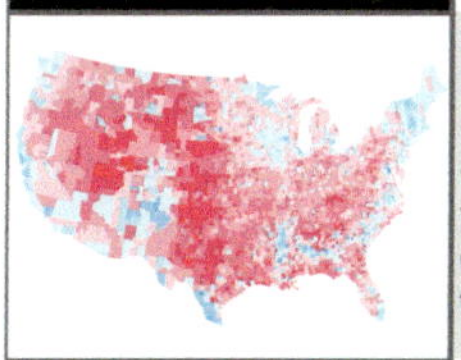

SPECIFICATION
Differentiate fine-grained differences within the dataset so users can visually compare and quantify detailed variations.

GENERALIZATION
Simplify the dataset so many small differences are merged or smoothed, emphasizing broad patterns over precise comparisons.

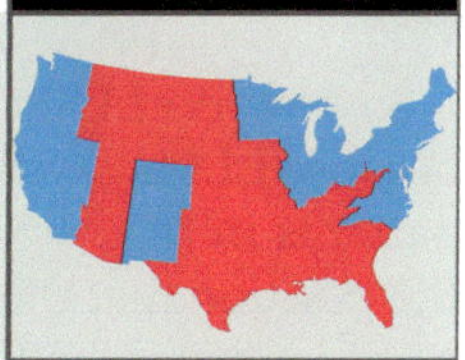

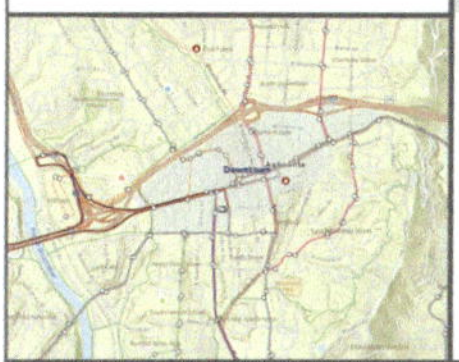

DETAILED BASEMAP
Use a detailed, highly accurate base map that preserves fine spatial geometry and supports precise spatial referencing.

MINIMALIST BASEMAP
Use a simplified or distorted base map that sacrifices geometric precision in favor of clarity, abstraction, or thematic emphasis.

Adapted from Muehlenhaus (2013). These different dimensions of visual literacy as outlined by Dondis (1973) were adapted for cartographic use by Muehlenhaus (2013).

Every map can be prompted to be have a design closer to one extreme or another to establish an aesthetic style and promote a particular message. This is visual literacy in action – natural language style.

FIGURE 8.3 **(Continued)**

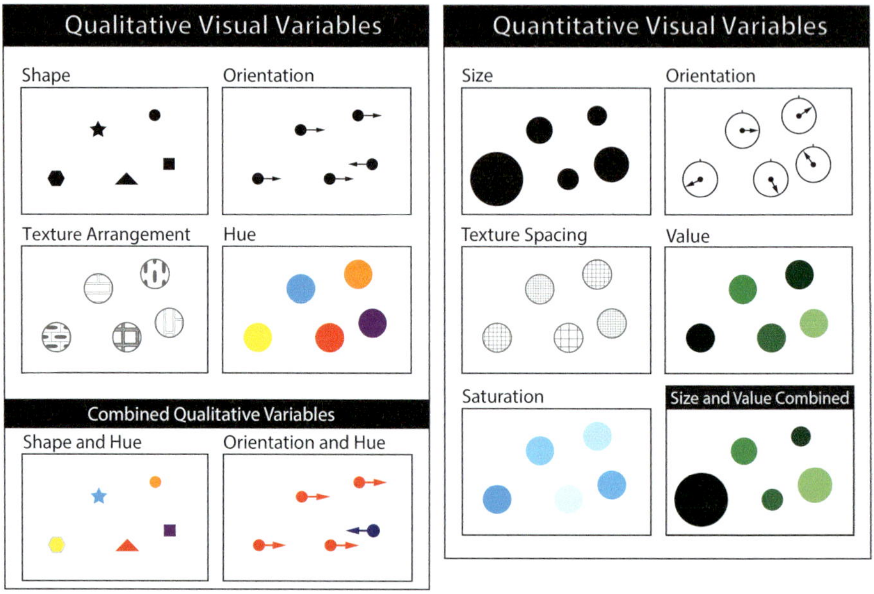

FIGURE 8.4
Bertin's visual variables for encoding qualitative and quantitative information in your map symbols.

So how do LLMs work? They have been mocked as "stochastic parrots," because LLMs do truly autocomplete text based on stylistic and algorithmic patterns in their training data. Though this author currently disagrees with the position that this differs greatly from how we humans learn and actually communicate, the verdict is still out, and we do know that specificity, not platitudes, drives LLM logic.

What this means for you is that everyday qualitative descriptors (i.e., adjectives and adverbs) such as "nice," "professional," "amazing," or "clear" are essentially useless to your LLM. Your words will be interpreted through statistical associations with other words, not design judgment. By contrast, a phrase such as "apply a high-value, low-intensity color scheme with strong directional flowlines" is far more likely to produce a coherent, intentional design. Still qualitative, but direct and unambiguous.

In the remainder of this chapter, we will use Dondis's framework as a scaffold for exploring generic map aesthetic styles, new form variables to use in prompting. No worries if you didn't major in cartography or have no experience in it. Once you can name the elements of perceptual design clearly, you will be able to instruct your preferred assistant to use them precisely – and, more importantly, you'll be better able to critique its output holistically, rather than via its choice of data visualization technique.

Aesthetic Styles in Cartography

If Dondis provides the lexicon of visual language, then aesthetic styles are the dialects – recognizable combinations of form that carry distinct cultural and rhetorical connotations. Just as spoken dialects signal region, class, or community, a map's aesthetic style quietly communicates genre, purpose, and attitude. Recognizing and deliberately invoking these styles is essential when working with LLMs, because stylistic cues are precisely the kind of patterns these LLMs are stellar at emulating – if you can name them.

> **THIS BOOK'S DIALECT AS DESIGN CHOICE**
>
> You've probably noticed that this book isn't quite as formal as a typical "textbook." This was a design decision on my part – and one that likely chafed my editor a bit. I've purposefully chosen a more softspoken than professorial voice (the latter of which I can pull off with aplomb when my kids aren't listening). This is a design dialect and a calculated choice. I know it will irritate many of my academic friends and colleagues. However, they aren't my target audience. You are! The young, new, fledgling, or mid-career, adapting to the changes that AI is bringing, cartographer. I am using the dialect, or voice, I use when I teach in the classroom or hang out with friends interested in the same topic. (In other words, my real voice, not the career professional one.)

Although there are innumerable micro-styles and hybrids, I outline some generic style templates (or dialects) below, not because these are official or even real. Rather, these are caricature templates to demonstrate the importance of visual style. Each was created using a single prompt. Feel free to experiment with these, add to and tweak, or mash-up them up as you see fit. Some of these styles can be defined in your agents' system prompts or JSON schemas. You can even refine styles enough to call them as "stylistic packages" in your map making. Check out the companion website for more system prompt ideas and schemas to implement.

Minimalist Modern styles emphasize clarity, whitespace, and visual restraint. Symbols are clean, color palettes are limited and often desaturated, and typography is unobtrusive. This style is favored in UI-driven applications where the map plays a supporting role, such as dashboards and mobile apps.

Prompt resulting in the map shown in Figure 8.5:

```
Design a highly readable interactive map of current
earthquake activity in California, styled like a modern
```

FIGURE 8.5
Minimalist example of earthquake activity in California.

```
digital newspaper graphic. Use restrained typography,
subtle color accents, and minimalist iconography to
convey magnitude and depth without visual clutter. The
interface should feel intuitive and quiet, with a sin-
gle unobtrusive filter pane allowing readers to toggle
magnitude ranges, depth categories, and recent time
periods. Aim for a refined, journalistic aesthetic that
foregrounds clarity and user exploration.
```

Data Visualization is a popular style dialect these days (particularly among the college-educated), popularized by online journalism and visualization blogs. It favors bold, saturated color schemes; bivariate or trivariate grids; prominent annotation; and large, expressive legends. Elements such as labels and callouts are often overprinted directly onto the map surface. This style communicates analytical sophistication and confidence – sometimes, I would argue, to a fault.

Prompt resulting in the map shown in Figure 8.6:

```
Produce a rigorously precise, data-driven seismic activ-
ity map of California modeled after professional out-
puts from geophysical research institutions. Represent
earthquake magnitude, depth, and temporal distribution
using standardized scientific symbology, quantita-
tive color scales, and statistically meaningful spa-
tial clustering. Prioritize accuracy, legibility, and
methodological transparency. Include a unified control
panel containing advanced filtering options for magni-
tude thresholds, depth intervals, temporal windows, and
geospatial bounding boxes. The final map should read as
a peer-review-ready visualization suitable for seis-
mologists and data scientists.
```

Though more intuitive for exploring and understanding complex overlapping datasets than wading through individual tables, maps, and charts alone, the relevance of such complex data visualization is waning. I imagine humans will no longer use these tools to explore spatial data as much as they did over the past 20 years. We can now ask AI tools to sift through the data, find the interesting and relevant outliers and correlations, and prompt the visualizations into a simple-to-read form.

Historical or Retro styles draw inspiration from hand-drawn or early printed maps. They make extensive use of textured backgrounds, parchment tones, crosshatching, or stippling; decorative cartouches; and period-appropriate typefaces. The example here is meant to appear analog-bureaucratic.

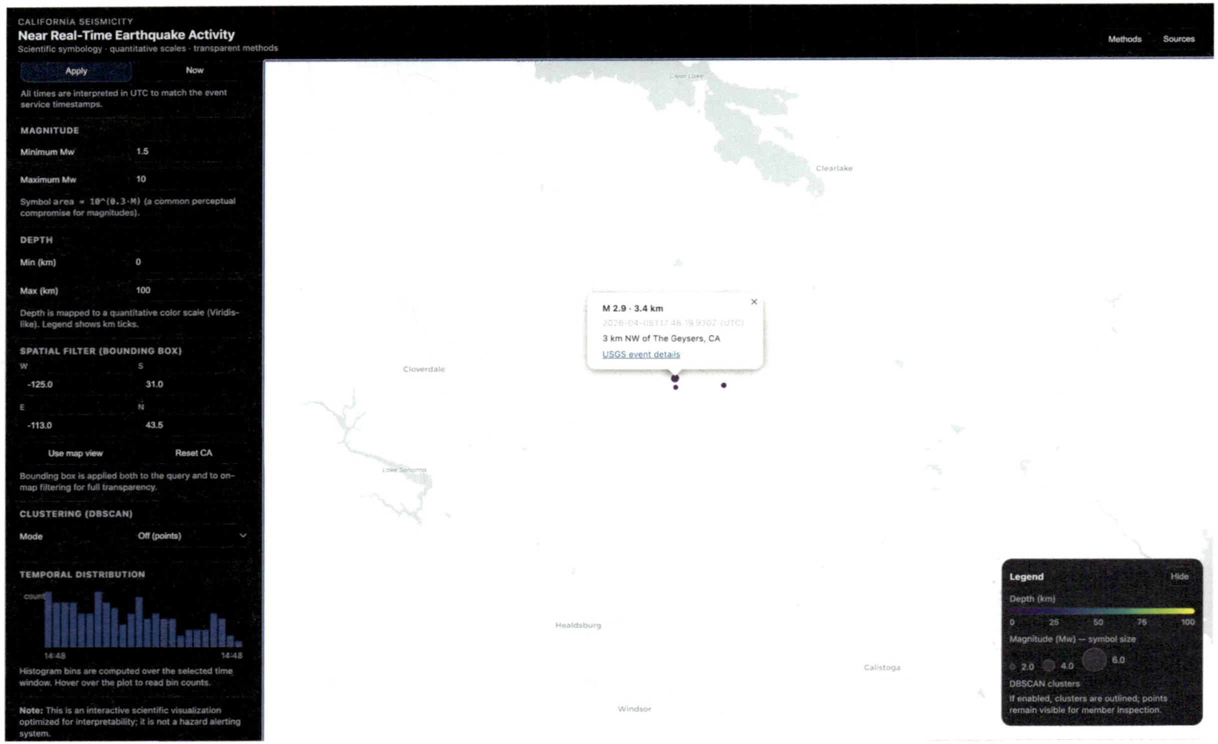

FIGURE 8.6
Data visualization aesthetic example of earthquake activity in California.

Prompt resulting in the map shown in Figure 8.7:

```
Create a mid-20th-century bureaucratic map of
California's current earthquake activity, styled as if
produced by an overworked federal agency circa 1957.
Use faded inks, typewriter labels, rubber-stamp sym-
bols, and slightly misaligned overlays to convey magni-
tude and depth. Embrace the aesthetic of aging paper,
imperfect registration, and official-but-dusty govern-
ment diagrams. Include a single filter pane designed
like a vintage form sheet, with checkboxes, dropdowns,
and fields reminiscent of old civil-service paperwork.
The map should feel authentically analog, procedural,
and charmingly institutional.
```

Realistic or Photographic styles emphasize terrain realism, aerial perspectives, or satellite imagery. They may overlay transparent thematic layers on photographic basemaps or use/simulate relief shading and contouring to give a tactile, "on-the-ground" impression. These styles are often associated with scientific authority or immersive storytelling.

Experiential or Artistic styles vary from conventional expectations. They may break standards and rules intentionally – using neon color palettes, non-Euclidean layouts, or dense textures to produce affective or persuasive effects, or perhaps even presenting a non-northerly viewpoint. While less common in mainstream GIS practice, these styles are powerful rhetorical tools when deliberately used.

Prompt resulting in the map shown in Figure 8.8:

```
Create a high-impact, visually arresting map that drama-
tizes real-time earthquake activity across California.
Present seismic depth and magnitude with striking sym-
bolic iconography that instantly conveys intensity and
urgency. Craft the map with the authoritative polish of
an official geological agency, but infused with dynamic
flair - glowing overlays, layered textures, and bold
visual hierarchies. Integrate a powerful, intuitive
single-pane control panel featuring smart, responsive
data filters that let users slice the seismic story
by magnitude, depth, region, and time window. Deliver
a map that feels alive, volatile, and impossible to
ignore.
```

I need to emphasize that these categories are examples – they are not truly defined or literature-backed categories. And none of these are mutually

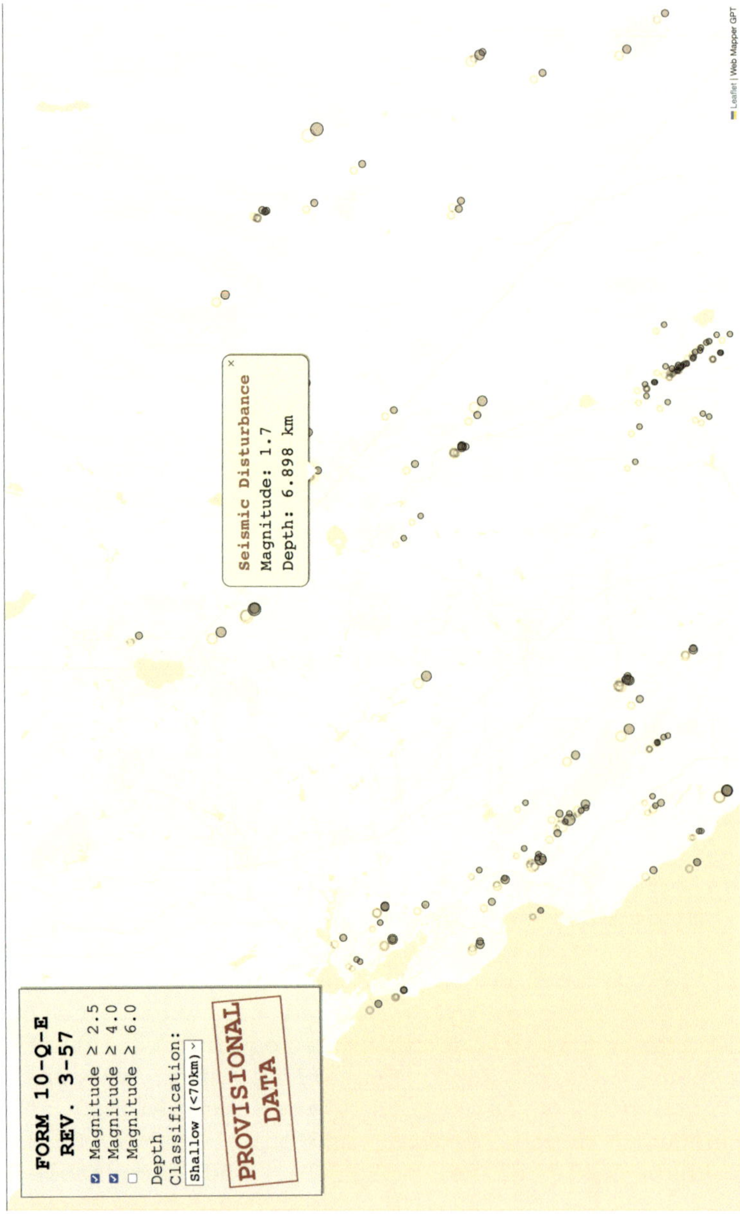

FIGURE 8.7
Retro-style map of earthquake activity in California.

FIGURE 8.8
Avant-garde-style map of earthquakes in California.

exclusive or exhaustive. Aesthetic styles function like dialects, not language. The point is not memorization, but precision in description to get what you want. LLMs excel at emulating stylistic patterns, but you must name and describe them clearly.

Map Form and Function

Aesthetic style is not merely decorative. It is integral to how a map communicates – how it guides the eye, establishes trust, evokes emotion, and frames interpretation. The style you choose (or fail to choose) can reinforce your map's purpose or quietly undermine it.

One of the most fundamental functions of style is to shape perception and attention. Consider the map examples above showing earthquakes. They all use the same dataset, and each was created using a single prompt. (Hence, as we will see in upcoming chapters, their map elements and layout could use some significant refinement.) Even before a map user reads the title or legend, however, their eyes will gravitate differently on each. The first map is whispering, inviting careful inspection. The last is shouting, drawing attention to the largest earthquakes but potentially dissuading a map user from analyzing the data in depth or too seriously.

The aesthetic and visual styles *you choose* set the emotional tone of a map. A retro parchment-toned basemap may evoke nostalgia or trust in traditional authority. A sleek minimalist interface signals contemporary neutrality. A neon artistic style may suggest playfulness or, in some contexts, alarm. These emotional cues are not arbitrary; they emerge from centuries of visual culture and are read subconsciously by audiences. Humans, like LLMs, are trained on years of data collection and processing throughout their lives. Unlike with LLMs, your map styles will evoke visceral and emotional reactions in people.

Equally important is credibility and trust. Audiences associate certain aesthetic styles with reliability – regardless of whether that association is justified. Editorial and atlas styles often confer a sense of seriousness and authority. Realistic basemaps imply scientific grounding. Conversely, garish or inconsistent styling can make even accurate data feel dubious. When prompting an LLM, style selection is therefore not merely a question of taste; it is a strategic communicative choice.

Finally, aesthetic style *must* be aligned with your target user's use case and contexts. A minimalist style might be ideal for exploratory dashboards on mobile devices, where cognitive load must be low. A scientific style might suit a static report where the map is read slowly and carefully. An artistic style might be deliberately chosen to *provoke* discussion rather than convey precise metrics. If the chosen style and the map's communicative function diverge, the design will feel dissonant.

Before prompt cartography, much of this alignment happened implicitly: a skilled designer chose colors, typography, and layout based on context and audience. In prompt cartography, the *designer must make this alignment explicit through language.* If you do not articulate your stylistic goals, the assistant will fill the void with defaults – often those of the platform's basemap and font stack, which may or may not serve your communicative intent. (Example: OpenStreetMap [OSM] offers the world incredible data. It's an amazing feat! OSM's default map tile design, however… let's just say, in my opinion, for most design uses it's a bit too jaundiced.)

Understanding why style matters allows you to use it strategically rather than accidentally. As we move into the next sections, we will break down map form into promptable variables – layout, color, typography, symbology – and discuss how to control them with precision. But these variables only make sense when tied back to a clear, overarching stylistic vision. Style is the connective tissue that turns a collection of design choices into a coherent visual language.

From Style to Promptable Variables

Once you have a clear stylistic vision, the next step is to translate that vision into specific promptable components. This is where abstract aesthetic language – "clean," "retro," "bold," "editorial" – must be broken down into structured, controllable variables that an LLM can act upon. Think of this as turning your design instincts into machine-readable prose.

The key categories of promptable form variables parallel long-standing cartographic design elements. Although the technology has changed, the levers are essentially the same:

- Map layout structure
- Color palette
- Typography
- Symbology
- Basemap design and selection

By addressing each explicitly in your prompt designs and schemas, you will create maps that are both aesthetically coherent and provide a comfortable user experience. The chapters that follow address these in depth. Here, the key insight is this: *style coherence emerges from early, explicit articulation, not late-stage correction.*

Style Drift and Aesthetic Defaults in LLMs

Even when prompts are well structured, a subtle but pervasive problem can arise in iterative workflows across numerous pipelines (e.g., data brief, visualization, etc.): *style drift*. LLMs, unlike human designers, *do not* maintain a persistent visual state. Each new instruction is interpreted in light of the current prompt input, not via an evolving aesthetic logic of the design. As a result, previously established styles can quietly erode with each new layer or modification.

Consider a situation where you prompt for "a modern, simple choropleth with restrained serif typography and an underlying muted gray basemap." The assistant returns a pleasing layout and design. You then add, "Overlay the attached provincial boundary geojson with 100% fill transparency and thin lines set to 80% black." In some systems, this seemingly minor addition can trigger unexpected shifts – labels revert to default sans serif, basemap colors reset, or legend styles change. These are not malicious errors; they are side effects of how many LLMs recompile styles on each instruction.

Aesthetic defaults exacerbate this issue. Most platforms have a baseline style – often something like a default minimalist gray base map, Roboto typeface, thin black outlines, and light blue water. If a prompt is ambiguous or incomplete, the assistant will lean back on these defaults. Over time, as your instructions (and prompts) accumulate, the map can drift back toward its default aesthetic even if you initially specified something very different.

Fortunately, there are several strategies to mitigate style drift throughout your map composition.

Declare Style Early

Explicitly declare your core stylistic identity at the start of prompting and refer back to this in your subsequent prompts. For example: "Maintain the previously specified modern choropleth style throughout the following changes. If you are unable to recall it, I can share it again." Reiterating key descriptors helps anchor the assistant. Like dealing with a perfectionist, giving it an "out clause" – if you forget, I can share it again – will also help prevent the LLM from lying to you that it remembers the original style.

Use Style Templates or Schema References

Many LLM systems allow you to define a JSON or style template once and refer back to it ("apply Template A"). This approach reduces the chance of re-interpretation at each step.

Chunk Your Prompts

Break your pipeline into discrete aesthetic stages: first establish layout and basemap, then color, then typography, then symbology. Avoid making large stylistic and structural changes simultaneously, as this increases the likelihood of cascading defaults.

Regularly Verify and Re-Specify Your Expectations

Treat the assistant's output like a design intern's draft. After each step, check whether any stylistic elements have silently changed. If they have, restate your stylistic constraints explicitly before proceeding.

Self-Critique

In your prompt, ask the LLM to justify how its current output design decisions follow your original declared style (or style template) and what compromises it had to make in 300 words or less. Forcing the LLM to self-critique is often a powerful guardrail. (Note, however, that sometimes LLM agents will be biased about their success! Remain critical.)

Style drift is not a sign of failure; it is a characteristic of working with any generative AI medium and will be discussed again later in this book. (In other words, you need to learn to work with it.) Traditional cartographers hold an aesthetic state in their minds and adjust individual variables without losing sight of the whole. LLMs, by contrast, typically respond to textual cues in isolation or by referring to the last few back-and-forth chats in your conversation. By understanding this difference and adopting a disciplined prompting strategy (with planning and practice), you will better preserve stylistic integrity across all the pipelines in your workflow.

On to Building Map Form Piece-by-Piece

With vocabulary established for describing visual form abstractly – Dondis's elements, aesthetic styles, promptable variables – the next task is making those abstractions operational. Abstract concepts like "visual hierarchy" and "restrained minimalism" only matter when they manifest in concrete map components: titles that frame interpretation, legends that clarify encoding, extents that determine what gets shown, and layouts that organize spatial grammar.

These elements don't design themselves. In prompt cartography, each one must be explicitly specified, or your LLM will fill the void with defaults – often borrowed from whatever basemap or template the platform happened to load. Chapter 9 addresses those concrete elements and why omission is not neutrality. It's delegation to someone else's aesthetic judgment.

If this chapter gave you the language to describe what maps should look and feel like, Chapter 9 teaches you how to ensure those intentions arrive on user screens.

Takeaways

- **You Can't Prompt What You Can't Name**

  In a prompt-cartography workflow, aesthetic quality depends on language. Vague descriptors produce generic maps; precise visual vocabulary enables intentional form.

- **Aesthetic Literacy Is a Design Skill**

  Style shapes perception, emotion, and trust. It is integral to how maps communicate, not something layered on after data has been visualized.

- **Dondis Provides a Usable Visual Grammar**

  Dondis's elements translate perceptual form into language, making them especially valuable for prompt-based cartography where intent must be expressed verbally.

- **Perceptual Form and Cartographic Variables Reinforce Each Other**

  Visual literacy and cartographic theory are complementary. Perceptual clarity supports effective data encoding rather than competing with it.

- **LLMs Respond to Specificity**

  Language models do not perceive maps. They infer design intent from textual patterns, making precise qualitative language far more effective than subjective adjectives.

- **Aesthetic Styles Function Like Dialects**

  Recognizable styles carry cultural and rhetorical meaning. Naming them explicitly allows LLMs to emulate coherent visual identities rather than default aesthetics.

- **Style Must Align with Purpose and Use**

  Aesthetic choices shape credibility, emotional tone, and usability. Misalignment between style and function undermines communication, even when data is accurate.

- **Form Becomes Promptable When Broken into Components**

  Layout, color, typography, symbology, and basemap design must be addressed explicitly to preserve stylistic coherence.
- **Style Drift Is Structural**

  Without deliberate re-specification, LLMs tend to revert toward platform defaults. Maintaining style requires intentional anchoring, review, and critique.
- **Visual Language Enables Critique and Control**

  When aesthetic intent is articulated clearly, outputs can be evaluated, revised, and improved holistically rather than reactively.

Conceptual and Applied Exercises

Finding the Missing Words

Find a map (print or web) whose overall look you admire but struggle to describe precisely. Which aspects of its form feel intuitive to you, and which feel difficult to articulate without resorting to vague terms like "clean" or "modern"?

Style as Constraint

This chapter argues that naming aesthetic intent early constrains downstream design decisions productively. Reflect on how explicitly naming a style might prevent certain default design choices rather than enable them. Name some of these defaults and explain how or why each might be omitted.

LLM Vocabulary Test

Upload a screen capture of a web map to an LLM:

```
Critically examine the visual style of the provided web
map image. Provide a description of its overall aes-
thetic and map form characteristics in plain language.
Then create a detailed JSON schema that I can use
with another LLM agent to mimic this map's visual and
aesthetic style.
```

Evaluate whether its response demonstrates a meaningful design vocabulary or relies on generic aesthetic clichés. Look at the JSON and determine if it accurately encodes the original maps aesthetic style. (Apologies in advance for all the curly brackets.)

Structured Prompt Exercise

Write a one- or two-sentence prompt that describes a map's overall form and aesthetic style without mentioning specific colors, symbols, or technologies. Then, ask an LLM to make a draft sketch representing the style. Refashion and iterate the prompt to see how the output changes – and hopefully improves.

For extended tutorials, hands-on labs, prompt templates, and updated examples related to this chapter, visit www.promptcartography.com.

9

Map Elements, Layouts, and the Curse of LLM Defaults

One of the most pernicious and misguided beliefs about prompt-based mapping is that the process somehow makes map design automatic – taking human creativity out of the loop. You ask for it, and a map appears. Or a legend exists. The layout settles into place and is perfect because large language models (LLMs) are all-knowing. Labels arrange themselves politely with minimal fuss. And because nothing is drawn by mouse or coded by typing, nothing is designed.

This belief is wrong – and it can be *conveniently* misleading. Particularly when it's used to dismiss people who are adapting. It's often meant to belittle people embracing new technology because it threatens established funding channels, business models, and people opposed to, or too tired for, more lifelong learning.

Not to ruin a perfectly good metaphor due to overuse, but no one would say that a film is by default rubbish just because the director didn't write the screenplay, do the acting, build the sets, produce the CGI, compose and play the music score, and edit the final film themself. The same goes for any art form – musicians who don't produce and record their own music; painters who don't manually crank out their own lithographs. And ironically, most of these outsourced processes I'm describing haven't even been done by other humans for decades either, but machines fed by human specialists (e.g., software does the sound mixing, Final Draft software the screenplay formatting). In this way, that's exactly what your team of LLM agents is – a machine, or more likely a bunch of machines, fed creative rules and guardrails by you to help produce original human creative output.

Map elements did not stop being designed by cartographers when people stopped drawing them explicitly with pens instead of using use a mouse or scripting. The same goes for prompting. Titles, legends, mapped extents, inset maps, interface controls, and visual hierarchy are still needed in prompt-built maps. The difference is that these elements, without specification, will now emerge from defaults, training data, and undocumented assumptions rather than from a cartographer's conscious design decisions. (Though, you might argue with pre-built API widgets, this is already the case for some web cartographers, unfortunately.) When prompt designers fail to specify what they expect, your LLM agents will fill the vacuum with best-guess defaults.

DOI: 10.1201/9781003797920-11

And defaults, as we all know, even when they aren't necessarily wrong, are the surest route to mediocrity in map design. Figure 9.1 highlights what happens when an LLM defaults to default mode.

In traditional cartography, map elements were tangible objects. You placed them deliberately. You negotiated for space. You made trade-offs when the page ran out of room. In prompt-driven cartography, those negotiations still occur – but silently, inside the model. The model decides how prominent the legend should be. It decides whether a title exists at all. It decides how much screen real estate the map deserves relative to interface controls, filters, or explanatory text. If you do not intervene, you have still made a design choice. You have simply delegated it – just like many novice cartographers might stick with defaults in GUI-based GIS software.

This is why map elements matter more – not less – in a prompt-mediated environment. They are no longer things you draw; they are things you must describe. Every meaningful map element must now exist twice: once as a conceptual intention in your mind and once as an explicit instruction in language. When those two diverge, the map almost always fails in subtle ways that are hard to diagnose after the fact.

Consider a simple example. You ask a model to "create an interactive map showing opioid overdose rates by counties in West Virginia." Without further guidance, the model will almost certainly include a legend, a title, and zoom controls. But what kind of legend? Does it explain classification? Does it foreground uncertainty? Does the title frame the data descriptively or rhetorically? Are users allowed to zoom in so far that they can see the Pizza Hut restaurant in Beckley but comparisons between counties are impossible? Each of these outcomes materially affects interpretation, yet none of them are guaranteed unless they are specified.

Prompt cartography, therefore, demands a shift in mindset. You are no longer designing map elements directly; you are designing the conditions under which map elements come into being. The role of the cartographer becomes less that of a draftsman and more that of an art director – someone who sets constraints, priorities, and intent so clearly that ambiguity has nowhere to hide.

This chapter begins from a simple premise: not only does prompting *not* absolve you of design responsibility, it *amplifies* every decision you make. If you cannot name a map element, you cannot control it. If you cannot control your agents, you are going to wind up in trouble. No one is going to assign responsibility to your LLM for being a bit daft, particularly when it was given insufficient instruction when your map fails. In the film industry, when a movie bombs at the box office, the director is typically the one who gets publicly lambasted – not the screenwriter, the set coordinator, or the boom mic assistant. It's the same here. If your map fails, people will be looking at you – the prompt director.

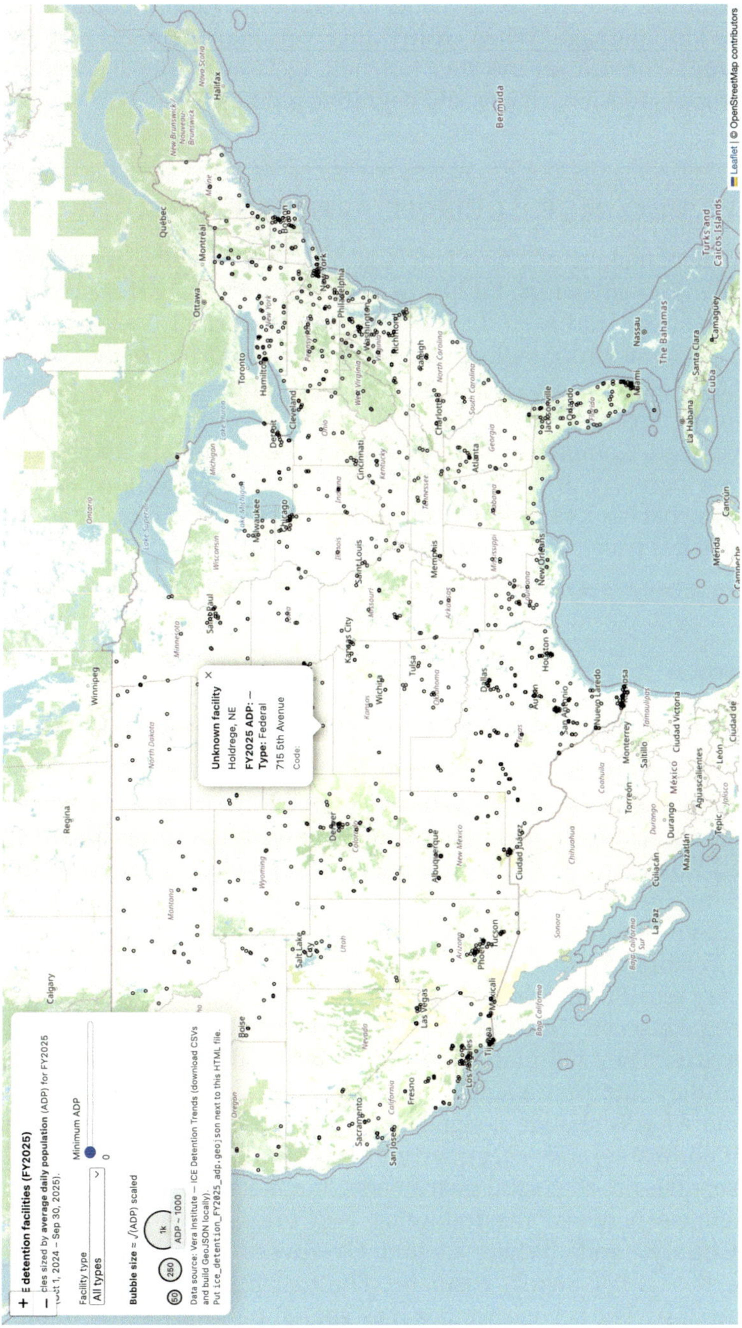

FIGURE 9.1
United States ICE detention facilities. Without specifications, unneeded map elements are included, often overlap and distract, and potentially make a map unusable.

The sections that follow revisit classical map elements not as relics of print cartography but as durable components of communication that must now be articulated in language. When treated this way, map elements stop feeling like vestigial formalities and start functioning as what they have always been: tools for persuasion, clarity, and cognitive guidance.

WE'VE COME FULL CIRCLE, ARTHUR ROBINSON

The thought of making maps via language only may be a bit scary to you still. To alleviate any stress, I'd like to point out something that should lighten the stress load a bit. One of the most important books in the canon of academic cartography published in the second half of the twentieth century had zero images, figures, or maps in it. *The Look of Maps*, by Arthur Robinson (1952), is one of the most elegant books on maps you'll ever read. Completely text based. When viewed in this light, what prompt cartography does is strip away all the distraction that we've added since that book came out. (Some of it is useful distraction, for sure, but none of it is necessary for making maps.)

Titles, Framing, and the Rhetoric of First Impressions

Titles are often the first casualty of prompt-driven mapmaking.

This is partly historical. Early web maps frequently relied on surrounding webpage context to explain what a map showed. Later, dashboard culture pushed titles into headers, tooltips, or collapsible panels. Now, with AI-assisted map generation, titles are often omitted entirely – either because the model assumes context implicitly, or because the prompt never mentions them at all. Fortunately, this is an easily preventable mistake.

As you may recall from *Web Cartography*, I am a proponent of map titles – when they make sense. Titles are *not* decorative labels. They are rhetorical framing devices. They tell map users what kind of thinking is required of them before they interpret a single symbol. A well-designed title does not merely name what is shown; it signals how the map should be read, what questions it addresses, and what interpretive posture the user should adopt. In this sense, titles are rhetorical instruments, not metadata.

The difference between the following two titles – "Opioid Overdose Rates by County" and "Where Opioid Overdoses Are Concentrated in the Midwest" – is *not* merely cosmetic. The first presents information as neutral inventory. The second frames it as spatial pattern worthy of attention. Neither is inherently dishonest; both guide cognition in very different ways, however. Prompt-based workflows amplify this distinction, because a map title often results in reactions from two very different audiences: one is your LLM agents (to give them context about what they are mapping) and the other is your human audience (to grab their attention and inform them). Titles can also be used to galvanize interest in your map, as shown in Figure 9.2.

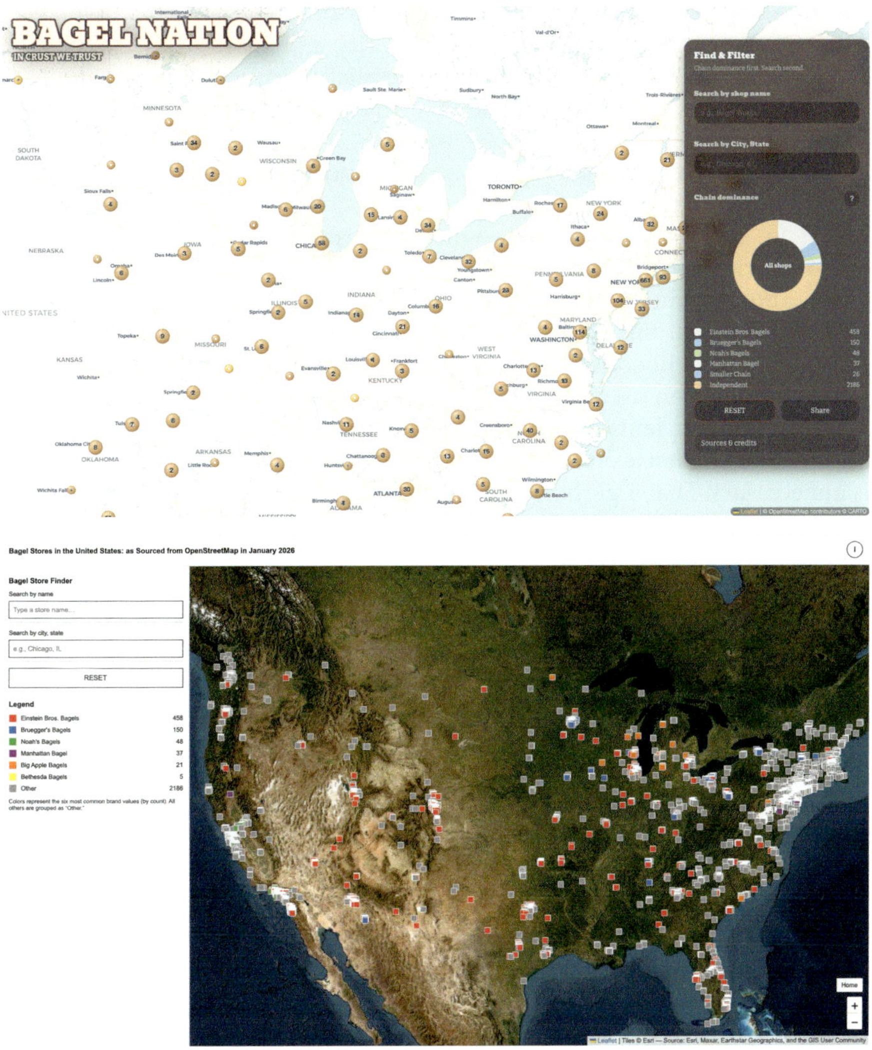

FIGURE 9.2
Two distinct title techniques for the same map showing where to buy a bagel in the US. One is more likely to grab and hold a map user's attention but may not guide map user as well.

LLMs respond strongly to framing language. A title embedded in a prompt – or echoed consistently across prompt components – acts as semantic ballast. It helps anchor classification choices, color schemes, legend phrasing, and even layout emphasis. When titles are absent, models tend to default to generic descriptors that flatten meaning and dilute intent. When titles are vague, the rest of the map often follows suit.

This dual role of titles introduces a useful distinction. Some titles are meant primarily for humans. Others are meant primarily for models. The most effective approach is to write both – even if one will never be displayed. A human-facing title might emphasize clarity and rhetorical tone, while a model-facing framing line may be more explicit, instructive, or constrained. Treating these as separate but related artifacts gives you far greater control over outcomes.

The same logic applies to splash screens and introductory frames. While splash screens have fallen out of favor in much of web design, they remain useful in cartographic contexts where interpretation matters more than speed. A brief introductory frame – whether visible to the user or used solely to condition the model as you build the map – can establish temporal scope, geographic focus, and analytical intent before interaction begins. When used sparingly and deliberately, such framing increases trust and reduces misinterpretation. When used gratuitously, it becomes friction.

This tension explains why titles and splash screens inspire strong opinions. Information architects often argue that any barrier between a user and the data is unnecessary. That logic holds when the task is retrieval – finding the nearest bus stop, restaurant, or restroom. It fails when the task is understanding. Maps that aim to explain, persuade, or contextualize often benefit from orientation. Titles provide that orientation in a single, cognitively efficient gesture.

In prompt cartography, omitting a title does not produce neutrality. It produces ambiguity. The model will still infer a frame. The user will still supply one. The only difference is that neither will necessarily align with your communicative goals. Explicit titles – whether rendered on the map, embedded in the prompt, or both – remain one of the simplest and most powerful ways to assert design intent at the very beginning of the mapping process.

First impressions matter. Titles are where you decide what those impressions will be.

Title and Splash Screen Design Tips

Titles are not decorative labels. They are orientation devices. They tell your reader what kind of map this is, what the map is *claiming*, and what kind of thinking it expects. In web maps, titles often get "outsourced" to surrounding page context, dashboard headers, or assumptions in the designer's head. Outsourcing becomes riskier with LLMs, however, because models will often generate a generic frame when you don't provide one.

A splash screen or brief intro frame can do the same job when interpretation matters more than speed. Used sparingly, it gives users just enough context to avoid misreading the map and time for larger datasets to load in the background. (Big tech companies sometimes use this as a distraction technique to cover up latency.) Used poorly, it becomes friction and self-sabotage. Just like long-winded paragraphs on a print map, many people won't read large chunks of expository text on the web – period. You don't want to annoy people before they even start using your map.

***Recommendations:* Do**

- Write titles that frame interpretation, not just inventory ("What pattern should I notice?").
- Consider a "model-facing" framing line even if the human-facing title stays short.
- Use splash screens only when they truly reduce misinterpretation (and make them dismissible).

***Cautions:* Don't**

- Assume context "from the page" will always follow the map (people embed, screenshot, and share maps constantly).
- Let the model title your map by accident; you'll usually get something bland, tautological, or both.
- Use a splash screen as an excuse to dump metadata on the user when other elements (e.g., a legend and or sources tab) could handle this information more cleanly.

Prompt Example

```
Specify three good human-facing titles accompanied by
a one-sentence framing line that anchors intent, scope,
and tone. Explain the rhetoric and your rationale for
each.
```

Mapped Area, Extent, and the Myth of Infinite Zoom

Another pernicious assumption carried over from modern web mapping is that every map should zoom endlessly. Users expect it. APIs encourage it. Models default to it. Somewhere along the way, zooming and panning became synonymous with interactivity itself. This default behavior deserves scrutiny, because (in my not so humble opinion in this case) it's completely absurd. Numerous studies on panscalar maps and zoom levels show that rather than offer clarity, default zooming and unrestricted panning lead to spatial vertigo and disorientation.

Every map has an extent, whether acknowledged or not. The mapped area defines the spatial bounds of what the map claims to represent and – just as importantly – what it does not. In print cartography, this boundary was explicit. A neat line marked the edge of the world being discussed. In digital maps, the boundary often disappears, replaced by the illusion of infinite space. But illusion is not the same thing as neutrality. When extent is left undefined, interpretation becomes unmoored.

Prompt cartography intensifies this problem. When a model is asked to "create an interactive map," it will almost always enable unrestricted zoom and pan unless instructed otherwise. This is not a design judgment on the model's part; it is a learned default. Most training examples privilege Google-Maps-like behavior, where zooming deeper is always possible and rarely questioned. The result is a map that appears flexible but often undermines its own message.

Consider a county-level thematic map. Allowing users to zoom to street level does not reveal new information; it dissolves meaning. Choropleth values fragment into arbitrary shapes, visual comparisons break down, and users may infer precision that does not exist. The map has not become more informative – it has become rhetorically sloppy. Yet without explicit constraints, this outcome is nearly guaranteed. The example in Figure 9.3 offers an example of a global-scaled issue.

Extent, then, is not a technical setting. It is an argument. By defining a mapped area deliberately, you signal to users what comparisons are valid and what questions the map is prepared to answer. A regional extent invites regional reasoning. A national extent foregrounds differences among states or provinces. A global extent invites ethical questions about projection, scale, and comparability. Zoom controls can support these arguments – or quietly sabotage them.

Prompt cartography forces this decision back into the open. Because models will happily enable interaction by default, restraint must be articulated in language. Statements such as "limit zoom to preserve county-level interpretation" or "constrain panning to the study region" are not implementation trivia; they are expressions of communicative intent. They tell the model – and later the user – where meaning holds and where it breaks down.

This does not mean zoom is inherently bad. Zoom can be powerful when it reveals genuinely different phenomena at different scales, or when it supports exploration without distorting interpretation. But zoom should be justified, not assumed. The same applies to panning. Infinite panning suggests infinite relevance. Most maps do not deserve that claim.

FIGURE 9.3
Unrestricted zoom undermines thematic coherence when users can zoom far beyond the scale at which data remains meaningful. When extent is left undefined, interpretation becomes unmoored, and maps that feel flexible can undermine their own message.

The myth of infinite zoom persists because it feels empowering. It suggests freedom, control, and depth. In practice, it often produces confusion, false precision, and interpretive drift. Good cartography has always involved saying "this far, and no farther." Prompt-based mapping does not remove that responsibility. It makes it harder to dodge.

When you define the mapped area clearly – both for the model and for the user – you replace illusion with intention. The map becomes honest about its scope. And honesty, in cartography, is a form of respect.

Mapped Area and Extent Design Tips

The mapped area is not just a canvas. It's a boundary around what the map claims is relevant. In print, that boundary was explicit. In web maps, it's easy to slide into "infinite world" mode – unrestricted pan, endless zoom, and the quiet implication that everything is equally important everywhere.

Extent is not merely a technical setting. It's a statement of scope and comparison. A county-level choropleth invites certain kinds of reasoning. Letting users zoom into street-level detail doesn't add meaning; it often dissolves meaning and implies precision you don't have.

Recommendations: Do

- Match extent and zoom range to the analytical resolution of the data.
- Constrain panning to the study region when your message depends on bounded comparison.
- Treat "default map extent" as a deliberate editorial decision, not a convenience.

Recommendations: Don't

- Allow zoom so deep that users can admire individual buildings when your data is aggregated to counties.
- Let users pan into irrelevant geography "just because the API can."
- Confuse interaction with insight. Sometimes a fixed extent is the most honest choice.

Prompt Example

```
Constrain zoom/pan to preserve valid comparisons; keep
the map anchored on the study region.
```

What About Neatlines and Map Navigation Tools?

Some old-school cartographers reading this, or those of you who read the first edition of this book, may remember something called a neatline (or even the concept of a frameline). I omitted any mention of these from this edition (outside of this paragraph). The reason: neatlines are implicit in every web map. It feels like discussing neatlines is a bit like using the word "Legend" as the title of your legend – unnecessary. However, this doesn't mean they don't matter. It means neatlines have ascended to omnipresence. Like everything else in prompt cartography, if you don't declare something, the LLM will decide for you. Fortunately, a neatline is almost always a browser edge, app space, or a frame in a multi-neatline comprised dashboard interface. Still important, but always there by default.

That brings us to map navigation tools like pan and zoom controls. These too are approximating discussing neatlines in redundancy. Users assume pinch-to-zoom, or scroll wheel on a mouse. Click-and-drag has been a thing in operating systems since the early 1990s. For accessibility, it may make sense at times to include the plus-and-minus buttons for zooming, but in general, they waste screen real estate, and few people are going to use them.

Map rotation is another thing altogether. It may make sense to include map rotation buttons, in addition to intuitive right-click-and-drag of the mouse or two-finger tap-and-twist on mobile methods. Rotation is a less commonly used feature by the public. It's also a far less commonly needed navigation technique than pan and zoom.

Pan and Zoom Controls Design Tips

Pan and zoom controls are familiar, which is both their strength and their limitation. Because users expect them, designers often include them reflexively, even when they undermine the map's message. Controls should exist to support a meaningful task, not to satisfy a genre convention.

When controls are appropriate, prioritize familiar, low-friction interactions. Click-and-drag (or touch-and-drag) panning is generally more intuitive than on-screen arrow buttons. Scroll-wheel and pinch-to-zoom are widely expected. Visible plus/minus controls can help novices, but they should stay visually secondary to the map itself.

Recommendations: Do

- Use common interaction patterns (drag, scroll, pinch) and keep controls unobtrusive.
- Limit zoom range to what the data can support.

- Provide redundancy when it helps accessibility (mouse + keyboard; touch + buttons).

***Cautions:* Don't**

- Treat the presence of controls as proof the map is "interactive."
- Add multiple overlapping controls that compete for attention.
- Hide meaning behind interaction – if zoom is required to understand the map, the default view probably needs redesign.

Prompt Translation

```
Use standard pan/zoom interactions; omit zoom and pan
GUI components; limit zoom to the data's meaningful
scales; prevent panning more than 4 degrees of longi-
tude or 3 degrees of latitude in any direction past the
maximum extent of all datasets.
```

Interface Elements via Language Prompts

Before we talk about visual hierarchy in full, we need to address which items in a map frequently ruin it: interface elements. In the first edition of *Web Cartography*, interface elements were discussed as graphical objects: buttons, sliders, menus, and controls that mediated interaction between the user and the map. These elements were tangible. You could point to them. You could count them. You could drag them a few pixels left or right until things felt balanced.

That framing no longer holds – at least not entirely.

In prompt-driven cartography, interface elements still exist, but they are no longer primarily widgets. They are descriptions of affordances. A search box, a layer toggle, or a temporal slider does not come into being because it was placed on a canvas; it comes into being because the prompt implies that users should be able to perform a certain kind of action. The interface is an answer to a linguistic request, not a direct construction.

This distinction matters. When interface elements are treated as widgets, design discussions tend to focus on placement, style, and visibility. When they are treated as language, the discussion shifts to intent. Why should the user be allowed to filter this data? What question does the slider help them ask? What cognitive burden does this control introduce, and what interpretive payoff does it offer in return?

In earlier web maps, interface complexity was often constrained by technical effort. Adding a new control required additional code, testing, and layout negotiation. In AI-assisted interface pipelines, the opposite problem emerges. Interface elements are cheap. Models will happily generate filters, toggles, legends, and controls in abundance. The danger is no longer scarcity, but excess.

Interface Element Design Tips

Interface elements are promises. A filter implies the dataset can support valid filtering. A temporal slider implies meaningful change over time. A layer toggle implies comparability across layers. In prompt cartography, interface creation is very cheap, which means the temptation is to generate a map that looks powerful but behaves like an obnoxiously loud toy.

The best rule is simple: every interface element should correspond to a real question the map is prepared to answer. If you can't name the question, you probably don't need the control. If the question is muddy, the control will be muddy. Either way, the interface becomes cognitive tax.

Recommendations: Do

- Add controls only when they enable a meaningful, defensible action.
- Keep interface secondary to the mapped data unless the interface is the point (dashboards).
- Treat accessibility as a premise: controls should be describable, navigable, and readable.

Cautions: Don't

- Generate "filter confetti" (toggles and widgets added because they look modern).
- Add interactions that imply analytic legitimacy the data can't support.
- Let panels cover the mapped area by default unless your design explicitly calls for it.

Prompt Example

```
Only include map controls tied to explicit user direc-
tives; keep controls unobtrusive and accessible.
```

This is why prompt cartography demands a different kind of restraint. Every interface element should correspond to a meaningful question that the map is prepared to answer. If the question is vague, the interface will be vague. If the question is unnecessary, the interface becomes noise. The map may still function, but its communicative clarity erodes.

Pan and zoom controls illustrate this well. In the first edition, these were discussed as fundamental interaction mechanisms – nearly universal components of web maps. That observation was accurate at the time and largely remains so. What has changed is not their prevalence, but the ease with which they are enabled without reflection. When interaction is free, intentionality becomes expensive.

Other interface elements follow the same pattern. Layer lists suggest comparability. Filters suggest analytical legitimacy. Temporal sliders suggest meaningful change over time. When these affordances are introduced without clear purpose, they invite users to perform actions the map cannot meaningfully support. The interface promises insight; the data cannot deliver it.

The solution is not to ban interface elements but to describe them precisely. Instead of asking for "an interactive map," you articulate the interactions that matter and omit those that do not. You describe not only what users can do but what they should do. The model will still generate interface components – but now they are constrained by purpose rather than habit.

This reframing also clarifies accessibility concerns. When interface elements are treated as language, accessibility becomes a design premise rather than a retrofit. If an interaction cannot be described clearly in words, it likely cannot be navigated clearly by users. Prompting forces this reckoning early, where it belongs.

Interface elements have always shaped how maps are read. What has changed is that they now emerge from linguistic intent rather than direct construction. When treated carelessly, they multiply and distract. When treated as language, they become powerful tools for guiding attention, inquiry, and understanding.

Visual Hierarchy Remains in Charge

Visual hierarchy remains the single most important organizing principle in map design. This was true in print cartography, it remained true during the rise of interactive web maps, and it remains true in prompt cartography. If a map fails, it is rarely because the data is wrong. It fails because the reader has not been guided on where to look.

In traditional cartography, hierarchy was constructed manually. Cartographers adjusted color value, symbol size, contrast, and placement until the visual order felt right. In prompt-based cartography, hierarchy emerges indirectly through language. The model interprets what you emphasize, what you mention first, what you describe in detail, and what you leave vague. Those linguistic signals are translated – sometimes faithfully, sometimes not – into visual prominence.

This translation is imperfect, and that imperfection is precisely why hierarchy deserves renewed attention (Figure 9.4). LLMs are excellent at generating many map elements. They are less reliable at deciding which elements should dominate. Without explicit guidance, models tend to flatten hierarchy. Legends become visually heavy. Interface controls compete with the mapped data. Titles shrink or disappear. Everything becomes "important," which is another way of saying nothing is.

This is not a bug. It's a reflection of how models learn. Training data is full of maps where hierarchy is implicit, inconsistent, or poorly articulated. When the prompt does not intervene, the model averages. Average hierarchy is almost always bad hierarchy.

Visual Hierarchy Design Tips

Visual hierarchy is the reading order of your map. It is how you tell users where to look first, what matters most, and what can wait. When hierarchy fails, the map doesn't feel wrong in a single obvious way – it just feels tiring. Everything competes, and the user quietly quits.

LLMs can generate many elements competently, but they often flatten hierarchy unless you explicitly assert priority. Legends get heavy. Panels compete. Titles vanish. The mapped data becomes merely one rectangle among many rectangles.

Recommendations: Do

- State what must dominate (usually the mapped data) and what must recede (often controls and secondary text).
- Use contrast, size, and placement intentionally, not uniformly.
- Treat hierarchy as editorial: it begins with what the map is *about*, not what it *contains*.

Cautions: Don't

- Give every element equal visual weight "to be fair." That's not fairness; it's confusion.
- Allow interface chrome to outshine the data.
- Trust defaults to establish hierarchy. Default is average. Average is rarely good.

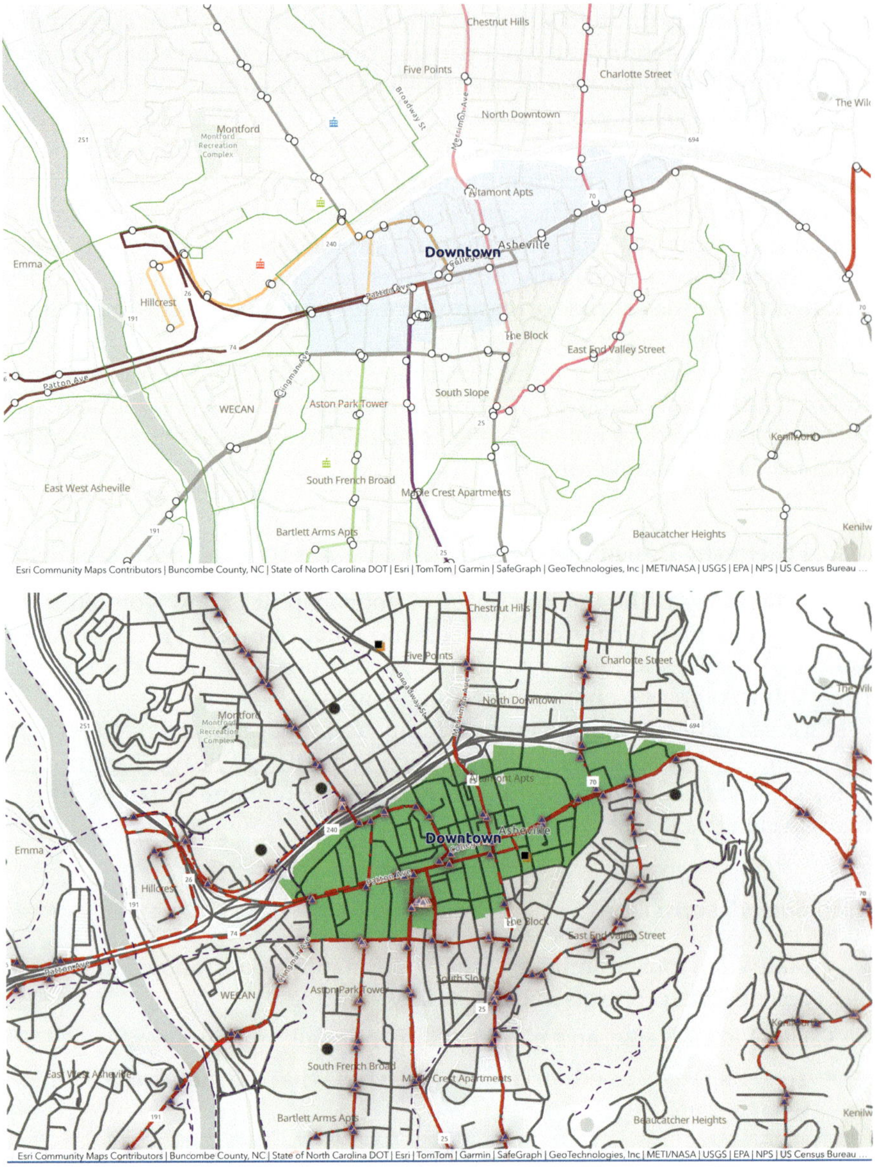

FIGURE 9.4
When hierarchy is not specified, secondary elements often overpower the mapped data, leading to cognitive confusion rather than clarity.

Prompt Example

```
Make the mapped data primary; set all other elements
as supportive, unobtrusive, and visually subordinate.
```

The solution is not to micromanage aesthetics but to articulate priority. Hierarchy begins with intent. Before any symbol is drawn or any layout is generated, the cartographer must know what the map is about. Not what it contains but what it communicates. That distinction mattered in print. It matters more now because ambiguity propagates.

Hierarchy also applies to elements beyond the mapped data. Titles, legends, interface components, annotations, and supplementary graphics all participate in the visual order. In many AI-generated maps, these secondary elements overwhelm the primary message. A beautifully symbolized dataset can be rendered cognitively invisible by an overbearing legend or an aggressively styled control panel. When this happens, the map has not failed technically – it has failed editorially.

If you are newer to cartography, or just looking for some guidance on visual hierarchy, Figure 9.5 may be of assistance. Building on Dent and Torguson's (2008) stellar visual hierarchy system for cartography, these hierarchy templates can be shared with your LLM assistants to help them establish base rules for visual hierarchy and anchor their hierarchy designs (Figure 9.5).

Prompt cartography forces hierarchy decisions upstream. Instead of adjusting opacity or font size after the fact, you decide – using language – what deserves attention and what should recede. You describe which elements should "dominate," which should "support," and which should "remain available but unobtrusive." These are not vague aesthetic preferences. They are operational instructions that shape the entire output.

It is worth stating this plainly: visual hierarchy does not emerge automatically from good data or clean code. It never has. This reality remains unchanged in prompt cartography. The cost of ignoring visual hierarchy is just higher. When visual hierarchy is left unspecified, the model will still produce one, and you'll rarely mistake it for your own.

Layout Systems: From Screen Real Estate to Spatial Grammar

Layout is where many web maps quietly succeed or fail.

In the first edition of *Web Cartography*, map layout was described as the organization and balance of all map elements on the screen. The goal was simple, if deceptively so: to arrange components in a coherent, intuitive

VISUAL HIERARCHY LEVELS FOR WEB MAP DESIGN

General Interest Web Maps		Thematic Web Maps		Animated Web Maps	
Level 1	Title / Splash Screen Map Symbology Key Reference Data Info Windows (opened)	Level 1	Title / Splash Screen Thematic Visualization Legend	Level 1	Title / Splash Screen Animation Symbology Map Symbology Temporal Legend / Interface
Level 2	Base Map Base Map Labels Navigation/Directions Tools	Level 2	Base Map (generalized) Info Windows (opened) Chart Graphics	Level 2	Base Map Legend Info Windows (opened) Locator Map
Level 3	Map Interactivity Pan / Zoom / Rotation Tools Print / Share Map Features	Level 3	Base Map Labels Map Interactivity Pan / Zoom / Rotation Tools Menus with Additional Tools	Level 3	Base Map Labels Map Interactivity Pan / Zoom / Rotation Tools Menus with Additional Tools
Level 4	Locator Maps Chart Graphics Multimedia Supplements	Level 4	Locator Maps Multimedia Supplements	Level 4	Multimedia Supplements Chart Graphics
Level 5	Supplemental Information Attribution and Copyright Neatlines / Grids / Graticules Tool Tips	Level 5	Supplemental Information Attribution and Copyright Neatlines / Grids / Graticules Tool Tips	Level 5	Supplemental Information Attribution and Copyright Neatlines / Grids / Graticules Tool Tips

FIGURE 9.5
Three visual hierarchy templates for interactive web maps to provide prompting guardrails. The first is for simple locator or reference maps. The second for thematic representation. The third for animated, narrative, or dynamic maps.

package that allows users to extract information easily and confidently. That definition still holds. What has changed is not the purpose of layout but the way layout decisions are made – and who makes them.

In print cartography, layout decisions were bounded by the physical page. In early web cartography, they were constrained by screen size, resolution, and browser limitations. Today, in prompt-driven design, layout is no longer something you "set" directly. It is something you describe into existence. The cartographer specifies relationships, priorities, and boundaries, and the system resolves them into a spatial arrangement. This shift elevates layout from a technical concern to a form of spatial grammar. Layout decisions are the syntax of your map's message.

Screen real estate remains finite. No amount of interactivity changes this fact. A map viewed on a wall-sized display, as opposed to one viewed on a smart device, must still negotiate competing demands for attention. The mapped area, legends, controls, text, charts, and narrative elements all vie for limited space. The difference now is that models tend to treat screen space optimistically. If not constrained, they assume that more can fit, that more should fit, and that more interaction is always beneficial. Such optimism is misplaced.

A layout that works is not one that includes everything but one that establishes clear zones of meaning. The mapped area must read as primary or secondary relative to other elements, not merely coexist with them. Supporting components must feel intentionally placed, not opportunistic. When layout fails, users are forced to decide what matters on their own – and many simply disengage.

One of the most durable distinctions introduced in the first edition was between fluid and compartmentalized layouts. Fluid layouts allow map elements to overlap, collapse, or dynamically adjust as users interact (Figure 9.6). Compartmentalized layouts (as shown in Figure 9.7) divide the interface into stable regions, each with a clear role. Neither approach is inherently superior. Each implies a different cognitive contract with the user.

Fluid layouts emphasize exploration and immersion. They work well when the map itself is the primary object of attention and when interface elements can afford to recede or temporarily disappear. Fragmented layouts emphasize clarity and persistence. They are better suited to analytical tasks, dashboards, and situations where users must compare information repeatedly without reorienting themselves.

Prompt-driven systems are, of course, capable of generating both – but as of this writing, they tend to be biased toward fluidity unless told otherwise. This bias reflects contemporary web aesthetics, not cartographic wisdom. Without explicit instruction, models often allow legends to float, panels to overlay the mapped area, and controls to appear and disappear in ways that feel modern but undermine stability. The result is a map that looks dynamic but feels unreliable.

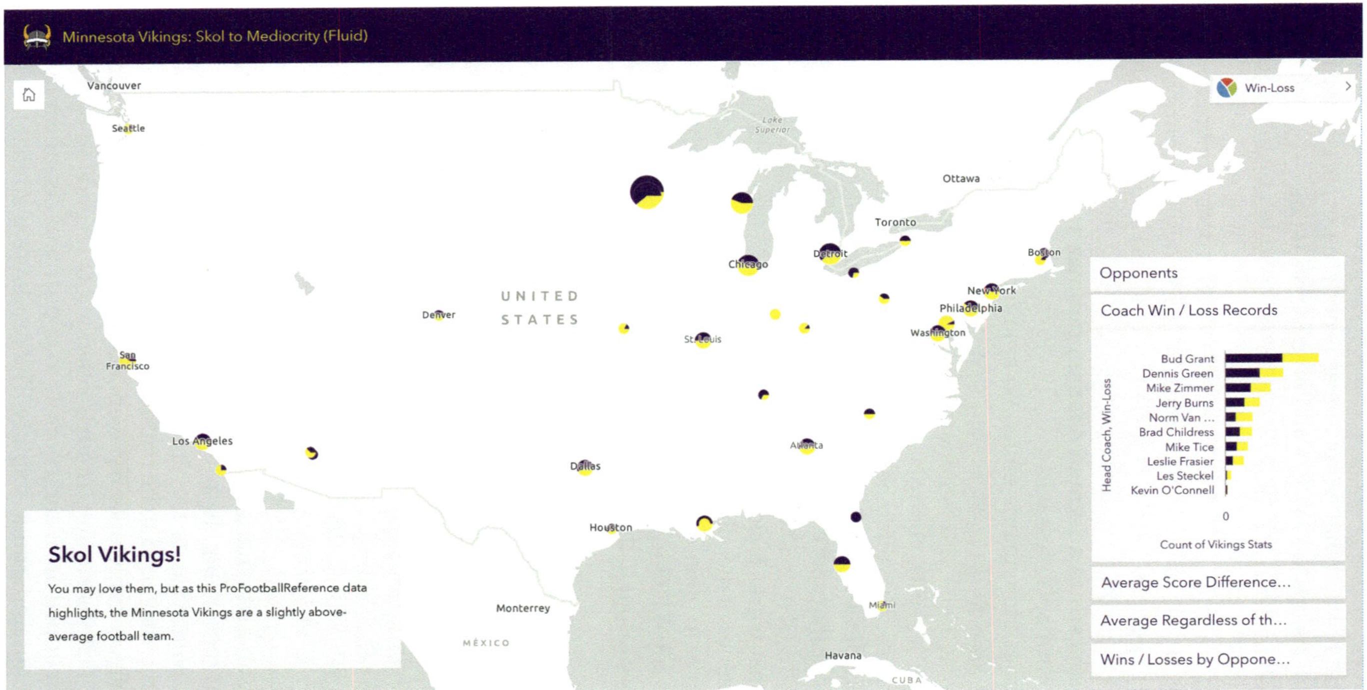

FIGURE 9.6
Fluid layouts often privilege map immersion.

FIGURE 9.7
Compartmentalized layouts imply analytics and multicomponent, linked interactive visualizations.

Layout Systems Design Tips

Layout is the spatial grammar of your interface. It defines stable zones, predictable reading order, and how much cognitive work users must do just to get oriented. In web maps, the common failure is not ugliness – it's aimlessness. Users don't know where to look for meaning, controls, context, or explanation, so they bounce.

Fluid layouts can be excellent when the map is the star and supporting elements can tuck away. Compartmentalized layouts can be excellent when stability and comparison matter. The problem is not choosing one. The problem is mixing both unintentionally, so things float, overlap, and shift without a rationale.

Recommendations: Do

- Decide whether your map is an immersive experience (fluid) or an analytical interface (compartmentalized).
- Define stable regions: map area, explanation, controls, legend, charts.
- Design for mobile explicitly; "responsive" does not automatically mean "usable."

Cautions: Don't

- Let panels cover the map by default unless the map is truly secondary.
- Allow key elements to jump around between states; it increases cognitive load fast.
- Treat empty space as wasted space. Empty space is often the reason a map feels readable.

Prompt Example

```
Specify a stable layout grammar (zones, positions, and
element priority) and explicitly describe mobile versus
desktop behavior and panel behavior.
```

This is where spatial grammar matters. Just as sentences rely on structure to convey meaning, layouts rely on predictable spatial relationships. Users learn where to look for explanations, controls, and context. When those locations shift unnecessarily, cognitive load increases. Prompt cartography gives you the opportunity – and the responsibility – to specify these relationships clearly: which elements remain fixed, which are secondary, and which may adapt without confusion.

Older discussions of layout often focused heavily on resolution, pixels-per-inch, and device-specific constraints. While those considerations are not irrelevant, they are no longer the conceptual center of the problem. Modern systems can respond to screen size dynamically. What they cannot infer reliably is intent. They cannot know whether a legend should dominate or whisper, whether a chart is explanatory or decorative, or whether the map is the star or part of an ensemble. Layout, then, isn't about filling empty space; it's about structuring the space to guide your map user's attention.

When you prompt for layout, you are not asking the model to arrange boxes. You are describing a reading order, a hierarchy of importance, and a rhythm of interaction. Get that right, and even a visually modest map can feel confident and intelligible. Get it wrong, and no amount of polish will save it.

Layout systems endure because human perception has not changed. Screens have multiplied. Tools have evolved. But users still seek orientation, balance, and visual logic. Your prompts must now provide those things explicitly – or accept whatever grammar the model invents on your behalf.

Layout and Balance

Tying back to visual literacy, I want to reemphasize that visual balance does not mean symmetrical balance. Map elements have weight that can offset one another. This means you can align all of your map elements so that the map feels balanced to a user while still being asymmetrical. Never conflate the idea of a balanced map to one that has elements on both sides of it. You may end up with a balanced yet unfathomably horrible user interface. (Just imagine if a company produced a piece of mapmaking software package with two important menus on opposite sides of the app. Ludicrous!)

Legends, Insets, and the Danger of Letting the Model Decide

Legends are among the most misunderstood elements in web maps. They are often treated as accessories – necessary, but secondary to the "real" work of mapping. In practice, legends are explanatory devices. They tell users how to translate marks into meaning. They may not always be needed – for example, when you have a single layer of features that are differentiated via symbology in any way. However, when legends are required, if they aren't implemented well, an otherwise great map may fail.

In *Web Cartography*, legends were described as interpretive tools rather than ornamental ones. That framing is even more important in prompt-driven cartography, because models are exceptionally good at producing legends that look correct while explaining very little. Colors are named. Symbols are listed. Units may even be present. And yet the legend may still fail to answer the most important questions a reader has.

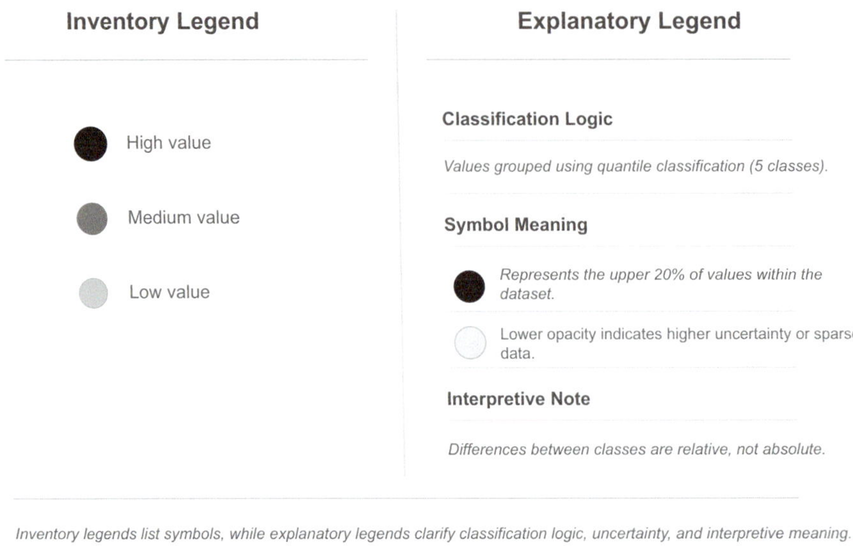

FIGURE 9.8
Inventory legends list symbols, while explanatory legends clarify classification logic, uncertainty, and interpretive meaning.

What does darker mean? Are these values comparable across space? What assumptions went into the classification? (You can sense the map-nerd level increasing with each question!)

Unless prompted otherwise, models tend to generate inventory legends – lists of symbols divorced from analytical purpose. These legends satisfy formal expectations but do little to support understanding. They may be correct in a narrow sense, but be careful, as they are often misleading from the broader perspective (Figure 9.8).

Inset maps, a "geography legend" so to speak, pose a related problem. Locator maps, context maps, and small multiples have long been used to situate a mapped area within a larger whole. In print cartography, such insets were precious. They consumed page real estate and required careful justification. In AI-generated maps, insets are cheap. Models will happily add them, resize them, stylize them, and reposition them without asking whether they serve a meaningful role. And much of the time, they really don't. Figure 9.9 shows a prompted inset map.

This *legendary* abundance introduces a new risk: the appearance of thoughtfulness without any substance. (The Oxford Dictionary definition of the word "enshittification" comes to mind here, but I promised my daughters I wouldn't be too crass in this edition of the book, so…)

A map inset that does not clarify scale, orientation, or context is visual clutter. A legend that does not explain interpretation is decorative text.

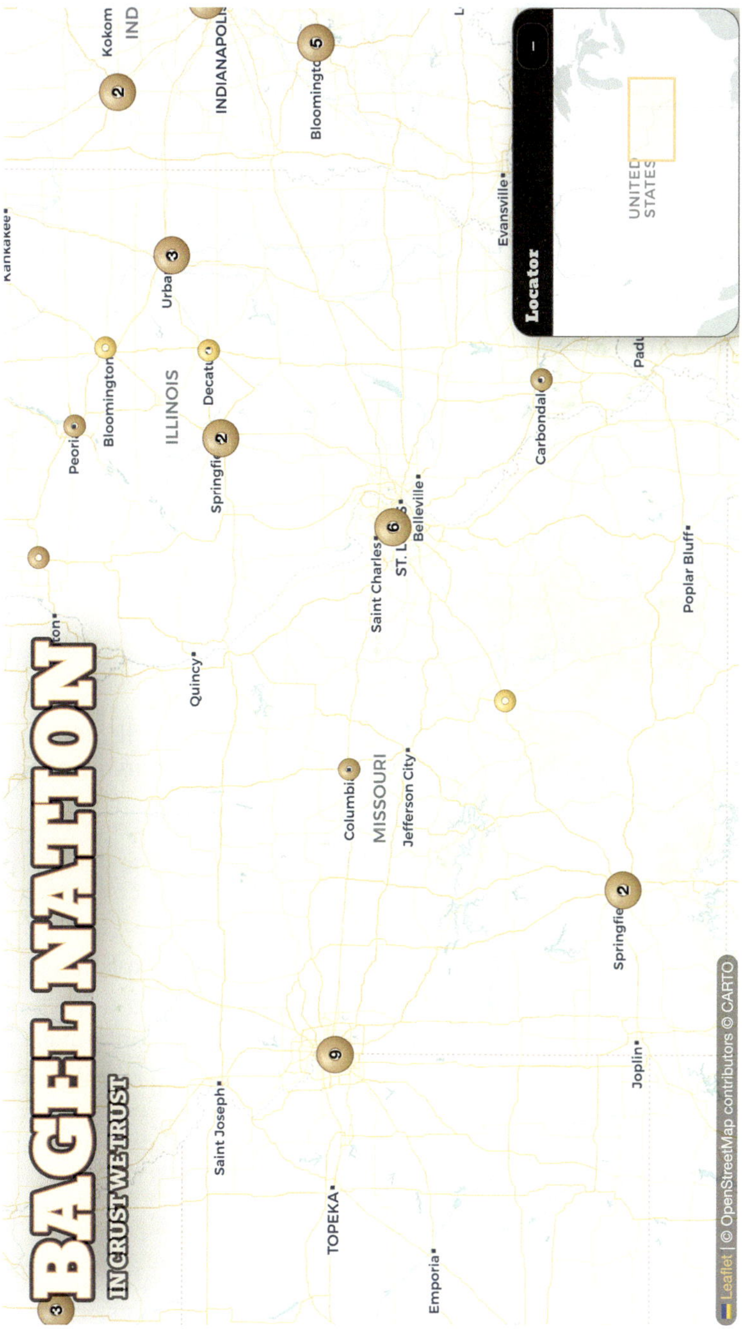

FIGURE 9.9
Inset maps are useful when a basemap does not provide a lot of context for where a user is in a larger geographic region. (Or when they are simply unfamiliar with fly-over country in the US as they might be here.)

In both cases, the model has made a decision that feels helpful but may actively undermine comprehension. The map looks complete. The user feels informed. But the cognitive work has not been done.

LEGENDS AND INSET MAPS DESIGN TIPS

Legends are explanatory devices, not decoration. A legend can be formally correct and still fail the reader. Many autogenerated legends become inventory lists: symbol, color, label – done. But the reader's real questions are interpretive: what do the colors mean, how were classes chosen, are values comparable, and what should I *conclude*?

Insets have the same failure mode. They can create the appearance of thoroughness while adding little. Locator maps are useful when location and context are truly unclear. Otherwise, an inset becomes visual clutter that steals attention from the main message.

Recommendations: Do

- Make legends explain interpretation: classification logic, units, uncertainty, and what "high/low" implies.
- Use insets only when they answer a question the main map cannot.
- Treat legend placement and prominence as part of hierarchy, not a leftover corner-filler.

Cautions: Don't

- Accept a legend that simply mirrors symbology without explaining it.
- Add insets "because maps have insets" and you drift toward the generic dashboard look – busy, safe, and strangely uninformative.
- Let the legend title itself "Legend" unless you're feeling nostalgic in a deeply unhelpful way.

Prompt Translation

```
Require an explanatory legend (not inventory) and jus-
tify any inset by the specific question it answers.
```

When legends and insets are not specified, the model fills the gap with convention. Convention is not inherently bad but frequently insufficient. A legend that simply mirrors symbol choices tells the reader what is shown, not how to read it. A locator inset that merely situates geography without reinforcing the map's purpose adds little beyond reassurance.

The remedy is explicitly prompting for intent. The cartographer must describe not only a legend's, but its function in the map as a system. (Otherwise, it will sit there like an unused appendix, just waiting to cause an issue.)

Should the legend emphasize classification logic? Should it foreground uncertainty? Should it remain unobtrusive until needed, or should it anchor interpretation from the outset? These are editorial decisions, not stylistic ones.

The same applies to insets. An inset should answer a specific question the main map cannot. "Where is this region relative to others? How does local detail relate to a broader pattern?" If no such question exists to answer, the inset does not belong – no matter how standard seems.

There is a temptation to trust the model here, because legends and insets feel auxiliary. They are not. They are often the only parts of the map that explain why a visualization looks the way it does. When left to defaults, they become artifacts of habit rather than instruments of clarity.

In prompt-driven cartography, explanatory elements deserve as much care as the mapped data itself. If you do not decide what they should explain, the model will decide what they should resemble. Those two outcomes are not the same.

From Design Rules to Prompt Literacy

By this point, one thing should be clear: prompt cartography is not about learning better tricks. It is about learning a new way to speak design. Once you can speak design, you will no longer need to spend hours on elbow grease things (HTML coding, python script writing, data cleaning). Yes, these skills will remain invaluable; no, they just won't be a core task of a cartographer's job anymore.

Everything discussed in this chapter – map elements, framing, extent, interface logic, hierarchy, layout systems, legends, and insets – existed long before LLMs entered the scene. None of these concepts are new. What is new is how they must now be expressed. The medium has changed from pixels and panels to language and intent. The rules survived, but the interface for building them continues to evolve.

A prompt-literate cartographer does not ask a model to "make a map" and hope for something usable. They articulate purpose, priority, and constraint with enough clarity that ambiguity has nowhere to hide. They understand that any omission is an instruction (use defaults), and every vague phrase is an invitation for the model to substitute its own (often harebrained) assumptions – usually borrowed from the wackiest parts of the modern web.

This is why the cartographer-as-director metaphor matters. You are no longer pushing symbols around a canvas. You are *staging a performance*. The map is the set. The data are the actors. The interface elements are supporting cast. The audience still arrives with expectations, habits, and limited patience. Your job is to ensure that what unfolds on the screen communicates what you intend – not merely what is possible. And you must manage a large team of LLM-based agents to do this.

Directing how your map communicates is also about building trust with your users. Outputting maps with poor layout decisions, subpar balance, and unnecessary widgets can lead to mistrust (i.e., it looks unprofessional) and skepticism (i.e., probably not a good thing if your maps live down to the stereotype of AI mediocrity). You now have as many helper agents at your disposal as you require, speeding up tasks and allowing you to do things you probably never could before. Use them to produce high-quality maps, not just fast ones.

LESSON LEARNED FROM PERSONAL EXPERIENCE

Understandably, sometimes you just need to get a job done. Perhaps the audience is small or you're testing new ideas. My colleague and I, for example, decided it would be a great idea to create a map every single day of the year using a fledgling LLM agent I had created.

The #365DaysOfMaps challenge sounded like great fun in December 2025. It took less than a week for the reality of the task to set in. Producing a web map every day – not during working hours – is a chore. Aside from a handful of gems, our output is fast and half-assed most days.

Though we forewarned people that most of the maps would be created in under an hour and that it was simply a proof-of-concept idea, the output won't help prompt cartography overcome biased stereotypes about map quality. I sincerely apologize for this.

To produce high-quality interactive web maps, always consider how these concepts translate directly into your prompt thinking:

- *Visual Hierarchy Becomes Linguistic Priority*. What you describe first, emphasize most strongly, or qualify most carefully will tend to dominate the output. If everything is described with equal weight, the map will look like it too.
- *Extent Becomes Narrative Boundary*. When you specify geographic scope – and just as importantly, limits on zoom and pan – you are telling the model where meaning holds and where it breaks down.
- *Layout Becomes Spatial Grammar*. When you describe which elements should remain fixed, which should recede, and which should support the mapped data without competing for attention, you are writing a reading order into the interface.
- *Legends and Insets Become Explanatory Commitments*. When you tell the model what these elements must explain, rather than merely that they must exist, you prevent the appearance of clarity without its substance.

Such decisions have immediate, observable consequences.

A prompt that says "Create an interactive choropleth map of county-level unemployment rates" will almost certainly produce a serviceable but forgettable map that perhaps includes infinite zooming, a generic legend (that has "Legend" as the title – gasp!), a title that restates the obvious, and an interface that competes politely with itself (look here… no, over there… no, wait, back *here*).

A prompt that instead specifies intent might say, in effect:

```
Design for county-level comparison; limit zoom to preserve interpretability; ensure the legend explains classification logic; keep interface controls unobtrusive; prioritize the mapped data over all secondary elements.
```

Notice what is happening here. You are not dictating implementation details. You are asserting design judgment. The model fills in the mechanics. You retain authorship.

This distinction separates prompt cartography from prompt tinkering. Anyone can iterate until something looks "fine." Prompt cartographers know why something works, can diagnose when it doesn't, and can correct it deliberately. They are not surprised by outcomes; they are occasionally disappointed; they are rarely confused.

Importantly, prompt literacy also changes how you will critique maps. When a generated map fails, the question is no longer "Why did the model do this?" It is "What did *I fail* to clearly specify?" That question is empowering.

It returns agency to you – the director. It allows critique to function as it should – iteration, not defense of your own ego.

Prompt cartography is not a break from traditional or web cartography; it is a compression of it. All the old rules still apply. In fact, if you are new to cartography, I recommend you also check out *Web Cartography* or any of the myriad stellar cartography fundamental books available (many of which are listed in the references and further readings section at the end of this book) to help you become more familiar with some of the key concepts of cartography body of knowledge.

Better still: don't be afraid to ask your LLM assistants to teach you about cartography – they are well read! I know my editor is groaning and shaking her head right now, but it's true. One LLM company was forced to pay me a settlement, because they allegedly pirated *Web Cartography*. I have built and will post links to some free LLM assistants that you can use as cartographic tutors on the companion website.

Takeaways

- **Map Elements Will Often Show Up Until You Say No**

  Titles, legends, layout, and interaction do not disappear in the prompt-driven design pipeline. They emerge from defaults unless explicitly designed.

- **Omission Is a Design Decision**

  When a map element is not specified, the model will still decide how it appears. Delegating silently is still delegation.

- **Titles Are Interpretive Instructions**

  Titles frame how both users and models understand a map before any symbols are read. Neutral-sounding titles still guide cognition.

- **Extent Is an Argument About Relevance**

  Zoom limits and mapped boundaries signal where meaning holds and where it breaks down. Infinite zoom implies confidence most datasets cannot support.

- **Interface Elements Imply Questions**

  Filters, sliders, and controls suggest analytical legitimacy. If the data cannot answer the implied questions, the interface misleads.

- **Interactivity Is Cheap but Intent Is Not**

  AI-assisted design make it easy to add interaction. The challenge is deciding which interactions deserve to exist.

- **Visual Hierarchy Does Not Emerge Automatically**

  Without explicit guidance, models flatten importance. When everything competes for attention, nothing communicates clearly.
- **Hierarchy Is Expressed Through Language**

  What you emphasize, describe first, or constrain carefully in a prompt tends to dominate the visual outcome.
- **Layout Functions as Spatial Grammar**

  Layout establishes reading order, stability, and trust. Predictable spatial relationships reduce cognitive load more than novelty.
- **Legends and Insets Explain Meaning, Not Appearance**

  A legend can be correct and still fail. Explanatory elements must clarify interpretation, not merely mirror symbols.
- **Prompt Literacy Makes Design Inspectable**

  When intent is articulated in language, failures can be diagnosed and corrected rather than guessed at.
- **Prompt Cartography Amplifies Responsibility**

  Delegating execution does not reduce authorship. It removes plausible deniability.

Conceptual and Applied Exercises

Default Autopsy

Find (or recall) a web map that feels "fine" but forgettable. Identify two elements that you suspect were default-driven (e.g., title phrasing, legend style, zoom range, control placement), and explain how each default subtly changes what the map *implies*.

Extent as an Argument

Imagine you are mapping county-level opioid overdose rates (or any county-level phenomenon). Write a short paragraph explaining what interpretive harm can occur if users are allowed to zoom to street level, and what a more honest zoom/extent policy would communicate about certainty and comparison.

LLM Defaults Stress Test

Ask an LLM: "Create an interactive map showing unemployment rates by county in {your chosen region}" *Don't add any constraints.*

Then ask the LLM: "Now explain what you assumed about title, legend purpose, zoom limits, interface controls, and visual hierarchy."

Compare the assumptions to what you would have chosen intentionally.

Structured Prompt Exercise

Draft a 2–3 sentence tool-agnostic (i.e., generic) *map element directive* that you could paste into any prompt-built mapping pipeline. It must explicitly:

1. state what should visually dominate;
2. constrain extent/zoom to preserve valid comparisons; and
3. require an explanatory legend (not an inventory list).

10

Color Theory and Palette Design

The city council chamber was packed. A "heat map" of household energy use glowed crimson across the projector, and the mayor frowned.

> "Why's my neighborhood *red*?! It's not dangerous," someone asked.

The cartographer – half terrified, half amused – explained that red only meant high kilowatt hours, not catastrophe or danger. But the damage was already done. The room had felt a crisis before anyone read a title or legend.

My first job after graduating with my undergraduate degree was as a regional planner. Though the previous scene is metaphorical, I assure you it's not off base. (Except in my case, I was showing someone's house in a hot pink color, and they were absolutely incensed.) This and similar situations can unfold in almost any municipality today. When presenting maps to the public, suspicions and tempers can flare. It's understandable, and if I could talk to my younger self, I would argue it's also largely preventable. (Hot pink? Really?!)

The takeaway here is that color must never be afterthoughts or shouldn't be selected simply because they look "kinda cool." Color is embedded with culturally distinct arguments. A bold red persuades before a single number is read (observe, for example, Figure 10.1).

This chapter is about directing the use of color in map design through language. Essentially, prompt cartography reframes our aesthetic choices as parameters of persuasion, whether we realize it or not. A good prompt makes explicit what cartographers have long handled implicitly: hue, value, contrast, texture, hierarchy. These are the same decisions designers have always made – now stated in plain English and rendered reproducibly in code or JSON.

Before we can teach a model to draw, we must first teach it to see what we mean. A large language model (LLM) does not perceive color the way humans do; it parses associations, words, and patterns about colors as embedded in its training and RAG knowledge sources. That is both our advantage and our challenge. By describing colors and palettes in precise but disciplined language, we give the model scaffolding for aesthetic judgment. The more intentional our adjectives and constraints, the more faithfully the model becomes a translator of color intent rather than a generator of stylistic noise it has learned from reading too many social-media posts by graphic design influencers.

DOI: 10.1201/9781003797920-12

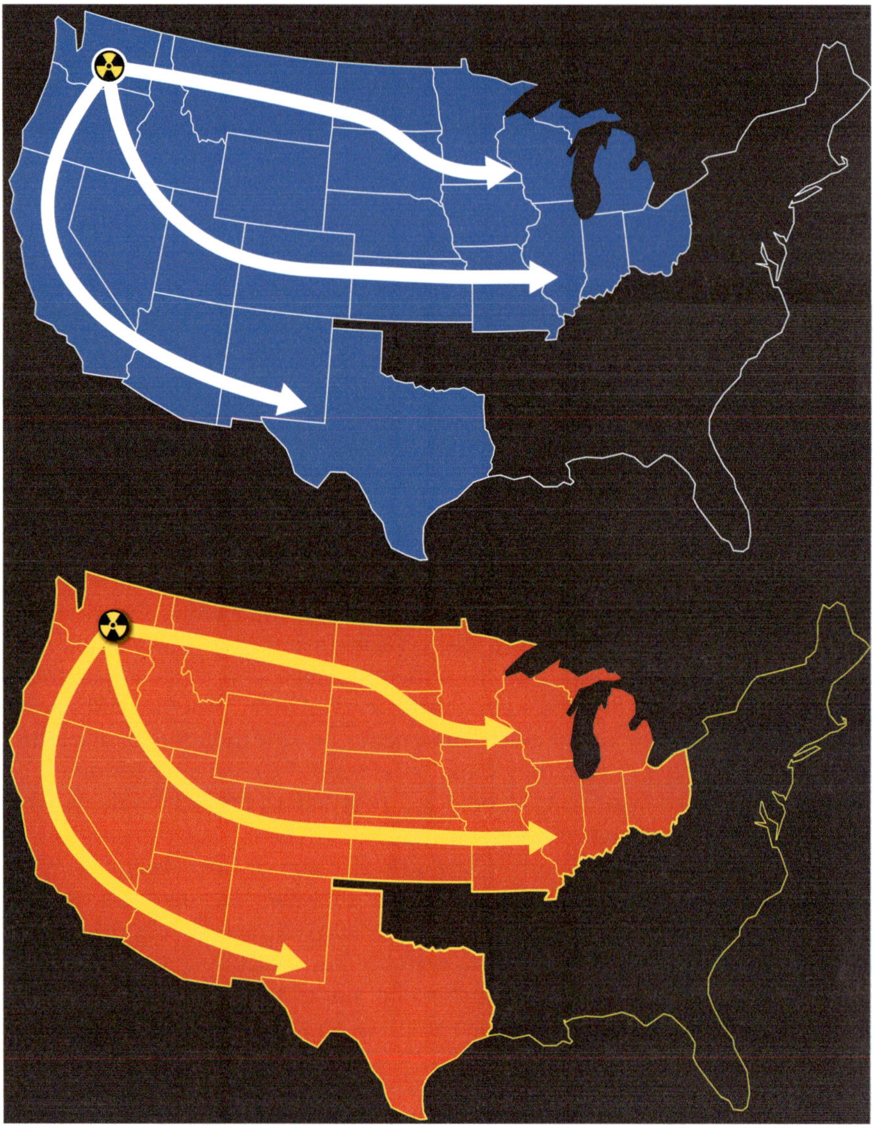

FIGURE 10.1
Two renderings of identical data: the first using alarmist reds and the second a neutral sequential palette. Same exact symbols but different story.

A Brief Color Literacy (So Prompting Isn't Guesswork)

First things first: you will need to establish a shared vocabulary about color with an LLM. What follows is not a deep dive into color science nor an attempt to turn you into a designer overnight. It is simply enough foundation to prevent color choices – human or machine – from becoming accidental arguments (as explained in the introduction).

For those of you who read the first edition of Web Cartography, a lot of this section may feel like a trip down memory lane. That's because much of it is near verbatim from the first edition. But if you are like me – someone who buys every *New York Times* bestselling book on quantum mechanics and rereads the same information in every book because it's never ceases to be fascinating – this section may well be worth reading for knowledge reinforcement. It's easy to forget details – without a built-in RAG… yet – so you may want to treat this section as a succinct reminder.

Color feels intuitive, but it is not simple. Most confusion arises because we talk about color emotionally ("warm," "calm," "urgent") while color systems describe it mathematically. Prompt cartography lives precisely on that boundary. Understanding a few core ideas makes your color prompts clearer, more predictable, and easier to critique.

Hue, Value, and Saturation

Nearly all practical color decisions in cartography can be described using three properties: hue, value, and saturation.

Hue is what most people mean when they say "color." Red, blue, green, orange – these are hues (Figure 10.2). When you tell a model to "use blue," you are specifying hue, but very little else. Hues are always embedded with cultural bias and meanings.

Color value refers to a hue's lightness or darkness (Figure 10.3). A pale yellow and a deep navy will feel radically different beyond just having different hues. Value is the primary driver of legibility and contrast, especially in

FIGURE 10.2
Hues are colors as described via nouns (blue) or compound nouns (royal blue).

FIGURE 10.3
Values are shades of colors, as described with adjectives (e.g., light and dark).

FIGURE 10.4
Saturation is brightness or conversely grayness of colors (e.g., muted, neon, etc.).

choropleth maps and font colors. Light valued black fonts fade, for example, because they are light gray.

Saturation describes a hue's intensity. Highly saturated colors feel vivid and energetic; desaturated colors feel muted, analytical, or subdued (Figure 10.4). (My teen daughter tends to call desaturated palettes "sad beige mom" colors. I blame the internet for this.)

These three color properties always work together. A bright red (high saturation, medium value) will feel urgent or alarming. A dusty red brown (low saturation, darker value) may feel historical or restrained. When prompts fail to provide the color desired, it is often because they specify hue but leave value and saturation to chance. On the other hand, when you ask an LLM to "limit luminance contrast" or "use a muted palette," you are implicitly manipulating value (don't have too much contrast between values) and saturation (desaturate those colors) – even though you may not use these terms directly.

Warm and Cool Colors (And Why They Matter)

Hues are often grouped loosely into *warm* (reds, oranges, yellows) and *cool* (blues, greens, violets). This distinction is not scientific, but it is culturally persistent. Warm colors tend to advance visually and attract attention, whereas cool colors tend to recede and feel calmer (Figure 10.5).

This matters, because at their core maps present numerous symbols in various colors, all fighting for emphasis and attention. A warm color placed on a map almost always claims importance, whether intended or not. When prompting for tone – e.g., *analytically calm* but *authoritatively urgent* – you are

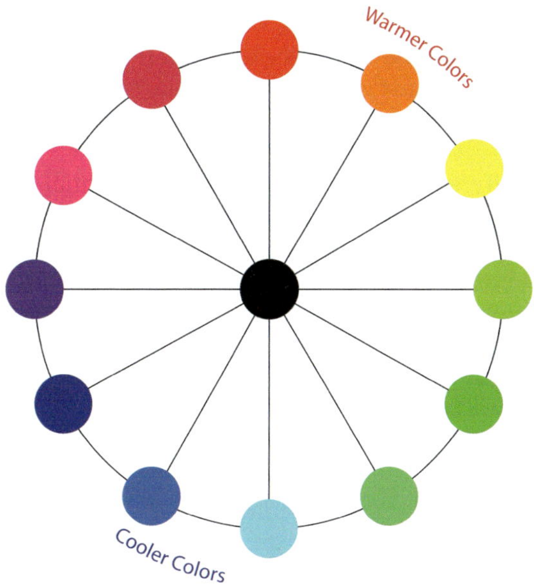

FIGURE 10.5
The color wheel organizes colors based on primordial interpretation – certain colors, due to their association with objects in the natural world, are associated with warmth and others with coolness. Effective prompts guide an LLM into choosing the correct constraints for color selection based on the nature of the data being represented.

implicitly guiding the LLM which colors to use from the warm–cool spectrum. If you don't guide your LLM, it may choose an unintelligible default.

LLMs inherit color associations from everything they have consumed regarding visual culture. They will happily lean into defaults, or simply refuse to take color selection into consideration unless you guide and constrain color use.

How Screens Actually Make Color

Digital maps rely on the RGB color model, in which red, green, and blue light are added together to produce other colors. White is the presence of all three; black is their absence. This is known as *additive color*.

Because RGB is numeric, colors are often expressed as hexadecimal values (e.g., #3b7fb6). These numbers are not aesthetic judgments; they represent coordinates in a color space. Whenever you prompt an LLM to return hex codes, you are asking it to be explicit rather than impressionistic.

One important caveat: the same hex color will not look identical on every screen. Displays vary in brightness, contrast, and calibration. This variability is unavoidable, which is why robust color design for screen devices relies on relative contrast and hierarchy, not on exact color matches as it does in print.

Why This Matters for Prompting

At this point, you know enough to see the trap. If you ask an LLM to "make a nice color palette," it must guess at hue, value, saturation, emotional tone, accessibility, and data logic all at once. That is not creativity – it is LLM roulette.

But if you specify even a few constraints – sequential versus diverging, muted versus saturated, high versus low contrast – you transform color from a stylistic afterthought into a controlled design decision. You are no longer asking the model to invent taste; you are asking it to execute intent.

With this minimal vocabulary in place, we can now treat color not as decoration but as rhetoric – described, constrained, justified, and reviewed through language. You're ready to begin prompting for the most emotionally charged of all visual variables.

Choosing the Right Color Scheme (and Making the LLM Explain It)

Not all color schemes make the same kind of argument. Before asking an LLM to generate a palette, it helps to be clear about what the data itself implies.

Most thematic maps fall into one of three broad scheme types:

- Sequential schemes are used for ordered, continuous values (e.g., temperature, income, density). They imply progression from low to high and rely heavily on value change.
- Diverging schemes are used when deviation from a meaningful midpoint matters (e.g., above or below average, gain versus loss). They imply balance, contrast, or departure from a norm.
- Categorical schemes are used for qualitative differences without inherent order (e.g., land use types, political parties). They imply distinction, not magnitude.

LLMs are often capable of selecting an appropriate scheme *if the dataset is described clearly*. But "often" is not "always," and subtle framing errors can push a model toward the wrong rhetorical choice. A diverging scheme implies judgment. A sequential scheme implies accumulation. A categorical scheme implies equivalence.

A good habit is to always ask the model to justify its decision before committing to a palette:

```
Given this dataset, which color scheme type is most
appropriate and why? What incorrect interpretations
might arise from using a different scheme?
```

This transforms scheme selection from an implicit assumption into an explicit design decision – one you can accept, revise, or override. The goal is not to catch the model in a mistake – no matter how good it might feel – but to surface its reasoning while there is still time for you to intervene.

Color by Prompt

Color is the first voice a map uses. Long before a reader decodes labels or symbols, hue and contrast whisper – or shout – what matters. In LLM-based design, that means color cannot be left to random sampling or the model's vague sense of what looks "professional." It must be described with the same precision we once reserved for code.

A single sentence can change an LLM's perception entirely. At least, that's the theory behind prompt engineering, and from my day-job experience as a prompt engineer at a GIS company, I assure you it generally holds up in practice. "Use a muted blue–gray palette to convey calm analysis" will produce a radically different map from "Apply bold primary colors for a civic-pride tone."

From Palette Generator to Palette Critic

When prompted carelessly, most models behave like excitable art students – enthusiastic but inconsistent. They return a handful of colors, often without context and never with justification. A better approach is to make the model defend its decisions.

Ask for its rationale.

```
Generate a five-color sequential palette for water-
temperature anomalies. For each color, explain its sym-
bolic role and confirm that all adjacent pairs meet
WCAG AA contrast.
```

The response may include hex codes, short explanations, even contrast checks – evidence of reasoning we can inspect. This shifts the LLM from palette generator to palette critic. Each run becomes a miniature design review rather than a lottery draw. Figure 10.6 shows the role of guidance in creating appropriate and useful color palettes.

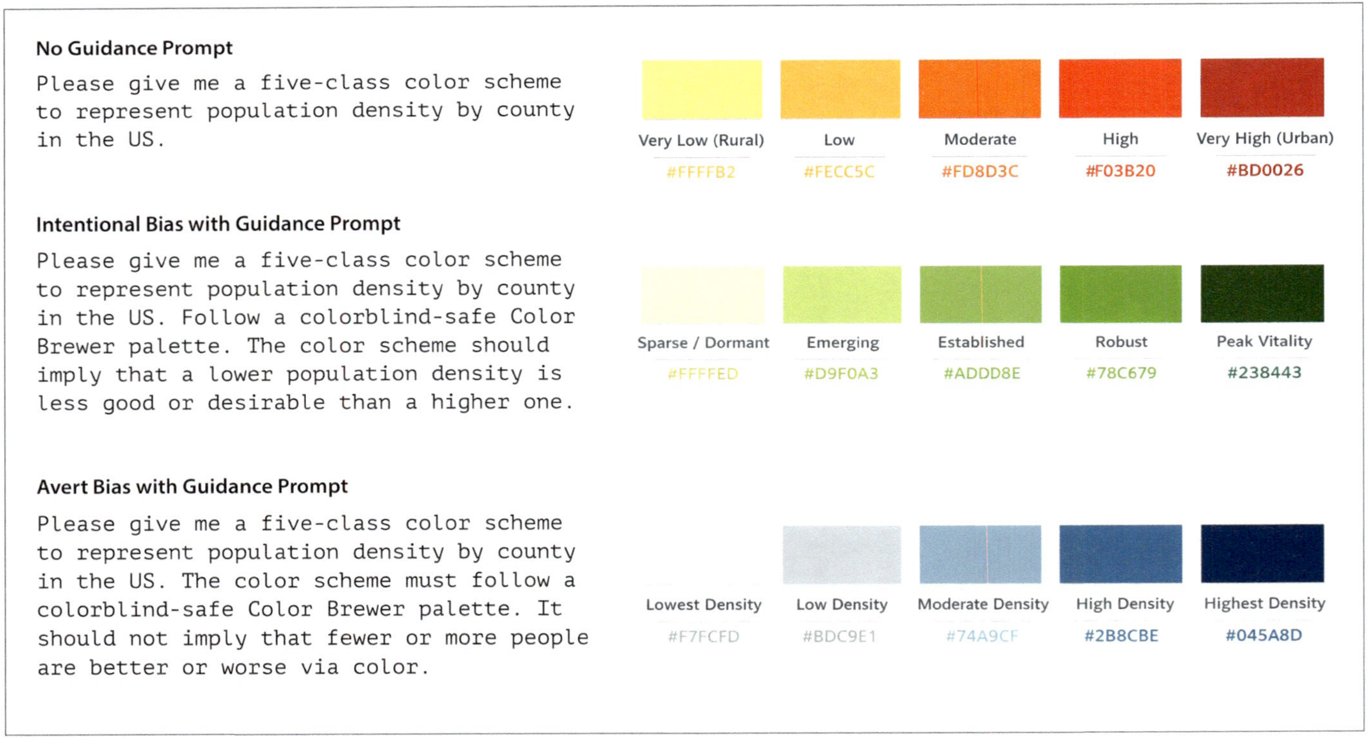

FIGURE 10.6
Three LLM-generated palettes for the same dataset. The first lacks guidance (ends up with a dangerous looking color scheme). The second shapes narrative intent and provides guidance. The output makes less densely populated areas look sickly and more populated ones vibrant. The last provides guidance and requests a neutral color scheme.

Semantic Color Prompts

Color language is slippery. Words like warm, trustworthy, or urgent carry emotional charge but lack numeric definition. Yet they remain valuable because they communicate tone. The trick is pairing emotion with structure:

```
Create a warm, trustworthy palette using low-saturation
earth tones between #b58e4f and #f2efe8. Limit lumi-
nance contrast to 2:1 for subtle transitions.
```

Here, adjectives guide intent while numbers anchor execution. Over time, readers learn to balance affective language with measurable constraints – much like musicians balance expression with tempo.

LLMs, trained on broad visual culture, often inherit common color metaphors: blue suggests trust, green implies growth, red signals alert. But these associations are not universal. In financial maps, red may signal profit in one context and loss in another. The model will follow your language literally, so define tone explicitly rather than assuming convention.

Harmony and Data Type

Colors placed next to one another create meaning. The meaning can be harmonious – think of your favorite color combo – or chaotic – orange and red. We don't try to establish color harmony in our maps for decorative reasons alone – though, sometimes I admit, I do... but I shouldn't! Color harmony gives your map symbols logic.

Scheme choice signals data logic. A sequential palette always implies progression. A diverging palette implies balance or deviation from norms. A categorical palette signals difference without order. Prompting is how we declare that intent, so the LLM produces palettes that are logical for your end users:

```
Use a smooth sequential palette for continuous tempera-
ture data. Use a diverging palette centered on zero to
highlight deviations above and below freezing.
```

By naming both data type and emotional register, we prevent aesthetic drift. The model knows whether to think in gradients or boundaries, calm or contrast, sequential or diverging. Figure 10.7 shows a comparison between color scheme outputs.

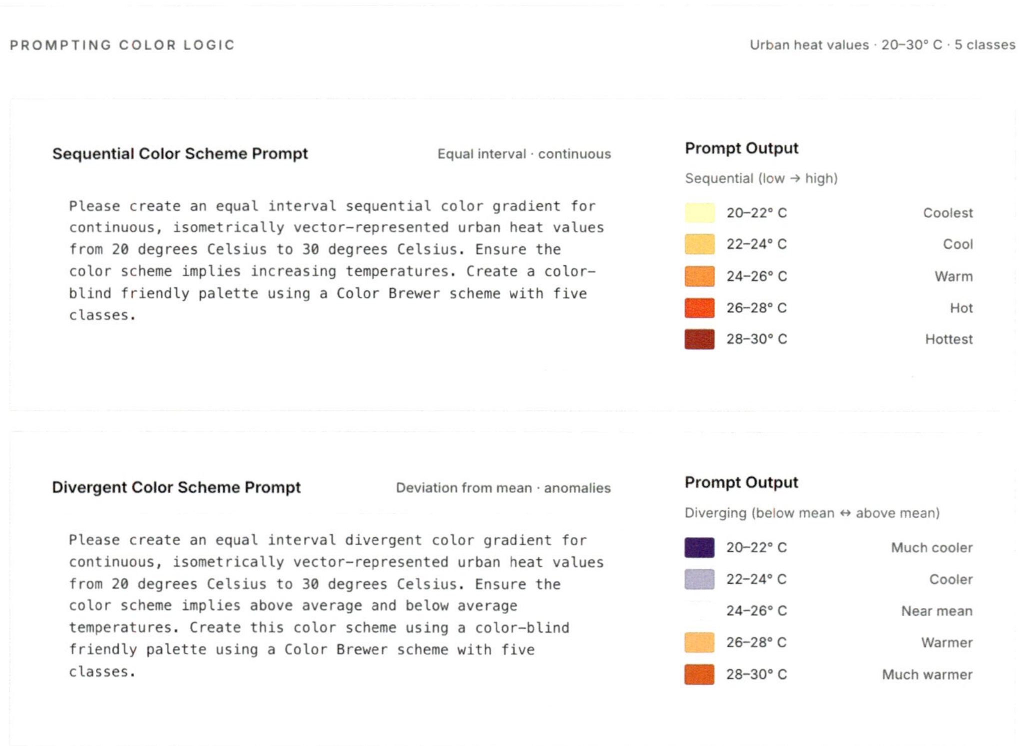

FIGURE 10.7
Prompt-based palette comparison: sequential gradient for urban-heat values versus diverging scheme for temperature anomalies. Different color schemes and decisions will tell very different stories and reveal different things about the same datasets. It's important to choose a color scheme that matches your map's intent.

Accessibility and Culture

Every palette is an invitation – or a barrier. Roughly eight percent of men and one percent of women experience some form of color-vision deficiency, yet many AI-generated maps still rely on red–green contrast. Good prompting can pre-empt exclusion:

```
Ensure adjacent hues remain distinguishable under deu-
teranopia and protanopia simulations. Use value con-
trast or pattern if color separation fails.
```

Accessibility is not charity; it is design integrity. The same applies to cultural interpretation. Red may mean danger in one context and celebration in another. The safest habit is to describe emotion and function together:

```
Ensure color palette conveys a positive trend without
necessarily implying moral judgment.
```

Color, then, is not just pigment. It is persuasion wrapped in math. The next two variables – type and symbol – complete this rhetorical triad. But before we add letters or icons, the palette must hold steady. It is the key that the rest of the design will play in.

> **HISTORICAL NOTE: COLORBREWER AND THE PALETTE REVOLUTION**
>
> In the early 2000s, Cynthia Brewer at Penn State University developed *ColorBrewer*, which instantly solved a problem cartographers and data visualization gurus had wrestled with for decades: how can we systematize color harmony *and* accessibility.
>
> Brewer's schemes encoded logic – sequential, diverging, categorical – long before today's AI assistants existed. Her color palettes are now integrated into many data visualization software packages. When you prompt for a ColorBrewer-style palette, you will be continuing the lineages of harmony, differentiation, and accessibility by default. Design ethics wrapped up in hex codes – wicked (pardon the hex pun).

Color as More than Looking Pretty

Color is the first argument your map makes, whether you intend it to be or not. Long before anyone reads a title, decodes a legend, or hovers over a tooltip, they've already absorbed your palette – and with it a set of assumptions about what matters, what's urgent, what's trustworthy, and what they should feel. In prompt cartography, that means color can never be an afterthought, and it definitely can't be left to whatever the model thinks looks "professional" based on its training diet of corporate PowerPoints and Instagram infographics.

What I hope you take away from this chapter is that color is not decoration. It is persuasion wrapped in hexadecimal codes. When you prompt an LLM to generate a palette, you're not asking it to pick pretty colors – you're asking it to translate intent into visual rhetoric. And that translation only works when you give the model enough structure to constrain its choices: data type (sequential, diverging, categorical), emotional register (urgent, calm, authoritative), accessibility requirements (contrast ratios, color-vision deficiency simulations), and cultural sensitivity (because red doesn't mean the same thing everywhere, and assuming it does is how you end up in a heated city council meeting trying to explain that crimson doesn't mean catastrophe).

The good news? Once you understand that hue, value, and saturation are parameters rather than mysteries, you can describe color with the same precision you'd use to describe anything else to an LLM. The challenge – and honestly, this never stops being a challenge – is remembering that every omission is also a specification. When you don't tell the model what kind of palette to use, it will still choose one. And that choice will carry meaning whether you endorse it or not. (Has this been hammered into your memory yet? Regardless, you'll be reading it again, as it can't be stated enough.)

So prompt deliberately. Ask the model to justify its decisions. Make it explain why it chose a diverging scheme over a sequential one, or why it thinks a muted earth-tone palette conveys "trustworthiness." Treat the LLM as a collaborator capable of design reasoning, not just a palette vending machine. Because when color choices become explicit – when they move from gut instinct or random sampling into structured, defensible language – they stop being accidents and start being arguments you can stand behind. Or at least next to, when an irate member of the public comes your way to complain about your hot pink color choice.

Takeaways

- **Color Is Both Rhetorical and Informative**

  Every palette makes an argument before a single symbol is read.

- **Hue Alone Is Insufficient**

  Value (lightness/darkness) and saturation (intensity) determine legibility, contrast, and emotional tone.

- **Warm Colors Advance, Cool Colors Recede**

  Use this hierarchy deliberately to guide attention and implication.
- **Scheme Type Signals Data Logic**

  Sequential implies progression; diverging implies deviation from a norm; categorical implies difference without order.
- **Omission Is Delegation (As Always)**

  When you don't specify palette constraints, the model chooses for you, and those choices carry unintended meaning.
- **RGB Is Additive, Screens Vary**

  Design for relative contrast and hierarchy, not exact color reproduction across devices, because it won't happen.
- **Accessibility Is Design Integrity**

  Roughly eight percent of men experience color-vision deficiency; value contrast and pattern redundancy prevent exclusion.
- **Semantic Color Prompts Pair Emotion with Structure**

  Words like "trustworthy" or "urgent" guide tone, but numeric constraints (hex ranges, luminance limits) anchor execution.
- **Make the Model Justify Its Palette**

  Asking an LLM to explain its color decisions transforms it from a generator into a critic you can inspect and revise.

Conceptual and Applied Exercises

Color as Persuasion

Think about a map you've encountered where color choices felt urgent, calm, or untrustworthy. What specific color decisions – hue, value, or saturation – drove that perception, and how might different choices have shifted the map's rhetorical effect?

Scheme Mismatch

Identify a real or hypothetical scenario where using a sequential color scheme instead of a diverging one (or vice versa) would fundamentally misrepresent the data. What interpretive harm would result from that mismatch?

LLM Palette Critique

Ask an LLM: "Generate a five-color sequential palette for mapping urban heat islands. Explain why you chose each color and confirm accessibility for color-vision deficiency." Evaluate whether its justification addresses rhetoric, accessibility, and data logic – or merely aesthetics.

Constraint Exercise

Write a color palette prompt for a map showing election results that deliberately avoids culturally loaded colors. Specify emotional tone, accessibility requirements, and data type without naming specific hues.

For extended tutorials, hands-on labs, prompt templates, and updated examples related to this chapter, visit www.promptcartography.com.

11

Prompting for Typography

If color is the first argument your map makes, typography is the voice that delivers it. And just like in real life, voice matters. A lot. You can have the most carefully researched data, the most thoughtfully chosen palette, and the most elegant symbology – but if your typography feels like it wandered in from a different conversation entirely, the whole thing falls apart. I've seen gorgeous maps undermined by Comic Sans labels (yes, really), and I've watched perfectly serviceable maps gain unearned credibility simply because someone chose Futura instead of Arial. Typography is sneaky that way. When it works, no one notices. When it doesn't, it's all anyone remembers.

In traditional web cartography, typography often got relegated to the final checklist – right after "Does the legend fit?" and "Can you actually read the pop-ups?" You'd set a font, fiddle with sizes until labels stopped colliding, and move on. But in prompt cartography, type can't be an afterthought, because the model needs to know what you want before it starts rendering anything.

And here's where things get interesting: you can't just tell a large language model (LLM) "make it look professional" and hope for the best. You have to describe what professional means in this context. Friendly but credible? Authoritative but accessible? Modern without feeling like it's trying too hard? These aren't vague aesthetic preferences – they're design specifications that translate directly into typographic decisions.

The challenge, as always, is that web maps don't give you the typographic control you'd have in print. Fonts fail. Devices substitute. What looks crisp and legible on your laptop might render as a blurry mess on someone's phone at 3 a.m. while they're trying to figure out whether to trust your unemployment data. That's why robust typography in prompt cartography isn't about finding the perfect font – it's about building resilient systems that preserve hierarchy, legibility, and tone even when your first three font choices aren't available. It's about encoding rules for label density, collision, and scale-dependent visibility so your map doesn't drown in competing text at the wrong zoom level.

What I'm about to walk you through isn't a comprehensive treatise on typography – people spend entire careers on that, and frankly, they're better at it than I am. This is a practical guide to making typography work in a prompt-driven visual design pipeline. We'll cover the basics you need to know (serif versus sans-serif, why legibility actually matters, how label

DOI: 10.1201/9781003797920-13

placement encodes meaning), the web-specific constraints you can't ignore (fallback stacks, rendering inconsistencies, the myth of control), and most importantly, how to describe typographic intent clearly enough that your LLM collaborators can implement it without turning your map into a ransom note.

Because at the end of the day, typography is just another design decision that needs to be made explicit. And once you can name what you want – once you can articulate the difference between friendly and frivolous, between authoritative and cold – you stop hoping the model guesses right and start directing it deliberately. Which, as it turns out, is the entire point of this book.

Don't Waste Your Time on the Typeface Versus Font Debate

Officially, a typeface is the design (the "look" of the letters), while a font refers to a typeface at a specific size. In practice, almost everyone uses the word *font* to mean the whole family, and modern computing has reinforced the habit. Operating systems have "font" folders. CSS uses the word font. Word processors use the word font. If you are one of those people who gets upset about this, I apologize in advance – clear communication matters more than semantics in this book.

For our purposes, we'll use:

- ***Font***: the typeface family name (e.g., "Roboto")
- ***Font style***: bold, italic, semibold, etc.
- ***Font size***: the size, which is still – *unfortunately* – difficult to standardize on screens

Serif Versus Sans-Serif (Or Why Web Maps Are Biased Against Serifs)

At the broadest level, fonts come in two common categories:

- Serif fonts have small finishing strokes on letterforms.
- Sans-serif fonts ("sans" = without) do not.

FIGURE 11.1
Screen-friendly fonts tend to have generous x-heights, open interior spaces, and consistent strokes. These font features matter more for map labels than the hipness of a font's brand name.

Web cartography developed a strong convention toward sans-serif fonts for a simple historical reason: early screens had mediocre resolution, and thin serifs could look jagged, blur, or disappear at small sizes on backlit displays. There are also cultural expectations in many Western contexts: sans-serif often reads as contemporary and "web-native," while serif can read as traditional or archival. Also, in the US, sans-serif fonts were typically used to represent human-made objects (e.g., countries, cities, buildings), and serif fonts more commonly referenced physical geography (e.g., mountains, lakes, seas, etc.).

The important update is that this old rule has softened. Screen resolution has improved dramatically over the past decade, which means serif fonts are perfectly usable – particularly for larger labels, titles, and headings – *if* you choose a serif designed for screens and keep an eye on legibility at multiple scales.

In other words, serif fonts are no longer forbidden fruit. (It's just fruit you should probably wash first.) Figure 11.1 highlights some key aspects of all fonts.

Legibility Isn't Just Taste: A Few Rules That Still Stand

Even though web maps are dynamic, many print-era typography principles remain steady:

- ***Use No More than Two Font Families***

  Only rarely do maps need more than two font families. Variety and differentiation should come from *weight, size, case, and spacing*, not from deploying five unrelated fonts like you're constructing some kind of ransom note.

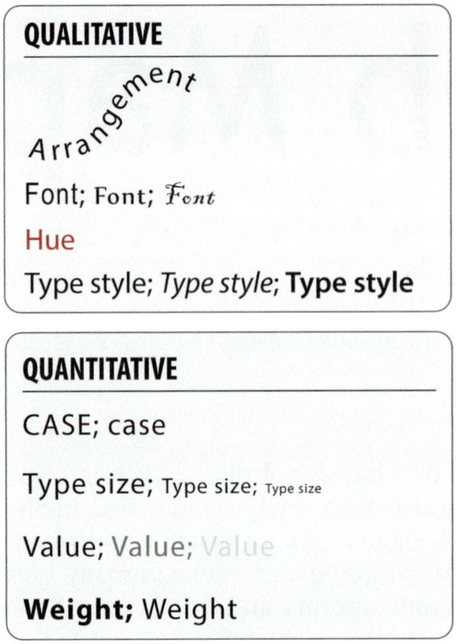

FIGURE 11.2
Type properties signal meaning. Prompt your LLMs to manipulate particular font properties to visually communicate more than just what the text actually says.

- ***Use Font Properties to Express Hierarchy***

  Size, weight, case, and style *always* communicate importance and category. You can use these properties to differentiate labels (cities versus counties) and sometimes even suggest quantitative importance (big cities versus small ones).

Figure 11.2 highlights the different ways that font styling can embed both qualitative and quantitative information on top of the information being communicated in natural language on your maps.

Label Placement Norms: The Quiet Rules Your Prompts Should Encode

Most readers don't consciously notice label placement – until it's wrong. Basic placement norms still matter in web maps because they reduce visual stress and increase trust. And here's the prompt-cartography twist: these

FIGURE 11.3
Label placement best practices for different spatial geometries.

norms are no longer trapped inside GIS label-engine dialogs. You can state them directly as constraints (Figure 11.3).

- Point labels generally read best in consistent positions relative to the point symbol.
- Line labels should be placed on smooth stretches, not on tight bends; a little offset helps them breathe.
- Area labels often benefit from letter spacing to fill space and clearly associate the label with the region.

Web Reality Check: You Don't Control Fonts Like You Can in Print

In web maps, typography is always at risk of "graceful degradation" turning into "typographic betrayal." Different devices and operating systems have different fonts installed. If a user doesn't have your chosen font, the browser will substitute something else, sometimes with wildly different spacing and tone. That's why robust typography includes fallback font stacks – an ordered list that tells the browser what to try next if the first choice fails. Always advise your agents to include a list of fallback fonts to ensure your

```
body {
    font-family: Calibri, Corbel, Verdana, Geneva, sans-serif;
}
```

FIGURE 11.4
Fallback font stacks in CSS. Fallback stacks preserve hierarchy and tone when preferred fonts are unavailable across devices and browsers.

styling doesn't go completely off the rails (Figure 11.4). Trust me, you don't want to produce a map in Georgia font.

This primer is not here so you can memorize typography trivia. It's here because prompting is design by language. When you describe a font system as "calm," "authoritative," "archival," or "friendly," the model is mapping those adjectives onto typographic conventions like serif versus sans-serif, weight, spacing, and hierarchy. The clearer your vocabulary, the less the model has to guess.

Now we can return to the main point: in prompt cartography, typography is no longer a quiet afterthought buried in CSS. It becomes an explicit, inspectable design decision – one you can demand the model to justify.

Prompting for Fonts and Font Styles

As duly noted above, before a single label is parsed, its shape and weight will already signal an attitude. A soft sans-serif font suggests approachability. A narrow serif font, on the other hand, implies tradition and scientific precision. Personally, sometimes I enjoy using a heavy geometric face to subliminally imply a map is authoritative. These instincts – long internalized by designers – translate surprisingly well into natural-language prompts, provided we are deliberate.

Until recently, web typography lived quietly in CSS files or label-engine dialogs. Prompt cartography pulls it into the open. Instead of writing {font-family: Roboto, Arial, Helvetica, sans-serif;}, you now prompt your agent to:

```
Use Google Fonts to find and choose a clear, modern
sans-serif for city labels, a restrained serif for
titles, and maintain an appropriate luminance contrast
between text and background. Include a list of several
similar fallback fonts that are commonly available on
both mobile and desktop devices.
```

Your LLM agent or model doesn't *feel* or sense typography. It reasons through the tone of your words, hierarchy cues, constraints, and its knowledge of how the fonts have been used elsewhere. By pairing adjectives with requirements, we guide it toward typographic logic rather than default Helvetica grayness.

Type Sets the Tone of Your Map

Every map has a voice. A flood-risk dashboard may need the calm precision of a news infographic; a heritage-trail app might benefit from the warmth of a travel poster. A handful of well-chosen adjectives – modern, approachable, archival, authoritative – anchor style far more effectively than brand names alone.

When prompted, the model maps those adjectives onto known font families or web-safe equivalents. The exact font matters less than the typographic role it plays (Figure 11.5).

Consider two near-identical prompts for the same parcel-ownership map:

```
Label property boundaries using a friendly, rounded
sans and soft gray labels.
```

```
Label property boundaries using a strong condensed
serif with high contrast to the parcel color.
```

Both are legible. Both are technically correct. Yet one invites curiosity and creativity; the other signals formality and authority. Your tone is your argument.

Fallbacks and Robustness

Maps circulate across browsers, devices, and operating systems. A pristine type hierarchy on your laptop can dissolve into chaos on a phone that lacks your chosen font. Prompting for fallbacks avoids this quiet failure.

```
Use Noto Sans as the primary font; if unavailable,
fall back to system-ui or Arial. Maintain consistent
x-height and relative weight across families.
```

This instruction may seem mundane, but it is a lifesaver. It tells both the model and the rendering pipeline to plan for imperfection. Robust typography is not about aesthetic purity; it is about graceful degradation.

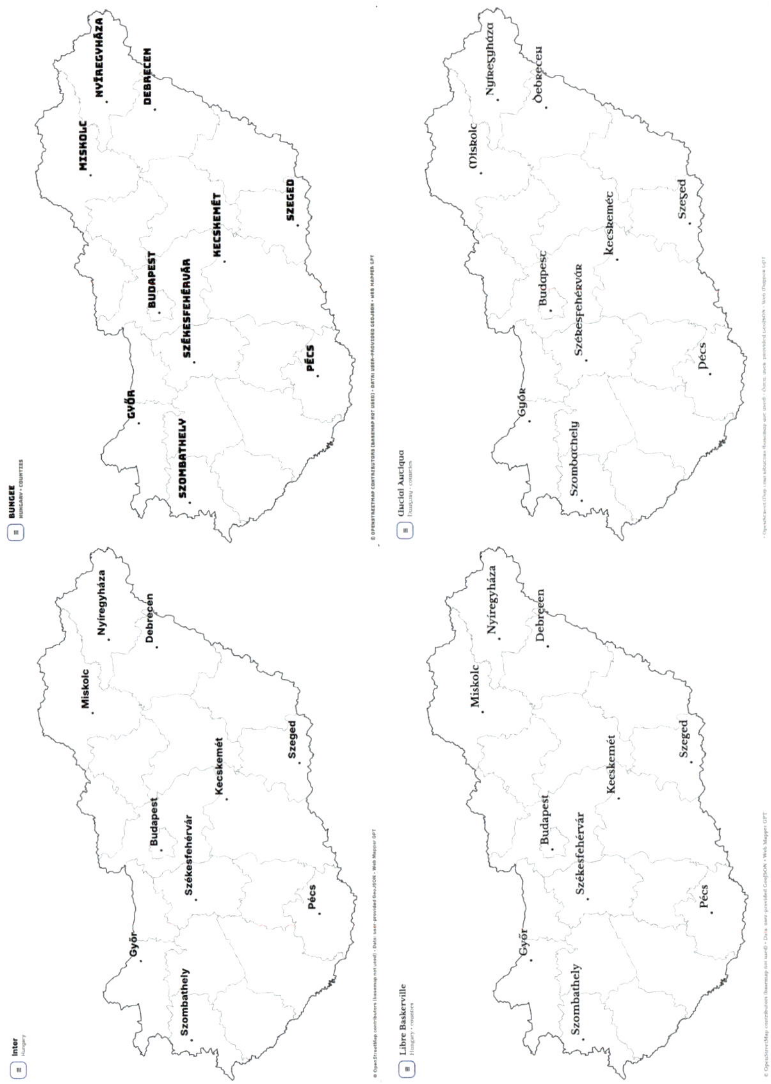

FIGURE 11.5
Type carries rhetoric. Four typographic systems applied to a simple base map of Hungary demonstrate just how much the tone or style of a map is shaped by font selection alone.

Density and Collision Rules

Good typography is not just about font choice; it is about restraint. In circumstances where you are labeling features on the map, for example, too many labels will compete for a map reader's cognitive oxygen. The visual hierarchy can quickly collapse. You can use prompts to encode label-engine like settings.

```
Prioritize county labels over city labels; reduce city
label size by 20 percent and apply alpha 0.7 when labels
overlap. Determine which zoom level will ensure minimal
label overlap and hide labels until the user has zoomed
to the appropriate level.
```

These two sentences teach the model how to negotiate space. It mirrors what experienced designers do instinctively in desktop or SaaS mapping apps – but now expressed in plain language.

Typography and Web Fonts as Key Component in Your Visual Design

Typography is the narrative spine of your map. It's the voice that whispers – or shouts – what kind of authority you're claiming, how seriously you expect to be taken, and whether the map is inviting exploration or demanding compliance. And here's the thing that still catches people off guard: type does all of this work invisibly. When typography is working well, no one notices it. When it's working poorly, people notice everything except the map you spent weeks building.

In prompt cartography, typography shifts from being a buried CSS artifact into an explicit, articulable design decision. You're no longer hunting through font stacks hoping something looks "about right." You're describing intent – friendly, authoritative, archival, modern – and letting the model map those adjectives onto typographic conventions: serif versus sans-serif, weight, spacing, hierarchy. The clearer your vocabulary, the less the model has to guess.

But here's the catch, and it's a big one: web maps don't give you the same typographic control that print does. Fonts fail. Devices substitute. Browsers make decisions on your behalf. That's why robust typography in prompt cartography isn't just about choosing a beautiful font – it's about building fallback stacks that preserve hierarchy and tone even when your first, second, and third choices aren't available. It's about encoding label density rules so your map doesn't drown in competing text. It's about recognizing that legibility isn't just taste – it's a set of constraints you can describe, test, and iterate on without resorting to trial-and-error tinkering.

I won't pretend this is easy. Typography has always been one of those design elements that separates the professionals from the enthusiastic amateurs, and prompt cartography doesn't change that. What it does change is where the work happens. Instead of spending hours adjusting font sizes and collision rules in a GUI, CSS, or other script files, you're spending that time articulating what you want the type to do – what role it plays, what hierarchy it enforces, what tone it sets. And then you're asking the model to implement those decisions consistently, so you can focus on whether the intent is right rather than whether the kerning is perfect.

Typography, then, isn't a finishing touch as it sometimes was in the past. It's a design decision that shapes how people experience your map at first glance. And in a world where maps are increasingly built through language rather than pixel-pushing, that means you need to be able to describe typographic intent as precisely as you describe anything else. Because if you can't name what you want – friendly but not frivolous, authoritative but not cold, modern but not trendy – the model will… wait for it… fill that silence with whatever it thinks looks professional. What an LLM thinks looks professional is rarely what your audience will.

So treat type like the voice of your map. Give it direction. Give it constraints. And for the love of all that is cartographically holy, never let it default to Comic Sans!

Takeaways

- **Typography Is Voice**

  Font choices signal authority, approachability, formality, or warmth before a word is read.

- **Legibility Isn't Just Taste**

  Hierarchy, spacing, contrast, and collision rules are functional constraints, not aesthetic preferences.

- **Use No More than Two Font Families**

  Variety should come from weight, size, case, and spacing, not from deploying unrelated fonts like a ransom note.

- **Font Properties Encode Hierarchy**

  Size, weight, case, and style differentiate importance and category (e.g., cities versus counties).

- **Web Fonts Either Degrade Gracefully *or* Disastrously**
  Fallback font stacks preserve tone and hierarchy when preferred fonts aren't available across devices.
- **Label Density Is Negotiated Space**
  Too many labels collapse visual hierarchy; prompts can encode collision rules and zoom-dependent visibility.
- **Type Sets the Map's Tone**
  A "friendly, rounded sans" invites curiosity; a "strong condensed serif in high contrast" signals formality and authority.
- **Prompting Makes Typography Inspectable**
  When type decisions are articulated in language, failures can be diagnosed and corrected rather than guessed at.
- **Robustness > Perfection**
  Design for resilience across devices, browsers, and unpredictable rendering environments, not pixel-perfect reproduction.

Conceptual and Applied Exercises

Typography Tone Audit

Examine a web map you use regularly. Describe how its typography sets a tone – formal, casual, authoritative, or approachable. Then identify one typographic decision that undermines or strengthens that tone.

Fallback Failure

Reflect on a scenario where a web font fails to load and a fallback font dramatically changes a map's credibility or tone. What typographic properties (weight, spacing, character width) would you prioritize in a fallback stack to preserve intent?

LLM Type Reasoning

Ask an LLM: "What typographic choices would make a disaster-response map trustworthy and legible under stress?" Critique whether its response prioritizes functional hierarchy and readability, or leans on aesthetic conventions.

Hierarchy Through Type

Write a prompt that specifies typographic hierarchy for a map with three label categories (major cities, minor towns, geographic features) without naming specific fonts. Focus on weight, size, spacing, and case to differentiate importance.

For extended tutorials, hands-on labs, prompt templates, and updated examples related to this chapter, visit www.promptcartography.com.

12

Map Symbology

Type sets the tone for your map; it's an invisible yet powerful component of visual communication. Though, typography establishes your map's voice, rarely can type carry your map's meaning alone – though, there are some fine typographic print maps out there that force me to refrain from using the word "never." You'll find that symbols are often what give your map's voice substance.

Symbols are where abstraction meets assertion. They declare presence, encode category, and signal importance. Where typography whispers, symbols more typically speak – sometimes loudly. And in prompt-driven cartography, symbols demand the same level of intentional specification as every other design element – perhaps more so, because models will happily fill the void with generic teardrops, cheerful pictographs, or decorative flair that undermines your map's argument before a single word is read.

This chapter shifts focus from the semantic weight of type to the visual logic of marks. You will learn how to prompt for symbols that clarify rather than decorate, how to encode meaning through size and shape without surrendering to defaults, and how to build critique loops that prevent your map from drifting into visual chaos. Symbols are not illustrations. They are arguments rendered visible. Prompt them accordingly.

Symbols Are Visual Arguments

Before we begin prompting for symbols, it helps to pause and clarify what symbols are used for.

Symbols are not illustrations. They are not decorations. They are visual arguments about presence, category, and importance. When they work well, they disappear into comprehension. When they fail, they quietly distort meaning or, worse, cause confusion.

Much of what people associate with web map symbols emerged not from design theory but from convenience. The familiar teardrop, pushpin, or thumbtack rose to prominence because it was easy to deploy, not because it communicated anything beyond location. These default symbols function as placeholders, not design decisions. They lower the barrier to making a map, but they rarely help a reader understand one (Figure 12.1).

DOI: 10.1201/9781003797920-14

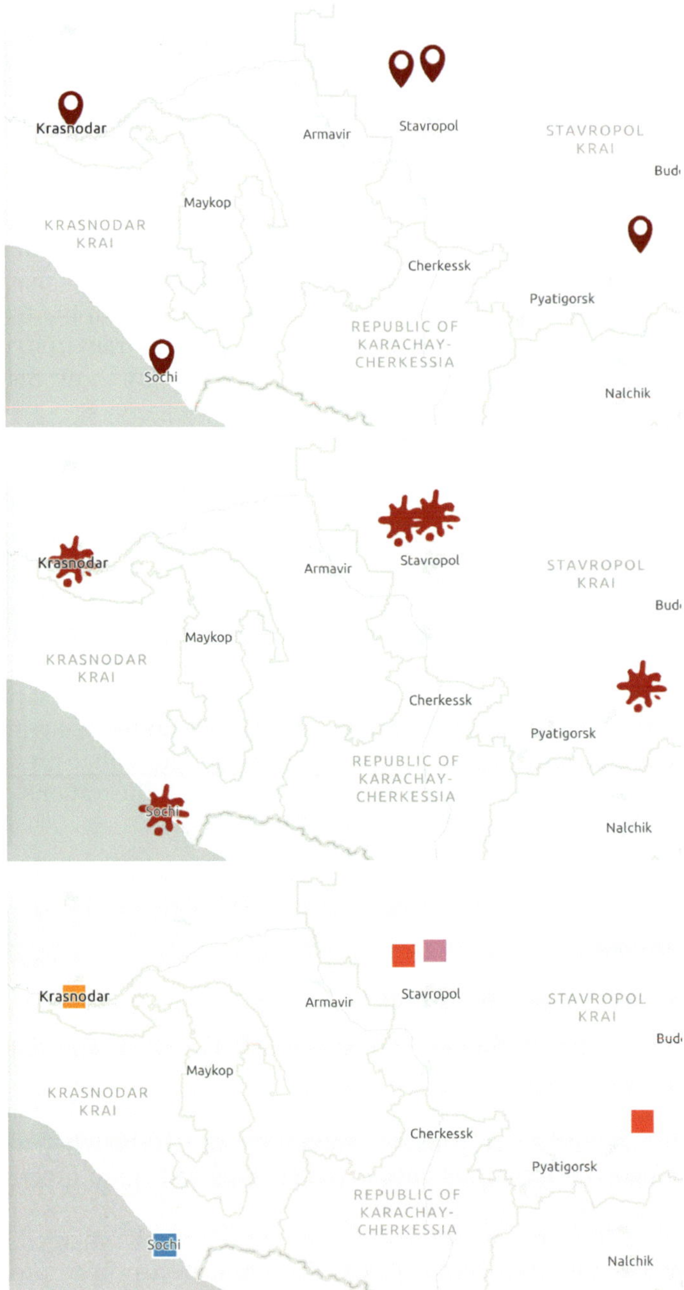

FIGURE 12.1
Default web map symbols function as location placeholders. They encode nothing beyond presence. Shape can communicate qualitative meaning quickly, and other visual variables can encode categorical or quantitative information.

There is nothing inherently wrong with defaults when communication is informal or exploratory. But for authored maps – especially those intended to persuade, explain, or support decisions – default symbols almost always under-specify meaning. Other than "something exists here," they tell us very little.

Effective symbology does the opposite. It allows the symbol to carry some of the interpretive burden on its own.

And here's a new prompt-cartography twist: when you don't specify symbolic logic, the model will fill the vacuum! Often with a completely uncalled for "helpful" flair. Perhaps more dangerously, with subtle rhetoric you didn't ask for and don't notice until it's too late. So your job is *never* to ask for "nice icons"; that's like telling your kid to "be good" when another family offers to take them to the cinema. Your job is to translate cartographic intent into constraints – "Be polite, offer to buy them popcorn, and make sure you say thank you."

Prompting for Symbology

If color sets mood and type sets voice, symbols provide the grammar in cartography. They are the nouns and verbs of any map – points that assert presence, lines that imply movement, polygons that claim territory. In prompt-based design, this grammar must be stated explicitly. Otherwise, the model fills the page with generic icons that are legible but semantically lazy.

Prompting for symbology means describing what a symbol does as clearly as what it looks like. Function comes first. Form must follow. When that order is reversed, symbols become decoration rather than evidence. Decoration quickly becomes noise – like 1980s wallpaper.

Below is a generic and default assistant prompt. You can tweak as necessary but read through it. It provides a skeleton of how you need to address a large language model (LLM) when designing symbols for you.

```
You are my cartographic design assistant. Follow these
rules:

  MAP GOAL
  - The map's argument is: [one sentence].
  - The user's task is: [compare / locate / detect pat-
    tern / explore].

  LAYER CONTEXT
  - Layer name + geometry type: [point/line/polygon]
  - Key field(s) and units: [..]
  - Data type: [raw count / rate / category / ordinal /
    continuous]
```

```
ENCODING RULES (NON-NEGOTIABLE)
- One primary encoding per layer.
- Shape = category. Size/value = quantity.
- If size varies, scale by AREA (not radius).
- Avoid pictorial icons unless explicitly justified.
- Add an overlap plan (opacity, outline, clustering,
  decluttering).
- Include an uncertainty plan if the data warrants it.

OUTPUT
1) Symbology plan in plain language
2) Legend labels (neutral tone)
3) likely failure modes and how you prevented them
```

Aside from color, there are two ways to embed your symbols with data-based information – shape and size. Shape allows you to encode symbols with qualitative or categorical information (e.g., apples versus bananas). Size allows for symbols that provide quantitative meaning (e.g., "look at the size of that thing (compared to that dinky thing)!"). We review both techniques now.

Shape for Categorical Difference, Not Storytelling Flair

Shape is best reserved for categorical distinction, not storytelling flair. A triangle does not mean "danger" unless you make it so. A circle does not mean "neutral" unless context reinforces it. Shapes acquire meaning through consistency and contrast, not cleverness.

This matters even more in an era of AI-assisted design. Left unguided, an LLM may reach for pictographic symbols (i.e., shapes that mimic real-world objects) because they appear recognizable and helpful. Recognition is not the same as comprehension, however.

Overly literal symbols can smuggle in unintended rhetoric – danger, blame, nostalgia, humor. They can also confuse users that do not recognize the symbolic connection – e.g., my children have no idea what a phone booth is other than the thing Doctor Who travels in.

Using shape is fine – indeed, quite useful – when it helps your narrative or pushes your communicative goals. It's *not* good when you don't request it – which again, LLMs are wont to do. Thus, prompting for abstraction first (i.e., geometric shapes) is not necessarily a stylistic preference but a useful safeguard. Once your map is functioning well, you can always come back and strategically change the icons to pictographic ones if you decide that works better rhetorically.

Example safeguards to include in your prompts.

```
Use simple geometric shapes with no interior detail.
Avoid culturally specific or era-dependent imagery.
Do not use "cute" or emotionally loaded pictographs.
```

When you do use shapes – and honestly, as much as I'm badmouthing them here, I'm a big fan of a well-placed blood splatter when mapping murders – there is one timeless principle to keep in mind. Generic caricatures *always* outperform detailed symbol illustrations (Figure 12.2). Abstract shapes not only widen applicability but also allow for more people to interpret the symbol without reading too much detail into it.

FIGURE 12.2
Generic symbols prevent accidental user bias. A dog symbol that looks like a Pit Bull may elicit strong emotional reactions from someone with a negative experience with a pit bull. A dog symbol that doesn't look like any breed is more neutral.

Size and Quantitative Logic (or Trust but Verify Your LLM's Output!)

Symbol size carries a promise of proportionality, and when you break that promise, you can quickly erode user trust. A classic mistake – one LLMs will repeat enthusiastically – is scaling symbol radius instead of area, exaggerating large values and compressing small ones (Figure 12.3). (A lot of amateur cartographers have made this mistake throughout the ages too. Whenever I see this, a part of me dies inside.) There is a simple fix for this as usual – linguistic precision. Always remind the LLM to use scaling. See the prompt guardrails below (and use them as needed).

```
Scale symbol AREA (not radius) to incident_count.
Use sqrt scaling to reduce dominance of outliers
  above 500.
Set minimum size for visibility and maximum size to
  prevent occlusion.
```

Tweaking the size of a point symbol based on a numeric field is technically a thematic mapping technique – proportional symbology (when unclassed) and graduated symbology (when classed). I briefly mention it here with symbols, however, because really that's what proportional scaling is – symbol

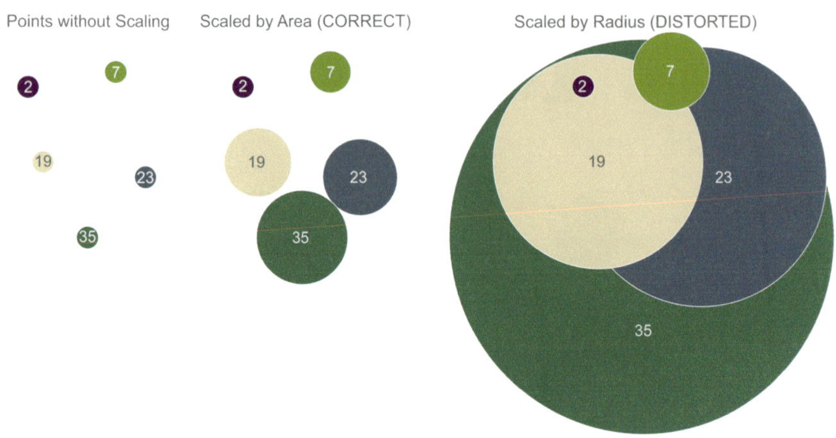

FIGURE 12.3
Proportional sizing error: always size on area, not radius. The distortion is anything but minimal. The values listed are represented, from left-to-right: no scaling, proportional scaling (correctly using area), proportional scaling (incorrectly using radius). Notice how ridiculously unbalanced the scaling becomes.

resizing based on a numeric field in the dataset be it wheat produced by province (polygons with distorted size) or oil shipping route tonnage (lines with distorted size).

Opacity: Hierarchy, Context, and Uncertainty (Without Lying)

Opacity is one of the most underused – and overcorrected – visual variables (Figure 12.4). Set too high, it obscures context. Set too low, it undermines confidence. Used intentionally, it becomes a powerful tool for hierarchy and uncertainty.

```
Render secondary overlays at 55-65% opacity to preserve
  basemap context.
Increase opacity only on hover/selection.
Do not animate opacity changes unless state changes.
Use reduced opacity to indicate low-confidence
  observations.
Explicitly label this in the legend as "confidence,"
  not "intensity."
```

Texture: Visual Redundancy Establishes Emphasis

Texture (shown in Figure 12.5) should rarely be the primary encoding variable. It's not as intuitive as other variables, and when overused can cause distraction. However, it excels as a secondary, or reinforcing, variable, ensuring that something stands out and is interpreted more accurately. It's particularly helpful for accessibility when a map's color count is reaching critical mass.

```
Apply subtle cross-hatch texture to overlapping pro-
  tected-area polygons.
Keep saturation low and alpha ~0.6 so boundaries beneath
  remain visible.
Avoid dense patterns that shimmer at typical zoom
  levels.
```

FIGURE 12.4
Transparency (or opacity) is often used to allow a user to maintain spatial context through polygon layers, imply density, or show lack of confidence in its validity. In this example, it allows the user to see where the different concentrations overlap.

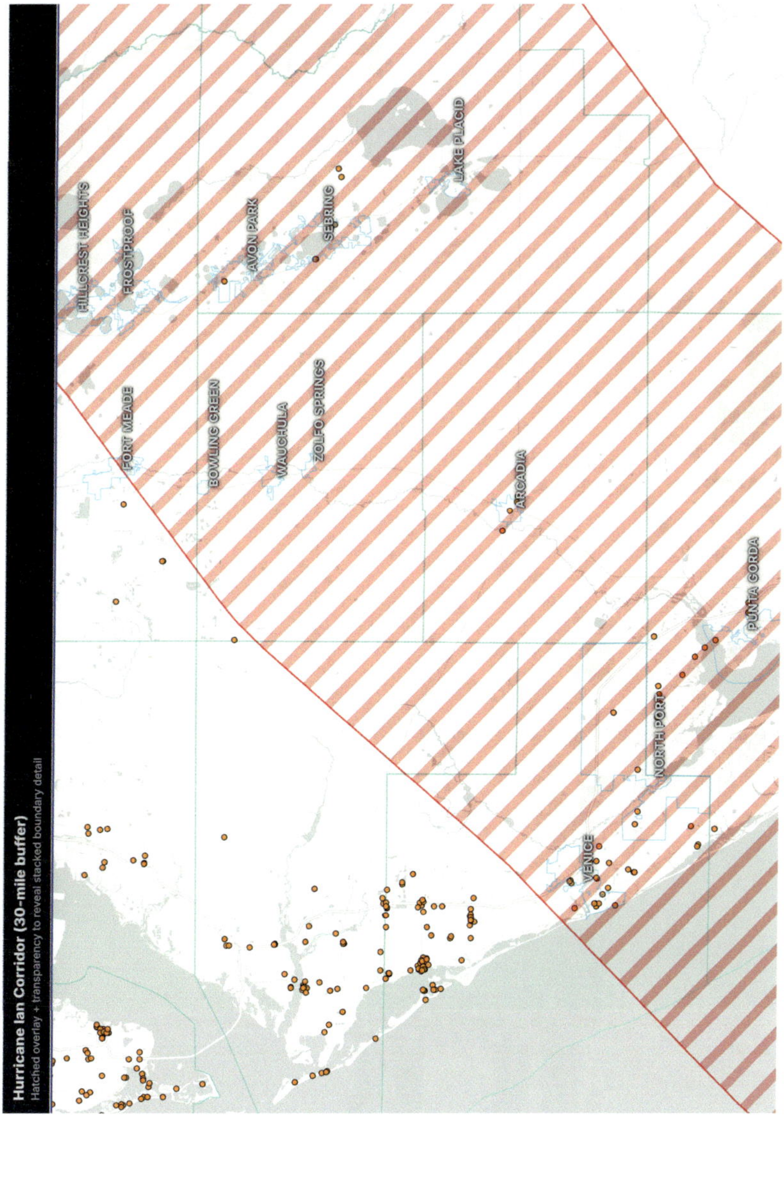

FIGURE 12.5
Hatch pattern and transparency used for the hurricane buffer to allow users to view base map context under Hurricane Ian and see boundary areas (cities and counties).

Using Map Symbol Libraries and Prompting Your Own

Many mapping systems provide icon libraries, and using them is often not a bad place to start – or even end up. Familiarity reduces cognitive load for your map users.

However, prompt cartographers can also ask LLMs to generate unique symbols – provided constraints are explicit. There are two types of symbols you will often ask your team to create: PNG (raster) and SVG (vector).

PNG Symbol Prompting

LLMs are very good at creating PNG images, allowing you to churn out symbology at a steady rate. At the time of writing, the PNG image file format remains the default image format for the web. It allows transparent backgrounds, which makes it very useful for map symbology. Many web mapping APIs allow their use.

The downside is that they are raster-based images. Thus, if you are scaling them, you will need to design them at multiple sizes to ensure that they don't begin to pixelate as they grow (Figure 12.6).

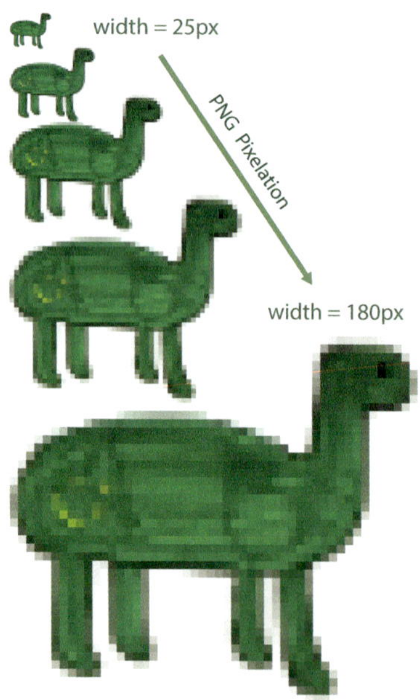

FIGURE 12.6
PNG image files are a stellar option for point symbol and easily incorporated by many APIs. However, they pixelate when scaled, and care must be taken when you are scaling them – even when mapping green dinosaurs without tails. Icon: Antja Muehlenhaus.

SVG Symbol Prompting

One thing I've found (and again, this may be obsolete by the time you're reading this book given the rate at which LLMs are changing) is that LLMs can be used to provide SVG ideas and schemas, but they are not always the best at execution.

However, SVGs are a stellar file type to use for symbols, particularly when scaling them. Why? The answer is in their name – scalable vector graphics (SVG). They scale without pixelation, as they are entirely vector-based. Better yet, SVGs are literally composed of nothing more than text (Figure 12.7). They can be embedded inside of your web map as part of the code.

If you find that an LLM is not producing a decent SVG for you directly, do not be dissuaded. There are tricks to get around this. Certain companies (e.g., Adobe) and open-source options allow you to upload a PNG and export it as an SVG.

When you do ask an agent to design an SVG icon or symbol for you, it will need lots of hints and guardrails. Below are some default ones that are advisable to use in one form or another.

```
Generate a minimalist SVG icon under 1 KB.
Stroke only (no fill), max stroke width 2px, rounded
  caps/joins.
```

FIGURE 12.7
Scalable vector graphics (SVG) images are literally text files, making them easily modifiable via code.

```
Must read clearly at 16px.
Avoid interior pictorial detail.
Return SVG + one-line intended meaning.
```

Prompting for SVGs is not about asking the LLM to show better taste. It's about writing more detailed rules for it to follow.

COMMON LLM SYMBOLOGY FAILURE MODES (AND QUICK FIXES)

- **Overly Literal Icons**: ask for geometric abstraction
- **Decorative Color Creep**: ensure every symbol color the LLM uses is tied to a semantic role (quantity, category, confidence), otherwise use only one color
- **Style Drift Across Layers**: re-declare style constraints periodically ("maintain the established map visual schema and ensure that SVG fits into the overall aesthetic context")

Symbols as Structured Decisions

By now, one principle should be unmistakable: symbols are never neutral. They assert presence, encode hierarchy, and quietly persuade – whether you intend them to or not. In prompt cartography, the risk is not that symbols will fail to appear. The risk is that they will appear thoughtlessly, generated from conventions you never endorsed and carrying rhetoric you never examined.

This chapter has equipped you with the vocabulary and a pipeline to prevent that outcome. You have learned to specify encoding logic before visual form, to critique symbols for their interpretive effects rather than their aesthetic appeal, and to recognize when abstraction serves meaning better than literal representation. These are not marginal skills. They are foundational to every map you will direct from this point forward.

Chapter 13 expands this logic beyond individual symbols to entire thematic systems – choropleth maps, proportional symbols, flow maps, and more. Each method carries its own default lies, and each requires prompts that acknowledge those lies rather than amplifying them. Symbology prepared you to think critically about marks. Thematic representation will teach you to think critically about the structural stories those marks collectively tell.

Takeaways

- **Symbols Are Visual Arguments**

  Effective symbols clarify presence, category, and importance rather than simply marking location.
- **Omission Is a Specification (Ad Nauseam)**

  When you do not direct symbol choices explicitly, models fill the gap with defaults that may undermine your map's intent.
- **Intent Determines Encoding**

  Specify what a symbol needs to communicate before describing what it should look like.
- **Shape Encodes Categories**

  Reserve shape for meaningful categorical distinctions rather than decorative or culturally specific flair.
- **Size Must Follow Proportional Logic**

  Scale symbols by area, not radius, to avoid distorting quantitative relationships.
- **Opacity Controls Hierarchy and Uncertainty**

  Use transparency intentionally to manage visual weight and communicate confidence levels.
- **Texture Functions as Redundancy**

  Texture works best as a secondary encoding variable, particularly for accessibility, rather than as a primary design feature.
- **Critique Symbols for Meaning**

  Evaluate whether symbols clarify interpretation or introduce unintended rhetoric.
- **Abstraction Increases Universality**

  Generic geometric shapes tend to communicate more broadly than detailed or era-dependent pictographs.
- **Prompt for SVGs with Constraints**

  When generating vector symbols, specify stroke weight, scale legibility, and semantic intent rather than relying on model defaults.

Conceptual and Applied Exercises

Symbol Rhetoric Check

Think of a map where symbols felt either too literal (pictographic overload) or too abstract (meaningless shapes). What balance between recognition and interpretation would have served the map's purpose better?

LLM Symbol Justification

Ask an LLM: "Design three symbols to represent low, medium, and high earthquake magnitudes. Explain your encoding choices for size, shape, and opacity." Evaluate whether it prioritizes perceptual accuracy or decorative appeal.

Abstraction Versus Detail

Write a prompt for symbols representing public transit stops (bus, train, ferry) that balances categorical distinction with universal recognizability. Specify constraints that prevent culturally specific or era-dependent imagery. Prompt for both PNG outputs and SVG versions separately. Notice differences and similarities in outputs.

For extended tutorials, hands-on labs, prompt templates, and updated examples related to this chapter, visit www.promptcartography.com.

13

Thematic Representation

Chapter 12 addressed individual symbols – marks that assert presence and encode meaning at the feature level. Here we scale that logic upward to thematic representation, i.e., the structural decisions that determine what kind of story a map is capable of telling.

Thematic maps do not merely display data. They impose interpretive frameworks. A choropleth suggests continuous fields. A dot map implies distribution. A flow map asserts movement. Each method carries assumptions about the data it represents, and each has predictable failure modes – places where the method's default logic quietly distorts meaning.

In prompt cartography, these distortions do not disappear simply because your map includes smooth transitions and polished interactions. If anything, they become more dangerous, because a well-executed lie is far more persuasive than a clumsy one.

This chapter provides a field guide to the most common thematic methods you will encounter and prompt for: choropleths, proportional symbols, dot maps, heat maps, isarithmic surfaces, flow maps, cartograms, and multivariate representations. For each, you will learn what the method does well, what it tends to lie about, and how to write prompts that make those lies visible – or prevent them entirely.

Thematic representation is where design meets data integrity. Prompt accordingly.

Data and Phenomenon: The Fundamental Distinction

Before diving into specific map types, it is crucial to understand a principle that will guide all your prompting decisions: you are never mapping phenomena directly. You are mapping data – measurements or samples taken from those phenomena.

Phenomena can be discrete or continuous, abrupt or smooth (as shown in Figure 13.1). Zoos are discrete and abrupt – they exist at specific locations and do not cluster predictably. Population density is continuous and smooth; it varies gradually across space. Exotic animal regulations are continuous but abrupt; they change instantly at state borders.

DOI: 10.1201/9781003797920-15

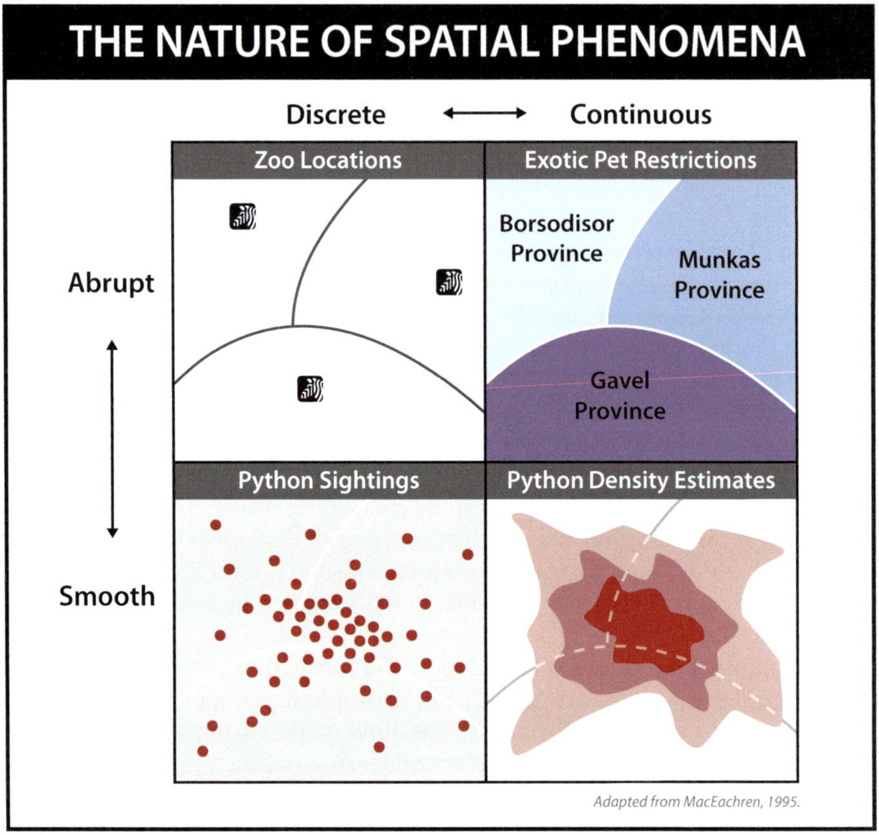

FIGURE 13.1
Depending on the nature of your dataset, certain visualization techniques facilitate user interpretation. You must pick an appropriate visualization type for the data you are mapping, or your map's communication will break down.

The critical insight for prompt cartography is this: how data is collected about a phenomenon impacts how you can – and should – map them. A phenomenon may be smooth (like python distribution in Florida), but if you only have county-level data, your representation will be abrupt-continuous. When the data's spatial nature diverges from the phenomenon's, your prompts must acknowledge this gap explicitly.

Prompt Strategy: When specifying a thematic map type, always declare the data's spatial characteristics and any known mismatch with the phenomenon:

```
Create a choropleth showing county-level python sighting estimates. Add a cautionary note that this represents aggregated reports by county, not actual python distribution, which is continuous and concentrated in specific habitat zones.
```

Choropleth Maps

Choropleth maps use color value, hue, or saturation to show data differences among enumeration units – provinces, counties, census tracts (Figure 13.2). They are the most familiar thematic map type, which means they carry the weight of user expectations and the danger of overuse.

When Choropleths Work

Choropleths excel at displaying standardized data – i.e., rates, percentages, densities – across enumeration units. When designed appropriately (lighter values representing less, darker values representing more), they are extremely intuitive. They effectively communicate continuous phenomena that vary smoothly across space. However, *like all thematic representations*, because they are massively simplifying data, they lie. It's important you understand how choropleths mislead your user so you don't accidentally mislead in the wrong ways.

The Choropleth's Default Lies

Raw Data Distortion

Mapping raw counts with choropleths fails to account for differently sized enumeration units. A state with a larger area will logically be assumed to have a higher population total, all things being equal geographically. Always standardize your data – i.e., choropleths may only be used to map rates (percentage population growth) or derived data (population divided by land area or another relevant value).

Internal Uniformity

By filling entire polygons with a single color, choropleths represent each unit as internally uniform. German federal state population density is not equal throughout each state, yet the choropleth reifies this fiction with its color-by-space approach.

Projection Sensitivity

Because choropleths fill entire enumeration units (spaces), areal distortion massively distorts both the nature of the data and its interpretation. Use equal-area projections. Always. (Well, almost always. If you can't, make sure that you bash your API for being so limited afterward, so then at least other people know *you know* you should have.)

Classification Decisions That Shape Meaning

Most choropleth maps classify data into groups (i.e., chunking like values into groups that are represented using the same color). Typically, one should

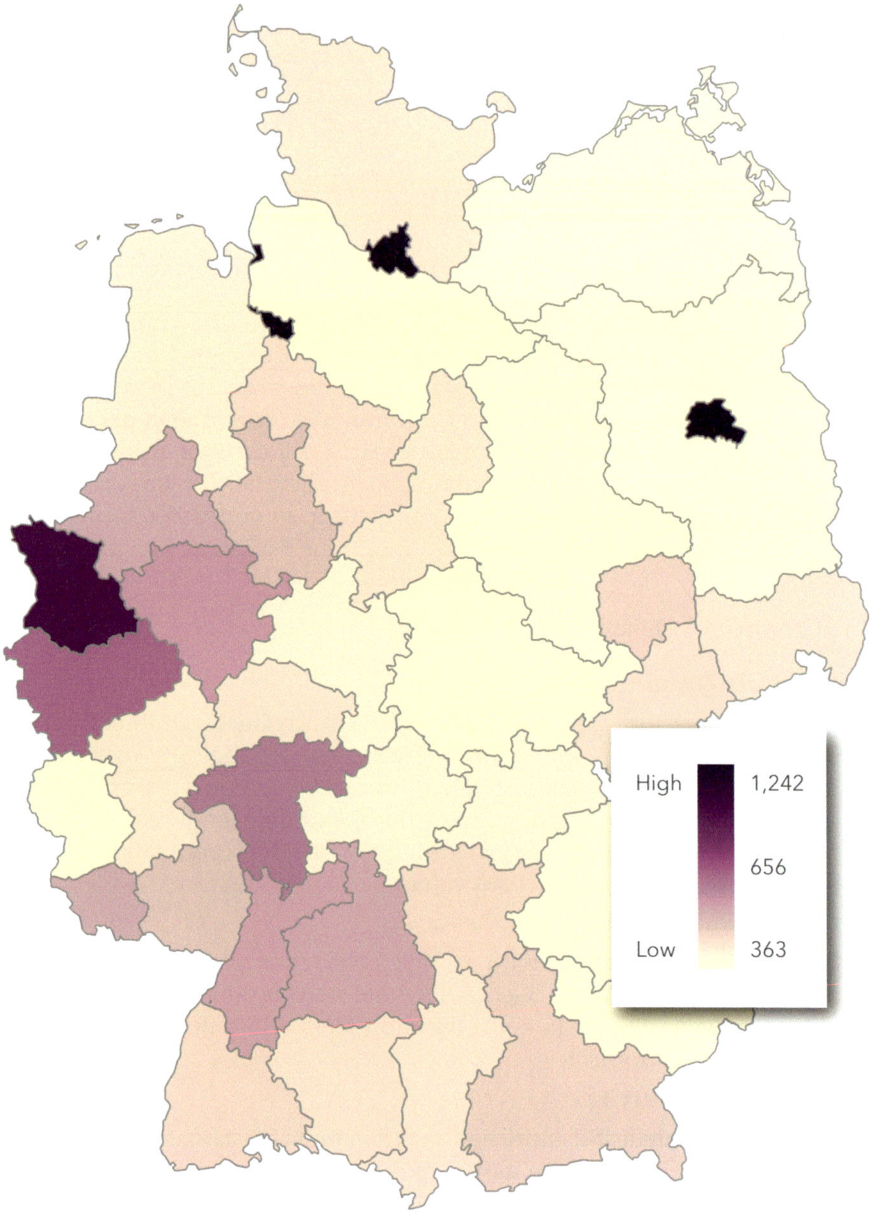

FIGURE 13.2
Choropleth maps are color-by-number maps. Each polygon area is colored using color value, saturation, and sometimes hue to communicate value from low to high. When "unclassed," such as the map in this example, the areas are colored based on where their values fall on the color gradient from low to high. This map shows German province population density in 2023.

have 3–6 classes to provide enough diversity to highlight a phenomenon but not so many as to overload your map user's ability to differentiate color schemes. How you classify your data will determine what your map says as much as any type of thematic representation. Choose your classification method carefully and be sure to prompt for it so your large language model (LLM) doesn't pick a default for you!

- **Natural breaks** identify the largest gaps in your dataset, useful for highlighting natural clustering and isolating outliers.
- **Quantiles** place equal numbers of data values in each class, appropriate for ordinal data but potentially misleading for interval data.
- **Equal intervals** divide the data range into equal steps, creating classes with potentially unequal numbers of values.
- **Standard deviations** highlight variation from the mean, optimal only for normal distributions (rare in geographic data).
- **Unclassed** assigns each unit a unique color value, reducing generalization but overwhelming cognitive capabilities beyond 6–7 distinctions.

The choice between these is not technical – it is rhetorical. Different classification schemes can produce dramatically different interpretations of the same dataset as shown in Figure 13.3.

Prompting for Choropleths

- `Design a choropleth of {layer name and field with ratio data} that highlights {your communicative intent}.`

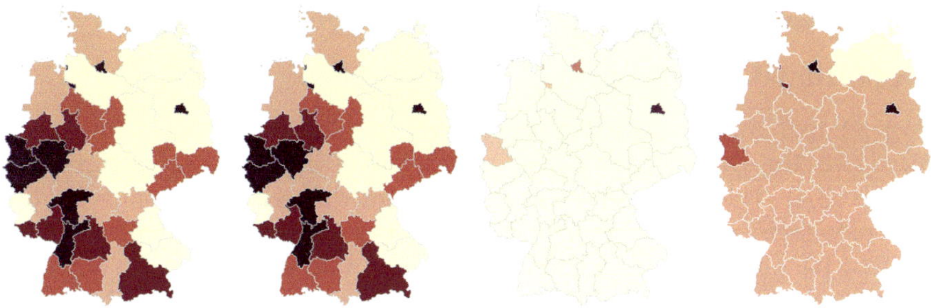

FIGURE 13.3
Same exact dataset as found in Figure 13.2 This time, however, I am using four different classification schemes (each with five classes). Clockwise from upper-left: natural breaks, quantiles, standard deviation, equal intervals.

- `Confirm field is standardized - if it's a raw count, STOP and propose alternative thematic representations.`
- `Use a colorblind-safe sequential palette (light=low, dark=high).`
- `Use 4-6 classes max - propose an ideal classing method and justify in 2 sentences.`
- `Specify appropriate equal-area projection to use given geographic extent of dataset (e.g., Albers Equal Area Conic for US data, Equal Area for global, etc.).`

Director Note: Never Forget the Choropleth Map's *Default* Lie. The phenomenon in each choropleth unit is *not* equally intense. If this fact harms your argument or narrative, switch to another thematic technique.

Prompt Variation Example

For exploring data patterns before deciding on final classification:

> *Generate an unclassed choropleth preview with a continuous color ramp to identify clustering patterns. Then propose three classification schemes (natural breaks, quantiles, equal intervals) with visualizations and 2-sentence justifications for each. Recommend one based on the data distribution and communication goal of highlighting regional disparities.*

Proportional and Graduated Symbol Maps

Proportional and graduated symbol maps use sized symbols – most commonly circles – to represent raw values tallied by enumeration unit (see Figure 13.4). They may be used to represent derived values as well, but typically, there are better thematic methods to use in these situations. One symbol per unit, scaled to show magnitude. Proportional and graduated symbols are remarkably effective, and in my humble but informed opinion, underused in general.

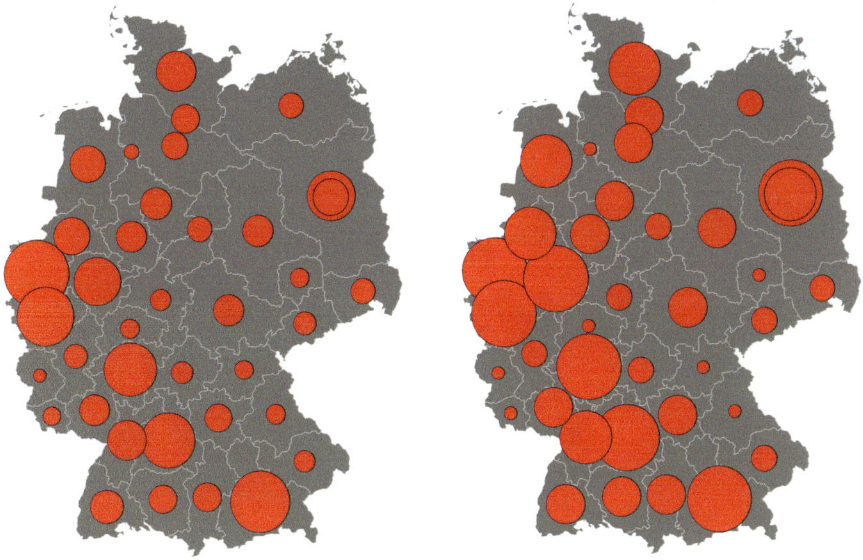

FIGURE 13.4
Proportional versus graduated symbol comparison. Each map is created using the same circle size parameters (smallest 20px and largest 100px). The map on the left uses proportional (unclassed) symbology. The map on the right uses graduated symbols – natural breaks was used to create five different symbol sizes.

The Critical Distinction

- **Proportional symbols** are sized exactly proportional to data values (e.g., Bavaria's symbol, representing 12,500,000 residents, will be exactly five times larger than Brandenburg with 2,500,000).
- **Graduated symbols** classify data into 3–5 groups (like choropleth maps) and assign discrete symbol sizes to each class.

Why Proportional Symbols Excel

Size difference is one of the most intuitive visual variables for humans, leading to innumerable fights over slices of cake among children (and the occasional adult) worldwide. Proportional symbol maps work for both raw and standardized data (though consensus favors raw counts). Unlike choropleths, proportional symbols can reveal distribution patterns across the landscape when enumeration units are compact. Bonus: symbols layer easily on

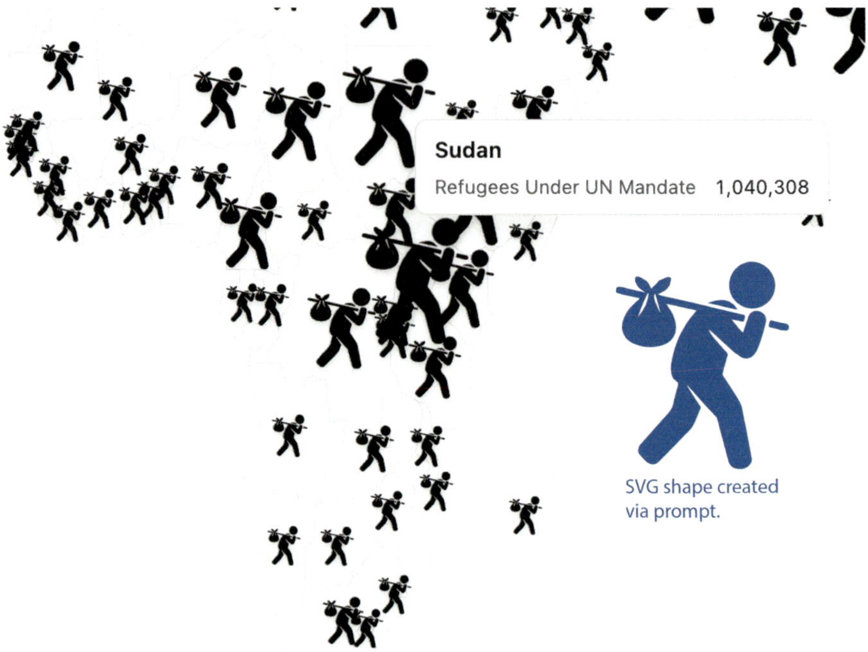

FIGURE 13.5
A proportional symbol map of international refugees using a generic refugee shape.

top of choropleth representations, allowing you to add a second data dimension to your visual communique.

When it makes sense rhetorically and aesthetically, you might, consider creating proportional symbols using pictographic icons. The icon used in Figure 13.5 was created with the following prompt:

```
Please create a single tone map icon, in simple silhou-
ette style of a refugee walking with a bag on a stick
over the shoulder. Generic map icon. Very simple art
with thin lines and use of negative space, designed to
be used in a proportional symbol map. Thanks!
```

The Look-Right-Is-Wrong Error

Unfortunately, LLMs will cheerfully scale symbols by radius instead of area unless you explicitly prohibit this. Radius scaling feels intuitive in language ("make it twice as big!") but is mathematically misleading and visually distorting. Always scale by area.

The Drawbacks of Proportional Symbols

- **Overlapping**: dense datasets often create symbol collisions requiring opacity, outlines, or decluttering strategies. Proportional symbols on a US map work without problem in western states (which are gargantuan in size) but cause headaches on the east coast when trying to squeeze in symbols over Delaware, Maryland, New Jersey, and Washington, DC, for example.
- **Area Estimation**: humans struggle to accurately estimate area differences, especially with circles (we do better with squares, but circles minimize symbol overlap).
- **Aesthetic Unfamiliarity**: proportional symbols are less common than choropleths, making them feel "wrong" to audiences accustomed to filled polygons. (Ironically, the same audience often readily accepts choropleth maps representing raw data, which is literally wrong, but...)

Prompting for Proportional Symbols

```
Create proportional symbols for [raw count] by unit.
Scale by AREA not radius - use sqrt() or log scaling if
  outliers dominate.
Set min/max sizes to prevent occlusion (e.g., min: 8px,
  max: 80px).
Add overlap strategy: opacity 0.7, 1px outline, cluster/
  declutter at small scales.
Legend: 4 reference sizes with example values (e.g.,
  1,000 | 10,000 | 50,000 | 100,000).
Use circles for compactness unless area estimation is
  critical (then use squares).
```

A False Believe (or Lie) about Proportional Symbols: you can read exact quantities from area. You cannot. Provide reference sizes in the legend and tooltips with actual values. Accept that visual approximation is the point.

Prompt for Graduated Symbols

When precise proportionality is less important than clear categorical groupings, request graduated symbols.

Create a graduated symbol map with 4 classes using natural breaks classification. Assign clearly distinct

```
sizes (12px, 24px, 42px, 64px diameter circles).
Label classes with value ranges. Include tooltips
showing exact values on hover.
```

Dot Maps

Dot maps use point symbols to show spatial distribution within enumeration units. Each dot represents either individual objects or a predetermined quantity (1 dot = 100 residents). They answer the question: where is it *within* the unit?

When Dot Maps Shine

Dot maps are excellent when a choropleth would flatten meaningful internal variation. They reveal distribution patterns that aggregated representations obscure. When you have geocoded point data (lat/long, addresses), they are often straightforward to create.

The Recurring Failure Mode

Dot maps frequently become maps of population density, reporting density, or sensor placement – regardless of what the dots supposedly represent (see Figure 13.6). UFO sightings mapped as dots simply show where people live. Crime dots show where police patrol. This does not make dot maps useless – it makes honesty essential.

Design Considerations

- **Symbol Choice**

  Use compact geometric marks (small circles, squares). Pushpins are visual noise and spatial anachronisms. Please, no pushpins.

- **Dot Value**

  If aggregating, clearly state the ratio (1 dot = N entities) and justify your choice.

- **Overlap Management**

  Use opacity, slight jittering, point clustering, or have the LLM resize the symbols at different zoom levels.

- **Distribution Honesty**

  Acknowledge whether patterns may reflect population/reporting density rather than phenomenon distribution.

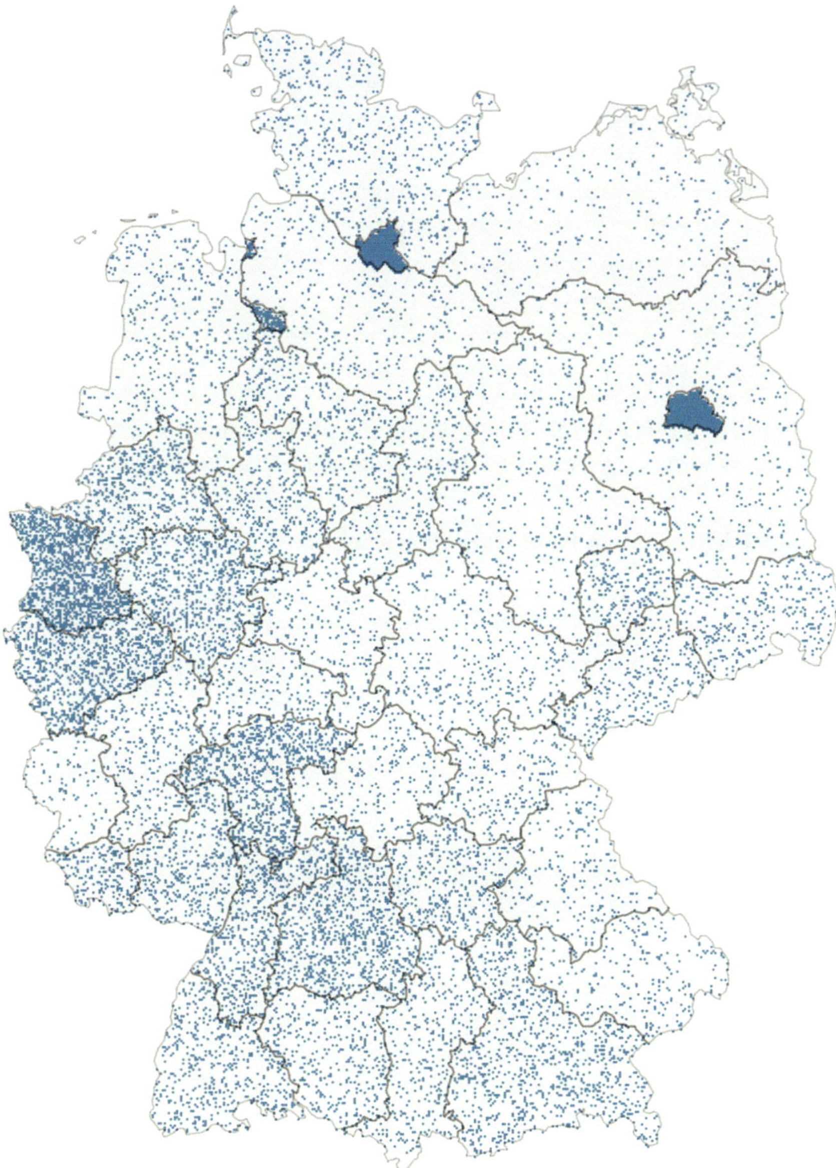

FIGURE 13.6
Dot maps can be used effectively to show distributions. They can be negatively impacted by the use of large areal units, however. This dataset is probably better designed as a proportional symbol or population density choropleth map.

Prompting for Dot Maps

- `Create a dot map for {phenomenon}.`
- `Use small geometric dots (3-5px circles) - no push-pins, no teardrops.`
- `If dots are aggregated, define "1 dot = N" and justify based on data density.`
- `Add overlap plan: opacity 0.6, slight random jitter (±5px) to reveal density.`
- `Include one sentence noting whether pattern may mirror population/reporting density.`

The Dot Map's Default Lie: this is where the phenomenon is, when it might be "this is where people are" or "this is where observation happens."

Prompt Variation: When Point Data Overwhelms

`Evaluate whether the point density makes individual dots interpretable. If overlap exceeds 40%, propose alternative representations (heat map, hexbin aggregation, or proportional symbols by census tract) and justify the recommendation.`

Heat Maps and Density Surfaces

Heat maps are hybrid forms of dot maps that display spatial intensity using color gradients. Unfortunately, they frequently still use a rainbow color scheme – typically blue (cool/sparse) through green (moderate) to red (warm/dense). It's unfortunate, because many humans do not naturally interpret this hue-based color scheme in a particular order. (In other words: whenever possible, avoid using a rainbow color scheme to represent quantitative data. It's illegible for many people, including this author.) Heat maps work by placing buffers around data points and changing these points' color as parts of it overlap with other point areas based on increased density.

When Heat Maps Work

Heat maps excel at showing general density patterns when exact point locations are less important than spatial concentrations (see Figure 13.7). They

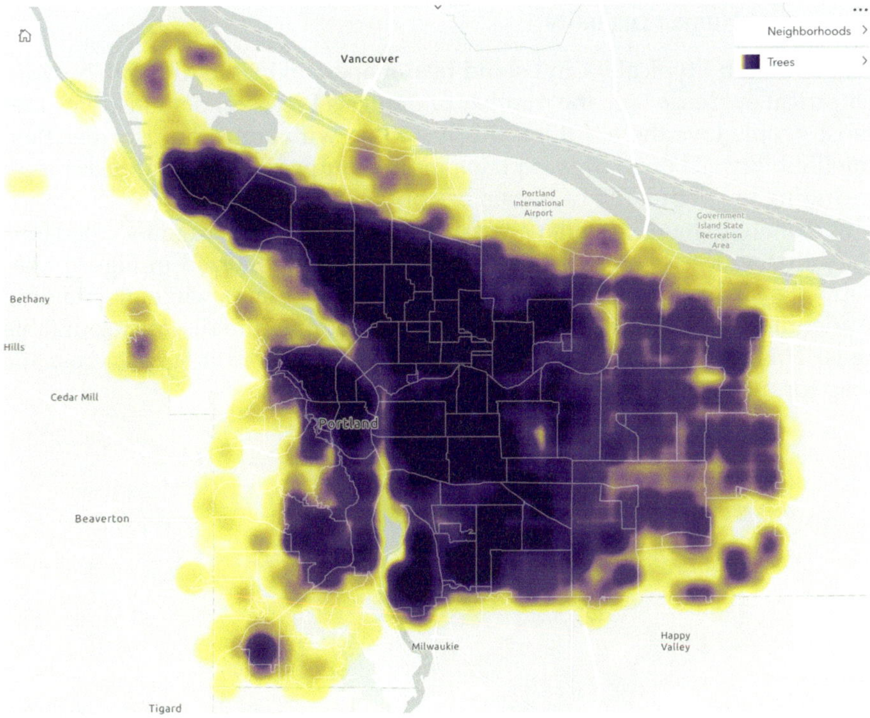

FIGURE 13.7
Heat map showing trees in Portland, Oregon. The information is interpolated and not specifically accurate in any one area. Attribution: Esri Default Gray Basemap.

work well for large point datasets where individual dots would create visual noise.

The Critical Drawbacks of Heat Maps

1. **Magnitude Ambiguity**

   It is impossible to visually determine how many points exist in a given area – legends help but never fully resolve this.

2. **Location Imprecision**

   Heat maps obscure exact point locations through smoothing.

3. **Semantic Confusion**

   Blue-to-red spectrums carry emotional associations (blue = depression/cold, red = danger/aggression) that may contradict your data's meaning. Rainbow color schemes are difficult to interpret. However, any sequential color scheme can be used in heat maps.

Binning for Human Legibility

I'll be honest: I typically can't stand heat maps. I'm biased, though, because empirical evidence (i.e., the number of requests I receive) demonstrates that most people love them. I think this has to do with appearance, rather than intelligibility. (Though, this explains more than people's love of heat maps, including the popularity of certain celebrities.)

Fortunately, there is a stellar alternative to heat maps. Hexagon-shaped bins (hexbins) and grid aggregations (square bins, as visualized in Figure 13.8) show density with greater precision than kernel surfaces while maintaining readability. Binning converts continuous smoothing into discrete, countable cells. This allows users to look up actual count data from areas across the map for quantitative comparison.

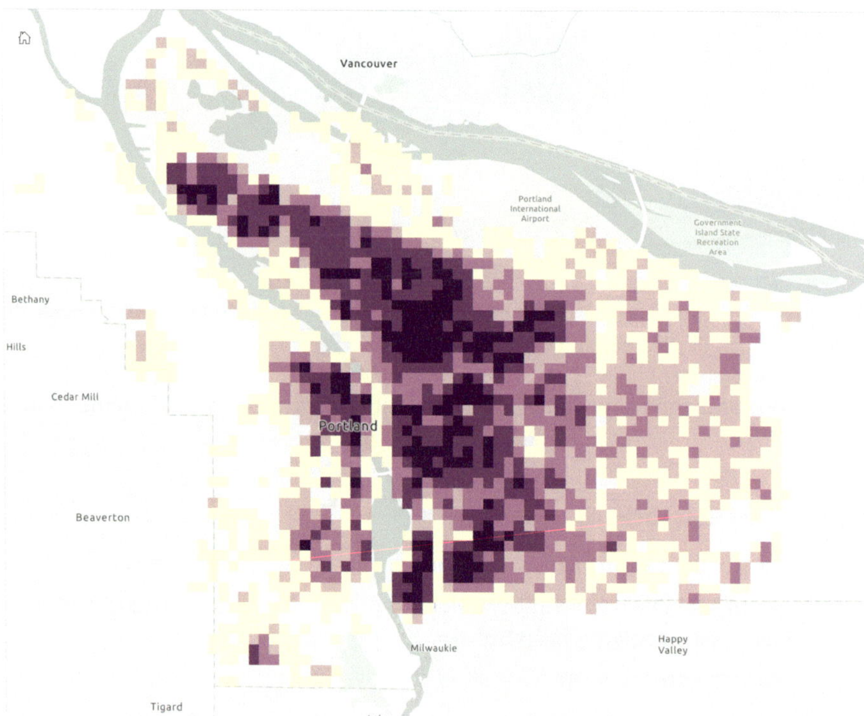

FIGURE 13.8
Bin map showing trees in Portland, Oregon (Esri Default Gray Basemap). Notice that bin maps allow for tool tips showing exact counts in a given bin area, allowing for numeric exploration if your map intent requires it.

Prompting for Heat Maps

```
Create a heat map showing density of {point phenomenon}.
Consider hexbin alternative if users need to estimate
  quantities. Aggregate into 100m hexagonal bins, use
  choropleth coloring. Otherwise…
Use kernel density estimation with {X-meter} radius.
Apply sequential color scheme: light yellow (low) →
  orange → dark red (high). Justify color choice for
  phenomenon. If you use a rainbow scheme, you're fired
  from the team. ;-)
If semantic confusion risk exists (e.g., crime data),
  use alternative palette (yellow-green-blue or single-
  hue sequential).
Add legend showing approximate point counts per color
  range.
```

How Heat Map Lie: smooth gradients imply precision where interpolation algorithms dominate. State smoothing parameters explicitly.

Isarithmic Maps

Isarithmic maps – contours and isolines – represent data using lines of constant value perpendicular to Earth's surface. Each line represents a consistent value at regular intervals (every 10 meters elevation, every 5°C). Most people know these from weather maps showing isotherms or topographic maps showing elevation contours.

Two Types Based on Data Collection

- **Isometric Maps**

  Created from point data (weather stations, depth soundings, stadium attendance).

- **Isoplethic Map**

  Created from enumeration unit data by placing control points (often centroids) and interpolating between them.

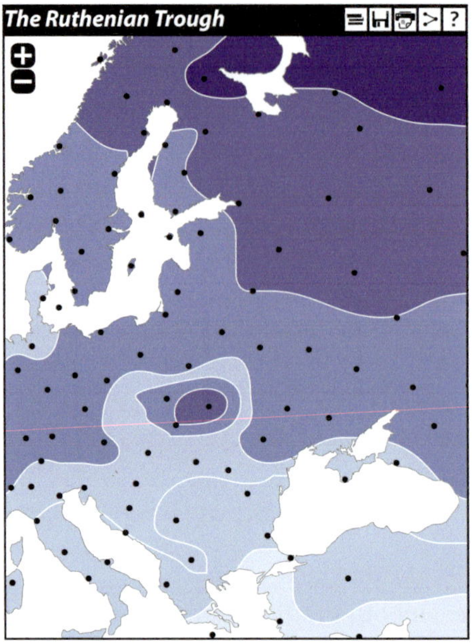

FIGURE 13.9
Isarithmic maps, like this one of European temperatures, are great for showing smooth, continuous phenomena, but depending on how close together the data points are that are used to create the map, they can imply far more accuracy than they really should.

When Isarithmic Maps Excel

Isarithmic maps shine for truly continuous phenomena that behave like surfaces: elevation, temperature, atmospheric pressure, pollution concentration (see Figure 13.9, for example). They naturally represent smooth transitions that choropleths render as abrupt boundaries.

The Beautiful Lie

A beautiful isarithmic surface can mask sparse sampling. Interpolation is mathematical, yes – and 100% guesstimate. Albeit a well-educated guess, like me estimating the time within 15 minutes right now, but still a guess. The map looks authoritative everywhere, but accuracy varies dramatically with sampling density.

Design Best Practices

- Choose appropriate isoline intervals – too many create clutter, too few obscure variation.

- Label only index lines (every 4th or 5th line) with slightly thicker strokes.
- Fill areas between isolines with appropriate color schemes (sequential for elevation, diverging for temperature anomalies).
- Ensure isolines look natural and fluid, not jagged.

Prompting for Isarithmic Maps

```
Create isarithmic contours from point measurements of
  [variable].
State interpolation method clearly (e.g., linear, krig-
  ing, spline) and assumptions (e.g., assumes gradual
  change between stations).
Choose contour interval to avoid clutter - label every
  5th as an index line (thicker, with value labels).
Add an uncertainty layer: use transparency or dashed
  lines in areas far from sample points (e.g., >5 km
  from nearest measurement).
```

The Isarithmic Map's Default Lie: the surface is accurate everywhere. Absolutely not, unless you collected measurements everywhere (which no one has, nor ever will, not even with Lidar).

Prompt Variation: Showing Data Collection Points

Include small point markers showing actual measurement locations. Use semi-transparent fill between isolines in areas with sparse sampling to visually indicate lower confidence regions.

Flow Maps

Flow maps represent movement – migration, trade, traffic, data transmission. They use lines (often with arrowheads) to show direction, and visual variables (thickness, color, animation) to encode quantitative and qualitative dimensions. They are my favorite thematic map types.

What Makes Flow Maps Powerful

They are one of the most dynamic and useful thematic representations available. They can show:

Direction

Arrowheads indicate movement; bidirectional flows use two arrowheads or none.

Magnitude

Line thickness scaled proportionally to volume (proportional flow maps) or classed (graduated flow maps).

Categories

Color hue differentiates flow types (ethnic groups migrating, product categories shipping).

Intensity

Color value/saturation adds a second quantitative variable (line thickness = total goods, saturation = percentage high-value goods).

The Aesthetic and Technical Challenge

Flow maps are visually striking and informationally dense (Figure 13.10) – and remarkably difficult to create well. Few algorithms produce aesthetically pleasing curves; most generate robotic straight lines. Manual drawing (digitizing Bézier curves) produces better results but requires significant effort.

Without filtering and bundling, flow maps become illegible hairballs. Interaction and selective emphasis are essential.

Prompting for Flow Maps

```
Create a flow map showing [movement type] between [origin] and [destination].

Use curved lines (Bézier or arc paths, not straight segments) for aesthetic fluidity.

Scale line thickness by {volume variable} - provide 3-4 reference thicknesses in legend.

Direction: add arrowheads for unidirectional flow; omit for bidirectional unless magnitude differs.

Filter: show only top {N} flows or flows exceeding {threshold} to prevent clutter.

Add interaction: tooltip on hover shows exact values (origin, destination, volume).

Consider flow bundling for dense networks:

  Apply edge bundling algorithm to reduce visual crossings while preserving general flow patterns.
```

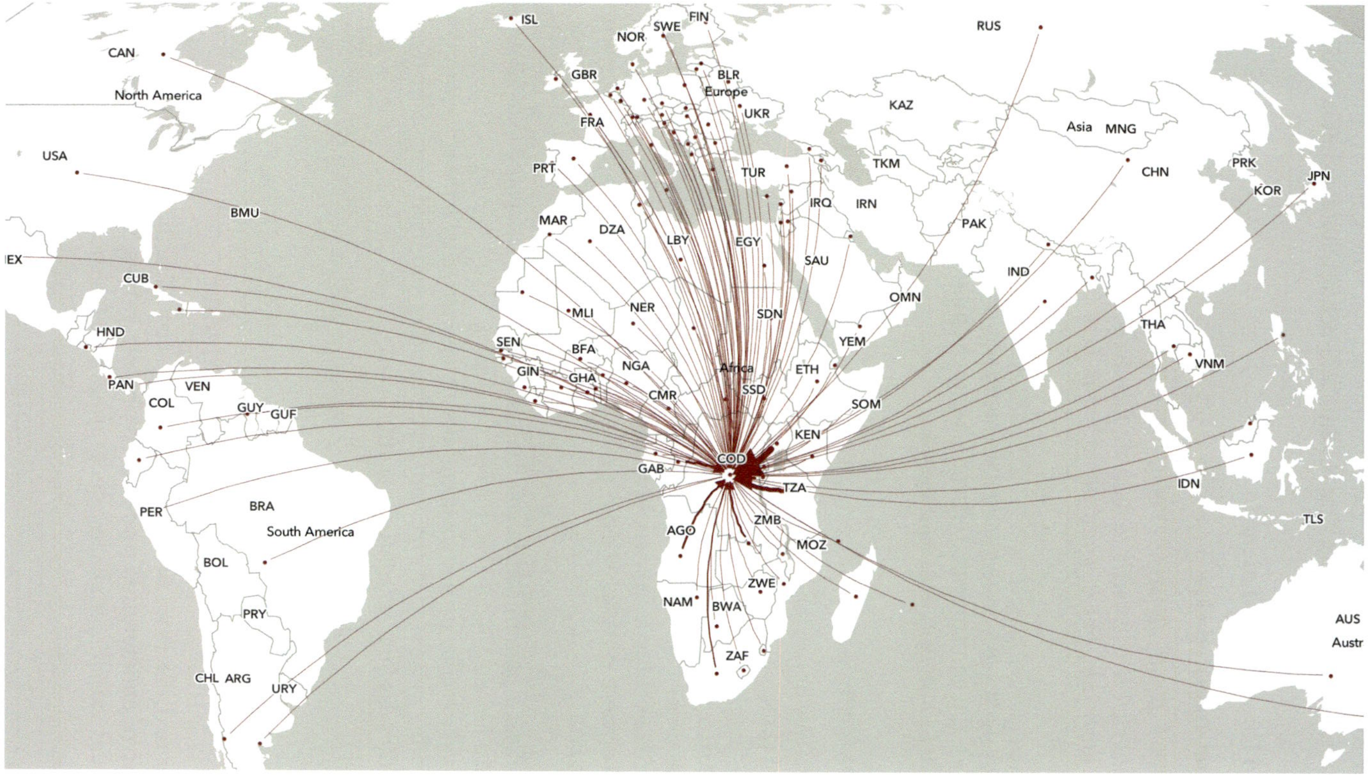

FIGURE 13.10
A simple hub and spoke flow map showing the number of refugees leaving the Democratic Republic of Congo circa 2020 and the countries to which they traveled.

The Flow Map's Default Lie: clean channels exist where messy, multidirectional movement occurs. Filtering is editorial – acknowledge what is omitted.

Advanced Flow Map Techniques

Prompt for Animated Flow

You will learn about animation in more detail in Chapter 14. However, I would like to point out that it is possible to animate flow lines. It should be done judiciously and with great care, but if you want to try it, here is a generic prompt to begin playing with:

> *Animate flow lines sequentially by magnitude (largest first) with 0.5-second intervals. Provide play/pause controls. Do not loop automatically. Include temporal legend showing animation progress.*

Heat Flow Maps

An alternative approach combines dot and heat map techniques using lines instead of points:

```
Create a heat flow map: each line represents {N units}.
Lines overlap as they converge on destinations, shift-
ing from light green (sparse) to dark purple (dense).
Add legend correlating color to approximate flow counts
per area.
```

This technique is powerful when you have so many lines overlapping that following single flows becomes impossible. Rather, the density of line overlap is the message. This technique is often used to show global shipping routes, with the highest traveled routes burning into the user's memory and the outlier route also being noticed, as it haphazardly sales from Kolkat to Dunedin.

Cartograms: The Rhetorical Sledgehammer

Cartograms differ from other thematic maps by directly distorting the size and shape of enumeration units based on data values. Size becomes the primary visual variable, replacing geographic area (as shown in Figure 13.11).

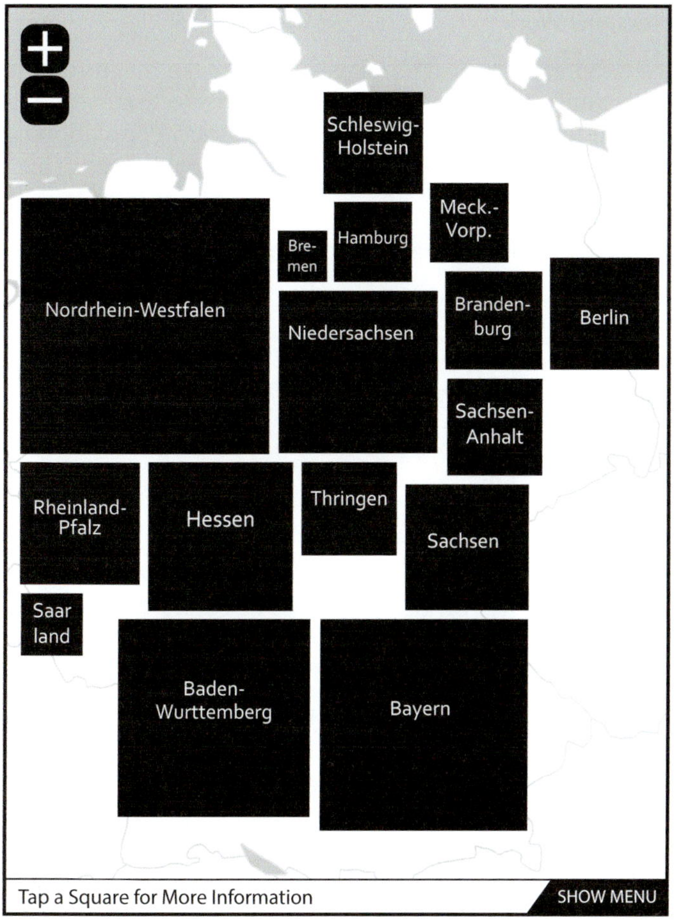

FIGURE 13.11
An example of a very simplistic cartogram. Cartograms are essentially proportional symbol maps without the underlying geography (i.e., polygon).

Two Approaches

- **Contiguous Cartograms**: maintain topology (units remain touching) but distort shape dramatically – difficult to recognize places.
- **Non-contiguous Cartograms**: preserve shape but separate units – easier to identify individual places but harder to judge proportions of the whole.

When Cartograms Work

Cartograms excel at rhetorical emphasis. They make magnitude impossible to ignore. Election maps showing states sized by electoral votes rather than land area reveal the true weight of different regions. Population cartograms clarify demographic realities hidden by conventional maps.

The Double-Edged Sword of Areal Distortion

Cartograms visually distort data so wildly that people get excited – this is both advantage and liability. Initial excitement can succumb to disorientation when users cannot locate or identify places. The novelty that draws attention also creates cognitive barriers.

Prompting for Cartograms

- `Create a {contiguous/non-contiguous} cartogram of {region} sized by {field / variable}.`
- `Resize enumeration units proportional to {variable} – justify whether contiguous or non-contiguous based on need for topology versus shape recognition.`
- `Use muted base colors to emphasize shape distortion over color coding.`
- `Provide a normal reference map toggle so users can compare distorted and geographic views.`
- `Add tooltips displaying both the sizing variable value and geographic area for comparison.`
- `In subtle tooltip pop-up include explanation of what the distortion means: "Size represents {variable}, not area."`

A Cartogram's Lie: distortion presents us with truth. *It does not*. Distortion equals emphasis – not truth, validity, or importance. Instead of emphasizing or respecting geographic area, shape, or direction, a cartogram emphasizes a particular field in the dataset. It may be a completely irreverent field. Make sure it displays what you intend.

Prompt Variation: Maintaining Orientation

`Include small inset map showing undistorted geography with highlighted current view region. When user taps on region, highlight it on the inset map.`

Multivariate Maps

Multivariate maps show more than one theme simultaneously – layered choropleths with proportional symbols, colored dot maps, proportional pie charts. Web mapping makes this easy, which increases the risk of cognitive overload. My rule of thumb, and it's not uncontested even though I believe it's right: less is *almost* always more. Sometimes, though, embedding more than one dimension of information into a single geographic layer can pay dividends.

When Multivariate Representation Succeeds

Multivariate maps work when spatial relationships between variables must be seen together – correlations, disparities, multivariate patterns. A choropleth showing income with proportional symbols for population reveals whether dense areas are wealthy or poor.

The Cognitive Load Problem

Each additional variable increases interpretive burden exponentially. Humans can track two to three visual channels simultaneously before comprehension degrades. There is a misnomer that we can track seven plus or minus two (i.e., 5–9) things visually, but this has been proven wrong many times. We can remember 5–9 things in our short-term memory, but not in our visual memory. Our brains typically can only perceive roughly three things at any given moment. The rest is filled in by our brain based on previous data and educated guesses. (Yes, even while you're driving or when a doctor is doing surgery on you.) So if you want to embed two or three bits of info into a map symbol (as shown in Figure 13.12), and it makes sense rhetorically, that will probably work well. Beyond that, linked views (e.g., bar charts that interact with the map) or small multiples will almost always communicate data more effectively.

Thematic Combinations

The most common approach to multivariate mapping is to place complementary thematic layers on top of one another. This works well, because it allows the human user to compartmentalize and connect the two layers being visualized as needed.

```
Create a choropleth showing population density (sequen-
tial blue palette) with proportional symbols showing
total GDP (scaled circles, 50% opacity, gold stroke).
Ensure symbols do not obscure choropleth patterns in
low-density areas.
```

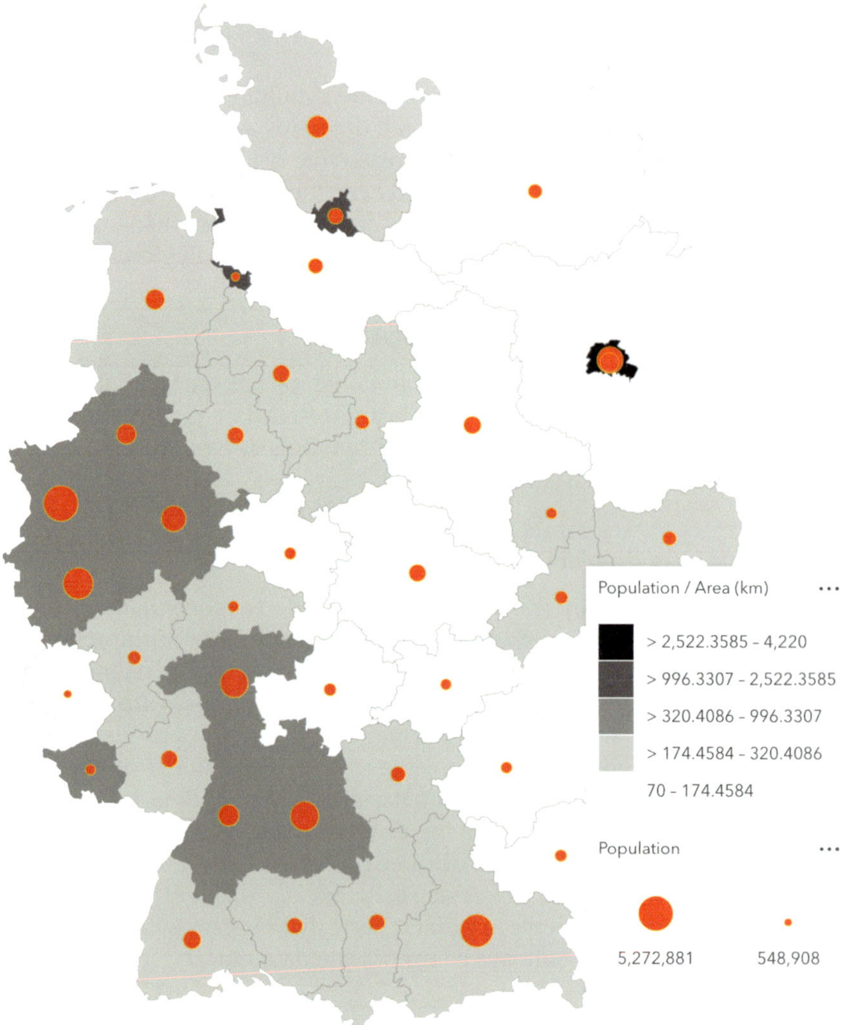

FIGURE 13.12
A simple multivariate representation in a single symbol. Hue and value represent population density, and size represents total population.

Colored Dots for Multiple Categories

Dot maps can show distribution of multiple categories using color or shape:

> *Create a dot map of election-night tweets: blue dots for Democratic Party mentions, red for Republican, gray for other parties. Use 60% opacity to reveal*

overlap density. Add legend and category filters to toggle visibility.

Chart Maps

Proportional charts (typically pie charts, donut charts, occasionally bar charts) encode multiple dimensions in a single symbol:

Create proportional pie charts for each state showing population (circle size) and age distribution (pie slices: 0–18, 19–64, 65+). Use muted colors (beige, gray-blue, dark gray) to avoid hue-based misinterpretation. Provide tooltips with exact percentages.

Pro-Tip: Bar charts work better in print; their vertical extent can be quite distracting in web maps. However, including them to highlight nuanced data within a pop-up or information window in a sparkline type fashion is often a useful solution.

When to Avoid Multivariate Mapping

Reserve multivariate techniques for cases where simultaneity is essential. Otherwise considering using:

- **Linked Views**: a user selection highlights that area's value across multiple maps and/or charts.
- **Small Multiples**: series of maps placed next to one another that (a) show different variables in the same geographic area at the same geographic scale for quick comparison, or (b) show the same variable in the same geographic area at the same geographic scale over some other variable (e.g., time or normalized by various other attributes).
- **Layered Interactivity**: allow users to toggle layers on/off.

Agent Prompt for Multivariate Restraint

```
Evaluate whether {variables} must be seen simultane-
ously to achieve the map's intent as stated in the
attached intent JSON. If yes, or if spatial correla-
tion is the goal of the map, create a multivariate map
with {required combination}. If comparison is the goal,
propose small multiples or linked views to the user
instead. Justify your recommendation based on cognitive
load and communication purpose.
```

The False Axiom Underlying the Use of Most Multivariate Maps: more information is better. *Not necessarily*. Clarity beats comprehensiveness. Just like searching for a phone number in your contacts app beats paging through a printed phonebook. The latter presents the user with a lot more information all at once and overloads and slows down information collection.

Thematic Mapping via Prompting: Integrating Method Selection

Never forget: the most important prompt is the one that asks whether you have chosen the right method at all. Use your team of LLM agents to help guide you. Yes, it's true, when left unguided they will make mistakes. They may often choose mediocre defaults when not asked the right questions. But if you want feedback and ask them specific questions (e.g., "Are there other, potentially more efficient ways, to visualize this dataset? What are they and list the pros and cons of each?"), they will help you avoid cartographic mistakes.

I recommend following the process below whenever you are designing thematic maps to make sure you design the map correctly and, most importantly, in a manner that meets your intent.

Method Selection Framework

Before specifying visual details, establish representational integrity:

> *I have {data description: type, scale, collection method}. My goal is to communicate {specific message}. Recommend three appropriate thematic representation methods with justifications. Identify the default lie each method tells and propose guardrails to make those lies visible.*

Iterative Refinement Pattern

1. **Generate Initial Representation**: "Create a {method} showing {variable}."
2. **Critique for Default Lies**: "Identify three ways this representation might mislead viewers."
3. **Add Guardrails**: "Add cautionary language, reference toggles, or supplementary views addressing those risks."
4. **Verify Alignment**: "Does this representation support the communication goal of {purpose}? If not, propose alternatives."

Prompt Chaining for Complex Decisions

When classification or aggregation decisions shape meaning significantly, try something like this:

> *Generate four versions of this choropleth using natural breaks, quantiles, equal intervals, and unclassed representations. For each, explain what story it emphasizes and what it obscures. Recommend one based on {stated communication goal}.*

Guardrails as Structural Elements

Effective prompts build honesty into the map structure, not as afterthoughts:

- `Add toggle between {method A} and {method B} to reveal representational differences.`
- `Provide tooltips showing raw values alongside standardized rates.`
- `Generate uncertainty layer showing {sampling density / interpolation confidence / aggregation level}.`

The Tip of the Iceberg

Every thematic map simplifies, and every simplification omits. Every omission is a lie of some kind – sometimes necessary, often excusable, but when done without consideration, it is frequently disqualifying. The question is never whether your map will tell the truth – it won't; it can't. You won't find any truth in mapping – anywhere. The question is whether your choice of which lies to serve to the user and how to show them clearly communicates useful information (i.e., your map's intent) or undermines it.

This chapter has walked you through (or perhaps more accurately, galloped) the most common thematic methods and their predictable failure modes. Choropleths pretend administrative units are uniform. Proportional symbols obscure exact magnitudes. Heat maps imply precision where smoothing algorithms dominate. Flow maps suggest clean channels where messy movement occurs. Cartograms distort geography for rhetorical emphasis. Each method has a personality, and each personality comes with baggage.

These common methods merely represent the tip of the iceberg. Once you are comfortable designing these, I encourage you to read up on other techniques and even experiment with designing your own. In no way am I suggesting the most common techniques are best. (I remind my daughter's regularly: "If the masses are all doing something, fortune and happiness typically lie in the opposite direction." I believe that is true in all things, cartography as well.) The common techniques above are merely good starting points for you to hone your skills before exploring additional techniques further.

In prompt cartography, your responsibility is not to eliminate thematic distortions but to acknowledge them, constrain them, and ensure that your audience understands what the map can and cannot claim.

Prompt guardrails, human critique, and explicit cautions are not signs of uncertainty. They are, as always, signs of intellectual honesty. Data type determines method appropriateness – rates and densities for choropleths, raw counts for proportional symbols, distribution patterns for dot maps, continuous surfaces for isarithmic representations, movement for flow maps, rhetorical emphasis for cartograms.

The choice between methods is never purely technical. It is always rhetorical. Different methods (including classification decisions) tell different stories with the same data. Your job as prompt director is to choose the method that serves *your* purpose while making its limitations visible.

Chapter 14 narrows focus to one of the most seductive – and most frequently misused – thematic techniques: animation. Motion promises to reveal change over time, but it often delivers visual noise and cognitive overstimulation instead. You will learn when animation earns its place, how to prompt for explanatory motion rather than decorative distraction, and why restraint is the most important variable in temporal representation.

Takeaways

- **Every Thematic Method Carries a Default Lie**

  Choropleths imply uniformity, proportional symbols obscure precision, and heat maps suggest false certainty about location. Memorize what these lies are. Don't start naively believing them.

- **Data Type Determines Method Appropriateness**

  Use choropleths for rates and densities, proportional symbols for raw counts, and dot maps when internal variation matters.

- **Phenomena ≠ Data**

  You map data (samples/measurements), not phenomena (reality). Acknowledge gaps between data's spatial nature and the phenomenon's.

- **Scaling Symbols by Area Prevents Visual Distortion**

  Proportional symbols must scale by area, not radius, to avoid exaggerating large values and compressing small ones.

- **Equal-Area Projections Strengthen Choropleth Integrity**

  When comparing magnitudes across space, areal units should be visually comparable.
- **Classification Is Rhetorical**

  Natural breaks, quantiles, equal intervals, and standard deviations each tell different stories. Choose deliberately and justify.
- **Color Value and Saturation Encode Quantity**

  Sequential data require perceptually ordered palettes – hue alone fails to communicate quantitative relationships intuitively.
- **Binning Improves Heat Map Legibility**

  Hexbins and grid aggregations allow users to interpret density with greater precision than continuous kernel surfaces.
- **Isarithmic Beauty Masks Interpolation Uncertainty**

  State sampling density and interpolation methods explicitly. Add uncertainty layers in sparse areas.
- **Flow Maps Demand Filtering and Hierarchy**

  Without limits and bundling, flow maps become illegible – interaction and selective emphasis are essential.
- **Use Cartograms for Rhetorical Oomph, Not Analytical Street Cred**

  Distortion emphasizes impact but often confuses spatial orientation – always provide a standard reference toggle.
- **Multivariate Maps Risk Cognitive Overload**

  Reserve multivariate techniques for cases where spatial relationships must be seen simultaneously – otherwise, use linked views or small multiples.
- **Prompt Guardrails Make Lies Visible**

  Explicit cautions about uniformity, interpolation, and smoothing prevent overconfident interpretation.
- **Method Selection Precedes Styling**

  Always ask whether you have chosen the right representation before specifying colors, sizes, or interactions.

Conceptual and Applied Exercises

Method Mismatch

Identify a dataset you are familiar with. Describe what interpretive distortion would occur if it were mapped using an inappropriate thematic method.

For example, use a choropleth map for raw data or a heat map when precisely located events are required.

Binning Consequences

Reflect on a choropleth map where classification choices (equal interval, quantile, natural breaks) would tell dramatically different stories. What ethical responsibility does the cartographer have in justifying their classification decision? What role might an LLM assistant play in a prompt-cartography scenario to help the cartographer decide on or justify this decision?

LLM Method Reasoning

Provide an LLM with a dataset description (type, scale, distribution) or upload a simple dataset (e.g., a CSV of province or country data) and ask it to recommend three thematic mapping methods with justifications. Evaluate whether its reasoning emphasizes communication clarity, technical feasibility, or default conventions. Attempt to prompt it to balance the three or promote one over the other.

Prompt Specification Exercise

Write a complete prompt for creating a thematic map that includes:

- data description
- method selection with justification
- classification scheme (if applicable)
- color palette with accessibility considerations
- projection specification
- legend requirements
- tooltip content
- A cautionary statement about the method's limitations to be included somewhere (perhaps in a data panel).

Compare your prompt to one generated by an LLM for the same dataset – which is more specific? Which anticipates more failure modes?

For extended tutorials, hands-on labs, prompt templates, and updated examples related to this chapter, visit www.promptcartography.com.

14

Change Over Time

We just examined thematic methods that organize data spatially – choropleths, proportional symbols, flow maps, and more. Each imposes a structural logic on how patterns are revealed. Now we will focus on a dimension that cuts across all space: time.

Anyone who has listened to U2's "Mysterious Ways" song knows, motion is seductive. Motion commands attention, implies dynamism, and makes maps feel alive. But liveliness is not the same as clarity, which is why motion *needs* strict rules.

In prompt cartography, animation is especially dangerous because models love to add it – symbols pulse, panels slide, layers fade with cinematic enthusiasm. Left unconstrained, animation becomes decoration. Worse, it becomes cognitive noise that distracts from the very patterns it claims to reveal.

The goal is to help you determine when animation earns its place in your map as opposed to being prohibited outright. You will learn the vocabulary for prompting temporal behavior – duration, rate of change, order, synchronization – and how to translate those concepts into specifications that produce explanatory motion rather than decorative distraction. Animation is not a feature you add at the end. It is a representational decision you make early, with clear rules about when motion stops. Prompt accordingly.

When Animation Earns Its Place

Animation becomes explanatory when it reveals temporal patterns, sequences, or relationships that static frames cannot communicate efficiently. DiBiase et al. (1992) identified three core visual variables for animation that remain essential for prompting: *duration* (how long something is shown, as highlighted in Figure 14.1), *rate of change* (how quickly transitions occur), and *order* (the sequence of presentation). A few years later, MacEachren (1995) added three more: *display date* (when something first appears), *frequency* (rate of events within a timespan), and *synchronization* (temporal correlation between datasets, illustrated in Figure 14.2).

DOI: 10.1201/9781003797920-16

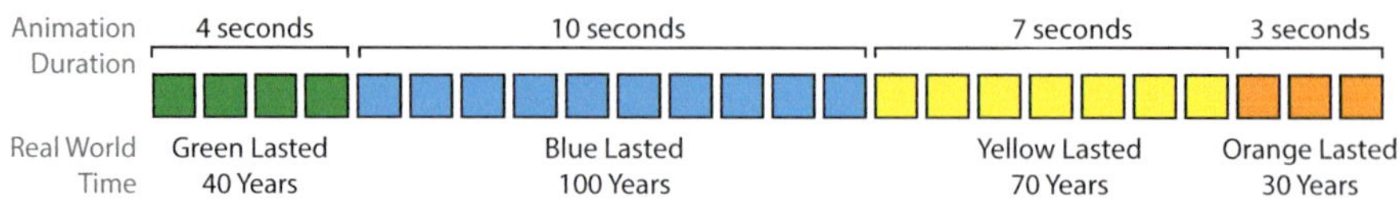

FIGURE 14.1
Duration as a visual variable showing empires displayed proportionally to how long they ruled.

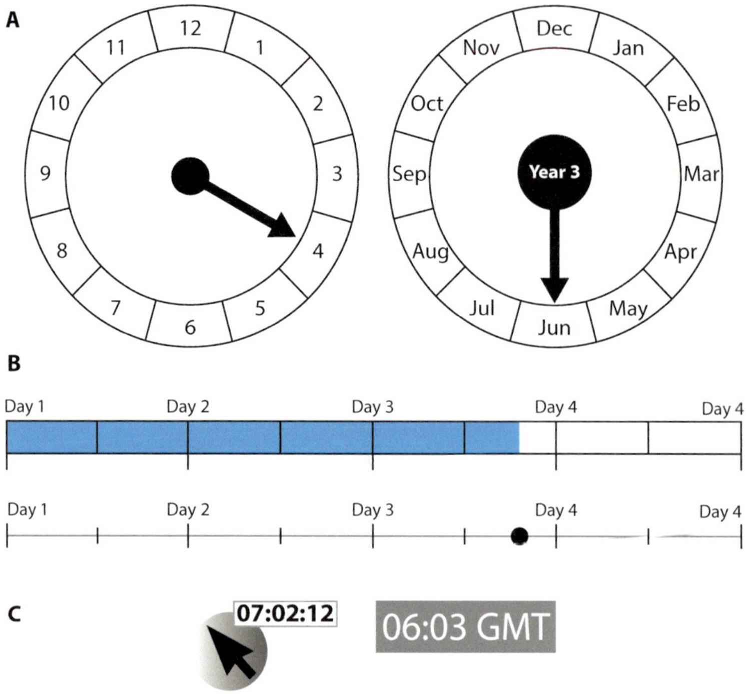

FIGURE 14.2
Synchronization example showing two animated datasets with correlated timing.

These variables give you a vocabulary for prompting intentional animation rather than decorative motion. Instead of vaguely requesting "animate the wildfire data," you can specify:

```
Animate wildfire perimeters chronologically by year
(1990–2025). Use a 250ms ease-out transition when the
year advances. Pause after one complete cycle – do
not loop. Display the current year prominently; update
the legend if classification thresholds shift between
years.
```

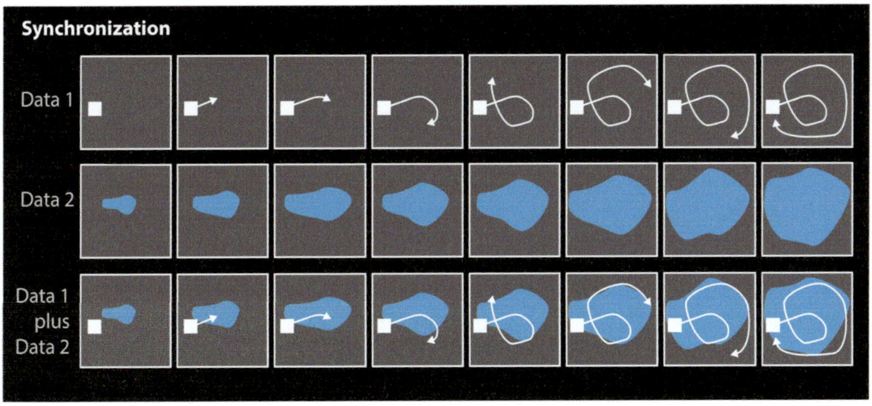

FIGURE 14.3
Rate of change examples showing different transition speeds for movement and size changes.

That prompt encodes *duration* (pause between years), *rate of change* (250 ms transition, as shown in Figure 14.3), *order* (chronological), and *frequency* (annual intervals). It also prohibits looping, which prevents the animation from becoming background noise.

Prompting Temporal Controls

Temporal animation requires user control. Research consistently shows that people struggle to retain information from passive animations (Harrower, 2007). Give them agency over playback, however, and their comprehension improves dramatically.

When prompting for animated maps, specify controls explicitly:

```
Add a time slider for yearly wildfire perimeters
  (1990–2025).
Rules:
- Animate ONLY when the year changes (slider drag or
  play button activated).
- Transition duration ~250ms ease-out; no looping.
- On play, pause after one full run; do not auto-repeat.
- Show the year label prominently above the slider;
  update legend if classes change over time.
- Respect user's reduced-motion preference (prefers-
  reduced-motion media query).
```

This prompt explicitly prevents auto-play (a common large language model [LLM] default) and respects accessibility preferences. The reduced-motion directive is critical – some users experience vestibular disorders triggered by animation. Honoring prefers-reduced-motion is both ethical and increasingly expected.

Temporal Legends: Make Time Visible

A temporal legend is not optional – it is structural. Without it, users cannot anchor what they see to when it occurred. Three formats dominate:

- **Temporal Bar Scales**: Horizontal timelines (like video players) showing elapsed and remaining time.
- **Temporal Circles**: Compact, ideal for cyclical data (seasons, 24-hour periods).
- **Clocks/Calendars**: Explicit date/time displays.

The best temporal legends sit close to – or within – the mapped area to minimize eye travel (Figure 14.4). Prompting this requires spatial specificity:

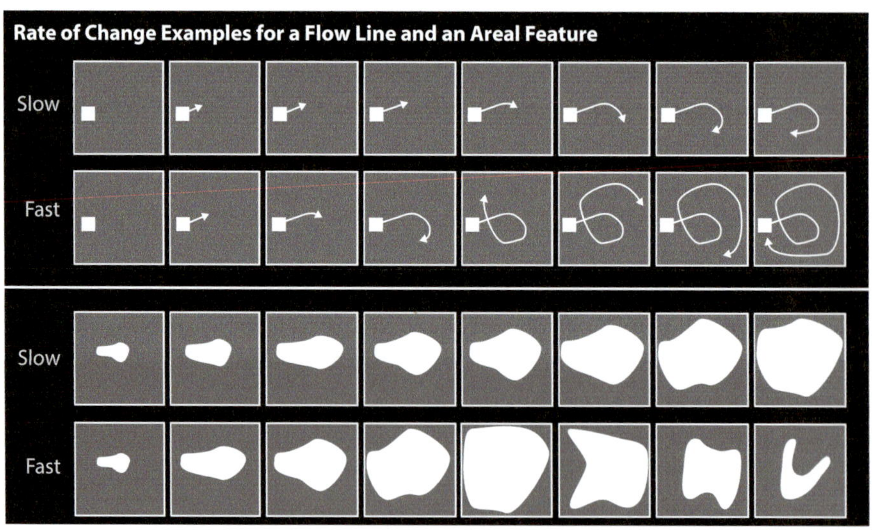

FIGURE 14.4
Different temporal legend examples showing circles, bar scales, and clock representations.

```
Place the temporal bar scale centered at the top of the
  map viewport, 20px below the title. Keep it visually
  subdued (neutral gray, 40% opacity) until the user
  interacts with it.
```

Distinguishing Animation Types in Prompts

Different animation types serve different purposes. Specify which you need: **Temporal animation** shows change over time.

```
Animate county population change by decade (1990–2020).
Use stop-frame transitions with 3-second hold per
decade.
```

Path animation emphasizes routes or journeys.

```
Animate the Lewis & Clark expedition route from May
1804 to September 1806. Progress the line gradually
westward; pause at major landmarks (Fort Mandan, Lemhi
Pass, Pacific Ocean) for 2 seconds with info panel.
```

Zoom animation guides scale changes.

```
When user clicks state, animate smooth zoom to state
extent over 800ms. Use ease-in-out easing.
```

Try to avoid requesting *fly-through* animations unless the oblique perspective serves a clear purpose. Users often become disoriented in 3D fly-throughs, and even when they don't, the cognitive load of the visualization spikes, potentially distracting them from the most important information. If you must use them, it often makes sense to include a synchronized 2D inset (i.e., locator or where-the-heck-am-I) map to help users maintain spatial orientation.

Motion Constraints: The Non-Negotiables

LLMs love to animate unless you prohibit it, so your prompts must encode restraint:

Motion constraints (non-negotiable):

- `Never loop animation automatically`
- `Do not animate by default on page load`
- `Animate only on explicit state change (time step, filter activation, user selection)`
- `Limit motion to one map element at a time`
- `Provide pause/stop controls`

A simple enforcement prompt:

```
Disable all auto-play animation. Require user interac-
tion to trigger any motion. Provide visible play/pause
controls. When prefers-reduced-motion is detected,
replace all transitions with instant state changes.
```

When to Choose Stop-Frame Versus Smooth Transitions

We have discussed discrete and continuous datasets twice now – during the dataset and thematic mapping pipelines. This time, we will use the concept to dictate when it is advisable to use stop-frame versus smooth (i.e., tweened) animations.

Stop-frame (discrete state changes) works best for:

- Data with temporal gaps (decennial census, Olympic medal counts)
- Raster time series (satellite imagery, weather radar)
- When change is categorical, not continuous

Stop-Frame Prompt

```
For decennial census data: use stop-frame transitions.
Hold each decade for 3 seconds, then cut instantly to
next. No interpolation between years.
```

Smooth transitions (tweening) work best for:

- Continuous temporal processes (storm tracks, troop movements)
- Morphing geometries (glacial retreat, urban expansion)
- When emphasizing gradual accumulation or decay

Smooth-Transition Prompt

```
For hurricane track: animate path with smooth 50ms-per-
frame progression. Add trailing fade (previous 6 hours
visible at 30% opacity) to show recent trajectory.
```

Simplicity and Duration

Keep animations short – 20 to 45 seconds maximum for thematic data. Longer animations overwhelm working memory and retention collapses. If your data spans decades, aggregate or sample rather than showing every year.

Prompt for simplification:

```
Animate only presidential election years (1960, 1964,
1968...) rather than all years. For each election, show
state winner by party color. Duration: 2 seconds per
election. Total runtime: ~32 seconds.
```

Simplification isn't compromise – it's respect for human perception. As already mentioned, we can focus on three (to maybe four if you play a lot of first-person videogames) visual elements simultaneously (Ware, 2008). Anything more becomes pattern noise. Animations have lots of noise by default – myriad moving parts clamoring for our attention. For an animation to work, it *must* be simple and clear.

Human in the Loop: Review Your Animation Output

After generating an animated map, it makes sense to quickly run through the following questions (by yourself, not with an LLM sidekick) to determine whether it's working well:

1. Does motion explain something a static map cannot?
2. Can a first-time user understand what changes without instruction?
3. Is the temporal legend visible and synchronized with animation state?
4. Does motion stop when it should, or does it loop endlessly?
5. Would this animation fatigue a user who views it three times?

Sorry, but if the answer to number 1 above is "no," *remove the animation*.

On the other hand, if the answer is no to other aspects of the list, you can probably modify to make the map more useful. For example, if question 4 is a bust, you can probably add loop constraints to fix it. If question 5 is "yes," perhaps shorten the duration or reduce motion intensity.

Use Animation for Visual Rhetoric and Temporal Explanation – Never Decoration

Animation, like every other design variable in this book, must justify its existence. It does not earn a place on your map simply because it is possible, or because it makes the output feel dynamic, or because users have come to expect it. Animation earns its place when – and only when – it reveals temporal patterns that static frames cannot communicate efficiently.

You now have the conceptual vocabulary and prompt structures to make that judgment deliberately. You have learned to distinguish stop-frame from smooth transitions, to prompt for user-controlled playback rather than auto-looping spectacles, and to treat temporal legends as structural necessities rather than optional accessories. You have also learned the most important constraint of all: motion is stoppable or someone is liable to get motion sickness all over your map!

In prompt cartography, animation is not a reward for good design. It is a representational commitment that carries risk. When prompted thoughtfully, it becomes explanatory. When prompted carelessly, it becomes the cognitive equivalent of a flashing banner ad – impossible to ignore, impossible to learn from. Your job as director is to ensure that every frame of motion serves interpretation rather than distraction.

Next, we shift from temporal representation back to interaction broadly conceived – how behaviors, states, and user flows shape the experience of your map as a living system. Animation was one form of motion. Interactivity is the grammar that organizes all motion, all disclosure, and all narrative progression in your maps.

Takeaways

- **Motion Must Explain Change to Justify Its Existence**

  Animation earns its place only when it reveals temporal patterns that static representations cannot communicate efficiently.

- **Duration, Rate, Order, and Synchronization Are Temporal Variables**

  These properties allow you to specify how animation unfolds rather than leaving behavior to model defaults.
- **Temporal Legends Are Crucial**

  Without visible time indicators, users cannot anchor what they see to when it occurred.
- **User Control Improves Comprehension and Retention**

  Passive animations overwhelm working memory; play/pause controls and sliders allow users to set their own pace.
- **Stop-Frame Transitions Suit Discrete Temporal Data**

  Use instantaneous state changes for decennial census data, raster time series, or categorical shifts.
- **Smooth Transitions Imply Continuous Processes**

  Tweening works best for storm tracks, glacial retreat, or gradual accumulation where continuity is meaningful.
- **Auto-Play and Looping Undermine Learning**

  Require explicit user interaction to trigger animation; disable auto-repeat to prevent motion from becoming background noise.
- **Respect Reduced-Motion Preferences for Accessibility**

  Honor prefers-reduced-motion media queries by replacing transitions with instant state changes.
- **Animation Length Should Respect Working Memory Limits**

  Keep animated sequences between 20 and 45 seconds; longer durations cause retention to collapse.
- **Path Animation Requires Pauses and Context**

  When animating routes or journeys, include holds at key moments and provide synchronized locator maps to maintain orientation.

Conceptual and Applied Exercises

Animation Justification

Think of a temporal dataset you've worked with or encountered. Articulate when animation would genuinely clarify temporal patterns versus when a small, multiple static display would communicate more effectively.

Motion Meaning

Reflect on the difference between smooth transitions and stop-frame animation. Describe a scenario where smooth transitions would falsely imply continuous change when the data actually represents discrete temporal snapshots.

LLM Temporal Design

Ask an LLM: "Design temporal controls for a map showing wildfire spread over 14 days. Should I use animation, a timeline slider, or both? Justify your choice." Evaluate whether its response prioritizes user control, interpretive clarity, or technical novelty.

Constraint Specification

Write a prompt for an animated map showing sea-level rise projections over 50 years. Specify pacing, temporal legend requirements, pause controls, and constraints on transition speed that prevent misinterpretation of uncertainty.

For extended tutorials, hands-on labs, prompt templates, and updated examples related to this chapter, visit www.promptcartography.com.

15

Bringing Your Visual Plan to Life via Interactivity

When you first hover over a live wildfire map, and the burn area pulses outward like a heartbeat, you aren't just seeing data – you're witnessing a design decision in motion. Someone decided that motion would signal recency, not alarm. That hovering would reveal context, not clutter. That the pulse itself would draw attention without demanding it.

Ten years ago, those decisions lived in JavaScript callbacks, stylesheets, UI libraries, and event handlers buried just beneath the browser's surface. The decisions were seen by map users, but never heard – unless there was a scripting error. Today, as a prompt cartographer, your design decisions still live beneath the surface in that same spot. Except now, your decisions are manifested for you by an agent following your natural-language direction. The agent listens to how you want your users' interactions to work and then implements it. When you write "update the legend symbology as the viewer zooms in," you are no longer implementing interface components. You are no longer commanding the computer via parameters and function calls. You are specifying behaviors based on descriptive, if sometimes vague, human interactions.

Designing interactivity becomes directed choreography, and your prompts comprise the score. In this shift, familiar map elements do not disappear. Legends, tooltips, panels, and transitions still matter. They are even built the same in the end, using the same scripts, styles, API calls, and DOM elements. What has changed is how they are conceived and developed. They are no longer just point and touch GUI widgets to be placed after the fact and configured to work with preexisting map elements (see Figure 15.1), but behaviors to be described deliberately – sometimes well in advance of the final map product itself.

From Wiring Events to Directing Experience

It might be argued that interactivity has always been the defining feature of web cartography. After all, it was interactivity – simple URL hyperlinks

DOI: 10.1201/9781003797920-17

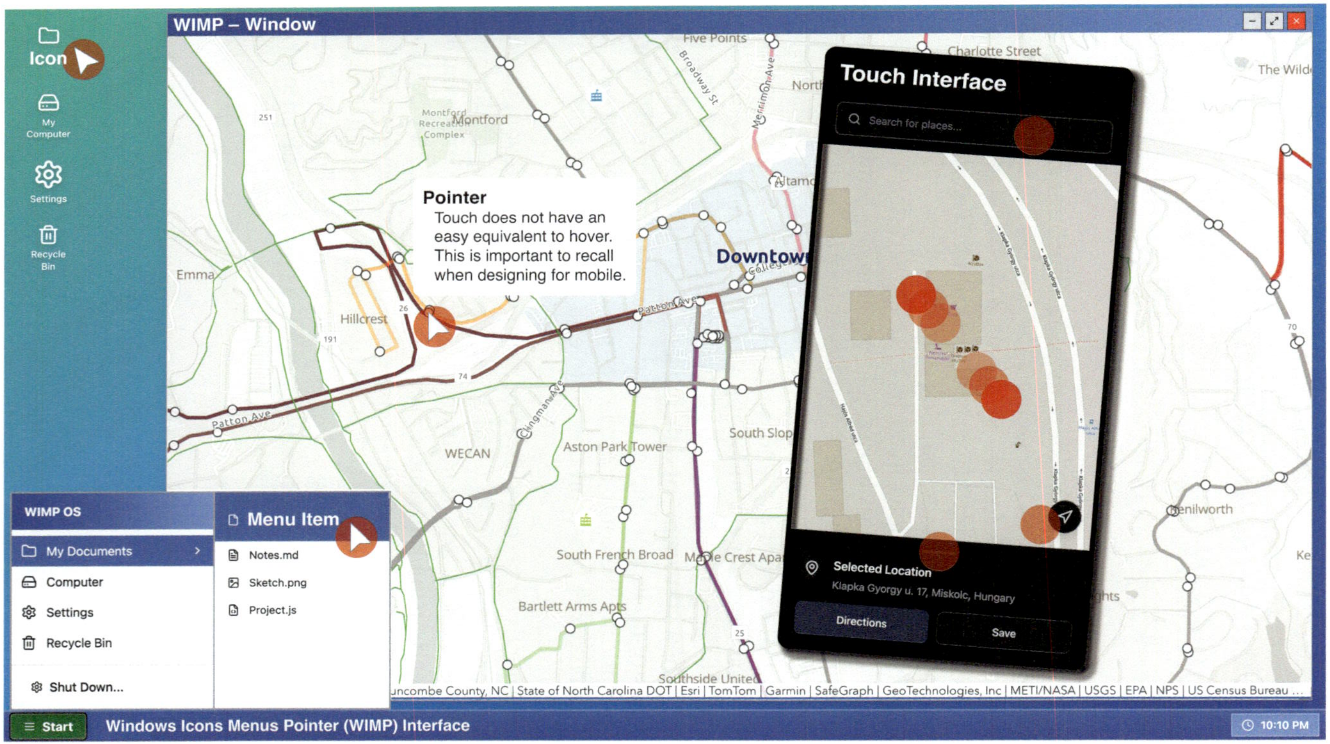

FIGURE 15.1
WIMP and touch interface elements. On the left: Windows, Icons, Menus, and Pointer (WIMP) interfaces were revolutionary inventions, allowing for the rise of digital cartography. On the right: the invention and rapid adoption of touch interfaces and tile-based web maps were another. These are being supplemented and likely supplanted by NLIs.

back in the 1990s – that transformed digital maps from static artifacts into navigable experiences. Over the past two decades, slippy web maps with interactive, draggable tiles have trained us to click, pan, zoom, and toggle our way around cartographic representations. These actions are not innate. They are learned behaviors that have become second nature worldwide, carefully cultivated by technology companies remarkably successful at social engineering how we move through digital space. Beyond maps, technology companies have been perhaps too successful at engineering us across the board, depending on whom you ask.

Prompt-driven design extends this lineage even further, but it also changes your role in a fundamental way. Instead of wiring events to interface elements, you now describe how the map should respond – when, why, and to what. You no longer need to build UI widgets from scratch or configure event listeners in code. You must, however, describe what the map will reveal, when it will withhold, and how it will guide attention. The designer's job becomes specifying inclusion, exclusion, and behavioral constraints. This is directorial work, not assembly work. You decide what the map will *not* do as much as what it will do.

Most interactive maps today can be understood through three overlapping interaction grammars, each serving different communicative goals:

Exploratory Interactions reward curiosity and encourage roaming – think of a map that invites users to click around, discovering patterns at their own pace.

Narrative Interactions guide attention and sequence meaning over time, directing users through a story or argument with intentional pacing.

Analytic Interactions prioritize precision, comparison, and control, offering tools for filtering, measuring, and dissecting data with rigor.

Prompt cartography allows these grammars to blend more easily than coding and revising code ever did. A single instruction regarding interactivity can invite the user to explore the map while quietly enforcing analytic constraints. Another instruction might allow for narrative sequencing while permitting exploratory detours.

However, explicitness matters. When you find yourself struggling to describe what an element should do – when it should appear, how it should change, or what it should withhold – pause. Have you thought this through well enough? Does this map element belong in the map at all?

The key point is this: *interactivity is never neutral*. Every hover, delay, filter, and transition nudges a user's interpretation, whether intentionally or not. One decision regarding interactivity often cascades into and impacts other user interactions. In prompt-driven pipelines, those nudges are no longer buried in code or implied by defaults. They're written, reviewed, revised, and questioned in plain language iteratively. And once you notice the power and danger of creating interactivity with natural language – well, as my kids might say, you can't unsee it.

Interaction Grammars

Interactivity in web cartography is not monolithic. Different maps call for different kinds of engagement, and understanding these distinctions helps you prompt with intention rather than habit. Interaction grammars provide a high-level taxonomy for thinking about how users engage with maps, and I will reference them throughout this chapter as we examine specific behavioral systems.

Exploratory interactions reward curiosity and roaming. These maps invite users to wander, click, and discover without prescriptive guidance. A well-designed exploratory map provides enough structure to prevent disorientation but enough openness to allow serendipity. Prompting for exploratory interaction means specifying what users can access freely and what constraints prevent them from getting lost. Instructions such as *allow free panning and zooming within the dataset extent* or *enable layer toggling but default to a coherent starting view* acknowledge that exploration requires both freedom and guardrails. Exploratory grammars work well for rich datasets where multiple valid interpretations exist, but they require careful attention to cognitive load. Without thoughtful disclosure systems, exploratory maps can overwhelm users with decision fatigue.

Narrative interactions guide attention and sequence meaning over time. These maps are argumentative. They lead users through a story, revealing information in a deliberate order to build understanding or make a case. Prompting for narrative interaction involves specifying progression, pacing, and emphasis. Instructions such as *introduce the map at a regional scale, then reveal local detail only after the user scrolls* or *guide the viewer through the dataset chronologically, pausing at key moments* turn interaction into directed experience. Narrative grammars are powerful for persuasion and education, but they demand restraint. Over-scripting can feel patronizing; under-scripting can feel aimless. The balance lies in giving users enough agency to feel engaged without losing the thread of your argument.

Analytic interactions prioritize precision, comparison, and control. These maps are tools for interrogation. Users expect to filter, measure, compare, and extract insights with rigor. Prompting for analytic interaction means specifying what operations are available and what constraints govern them. Instructions such as *allow stacking filters within category groups but not across them* or *provide a comparison mode that highlights differences between selected* *features* define the boundaries of analysis. Analytic grammars require clarity above all else. Ambiguous controls, inconsistent states, or unexplained defaults erode trust quickly. Users of analytic maps are often experts or stakeholders who will notice when behavior feels arbitrary.

The goal is to use prompting to deliberately blend these grammars so they show up at the right time. A map can be exploratory at the outset, narrative in its guided moments, and analytic when users drill down into specifics. The key is making those transitions explicit in your prompts. When grammars shift without clear signaling, users will probably feel disoriented by your map. When you shift them with purpose, users won't notice how much work you put into the UX.

Interactivity Versus Animation: A Quick Clarification

What's the difference between interactivity and animation? It can be confusing to keep the two straight, but there's a fairly straightforward method for doing so.

Interactivity depends on user input. It is motion or change that reacts to a user action – a hover, a click, a tap, a drag. In natural language, it's the result of a direct action by the user on the map. Animation, on the other hand, may begin or end due to a user interaction – pressing a play or stop button, for instance – but the motion itself is time-dependent and user-independent. It unfolds whether anyone is paying attention or not.

In prompt-driven design, the two often meet halfway. When you ask an element to *animate on click*, you are turning time into participation. A user initiates the sequence, but the map controls the pacing. Used well, animation can explain change and soften latency. A smooth transition between map states can help users understand spatial or temporal relationships. A brief fade-in can signal that new information has arrived without jarring the viewer.

Implemented poorly, however, animation becomes irritating noise – the interactive equivalent of Tufte's chart junk. Motion that exists for its own sake, or worse, motion that obscures meaning, undermines the map's communicative purpose.

Animation earns its place by explaining change or reducing friction. If a dataset updates and symbols shift positions, animating that movement can help users track continuity. If a panel slides open to reveal controls, a quick transition can feel more fluid than an abrupt appearance. But if symbols pulse for no reason, if transitions slow down simple tasks, or if every interaction triggers unnecessary flourish, animation becomes a liability. The test is simple: does this motion help the user understand something they wouldn't grasp otherwise? If the answer is no, *leave it out*.

Prompting for animation requires specificity. Vague instructions like *make it smooth* or *add transitions* will produce generic results that may or may not serve your purpose. Instead, describe the motion's role: "animate the transition between time steps to show spatial continuity or fade in new features after filtering to signal change without distraction." Specify duration and easing if it matters: "use a brief 200ms fade or ease in slowly to emphasize the shift." And always specify when motion should stop: "animate once on load, then allow static interaction or loop only while data is loading."

The distinction between interactivity and animation will matter throughout this chapter. As we examine behavioral systems – disclosure, commitment, memory, control, and motion – you'll see how user-initiated actions and time-based changes intersect. Understanding the difference helps you prompt with clarity, ensuring that motion serves communication rather than competing with it.

Disclosure Systems: Legend + Filters + State

Every interactive map quietly answers the same question, again and again: what should the user see right now?

Legends and filters manage that answer – not all at once, and not for everyone in the same way. In earlier web maps, legends were often treated as static explanatory artifacts, visual keys parked beside the map, available but inert. This is still the default in many mapping APIs. Filters, when they appeared, were framed as analytic tools layered on top of that explanation. In modern web cartography, however, this separation is beginning to collapse. Legends and filters are best thought of as disclosure systems, responsible for managing the user's cognitive load by controlling when and how information is revealed.

What Disclosure Controls

From a behavioral perspective, the most important property of a legend is not how it looks but what state it occupies. Is it expanded or collapsed? Does it react to filtering? Does it update as the map changes, or does it continue to describe a world that is not currently viewable by the user? These are not cosmetic questions – they are rhetorical ones.

A legend that always shows everything implies completeness, suggesting the map has nothing to hide, and users can see the full picture at a glance. A legend that reveals itself gradually implies guidance, suggesting the designer is leading users through complexity rather than dumping it on them. Neither approach is neutral, and both invoke and erode trust differently depending on the context and purpose of your map.

Prompting makes these implications explicit. There is a meaningful difference between asking for *a legend* and describing its behavior. The former produces a component; the latter produces an experience. Instructions such as "collapse the legend by default, expand only active categories," or "update legend entries dynamically as filters change" define how much, and how quickly, the map will orient its audience. Dataset and extent filters operate under the same logic. A filter is not merely a control – it is a statement about what comparisons you consider meaningful.

Prompts that specify defaults – which filters are active on load, which are mutually exclusive, which can be combined – shape how patterns are discovered and which comparisons are foregrounded. If a map opens with all filters cleared, it communicates that users have options to explore freely. When a map opens with certain filters preselected, it communicates emphasis and direction, like opening a film on a particular scene rather than a blank screen.

State as a Rhetorical Decision

State must be named deliberately. Words like *collapsed, expanded, active, disabled*, and *reset* are not implementation trivia – they are design decisions with rhetorical weight. When state is left unspecified, your agents will decide for you – and as you've had hammered into you, it's unlikely to be a stellar decision.

It's up to you to communicate how much to show or hide at certain stages, or after a user interacts with the map. The default states you choose will impact emphasis and imply your intentions. An overview state that shows everything suggests a neutral starting point; preselected filters suggest you've already made an argument about what matters most.

This is where earlier design decisions quietly reassert themselves. Legends inherit visual hierarchy from color and type choices. Filter groupings reflect semantic categories defined during data preparation in Chapter 7. Layout constraints from Chapter 9 determine where these elements can appear without competing with the mapped space. Prompting for disclosure does not replace those decisions – it activates them.

State also drifts when it's not monitored or maintained across user activity. A legend that describes all categories while the map shows only a filtered subset creates dissonance. Users notice when interface elements promise one thing, and the map delivers another. Synchronization rules – for example, "update legend entries dynamically as filters change or reset legend to full view when all filters are cleared" – prevent this drift and keep the interface honest (as shown in Figure 15.2).

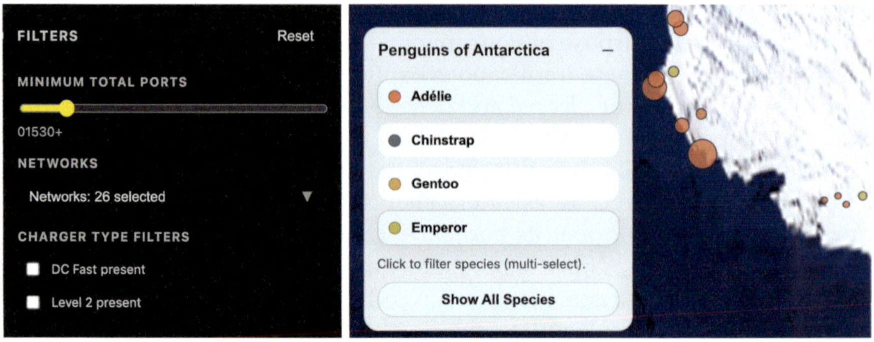

FIGURE 15.2
Filters should be limited to only allow certain behaviors that guide user experience. In this filter panel, users can filter EV stations by network, number of chargers, and what type of fast charging is available. In the example on the right, users click on the legend itself to quickly select which species to keep displaying on the map and which to remove. Simple, intuitive interactivity keeps cognitive load to a minimum.

Common Disclosure Failures

Legends tend to fail in predictable ways.

- Overloaded legends that display every category, subcategory, and attribute at once distract from the map itself, turning what should be a quick reference into a wall of information that competes for attention.
- Static legends that describe a world not currently viewable create confusion, especially when users have applied filters or zoomed to a limited extent.
- Unspecified defaults lead agents to make arbitrary choices – perhaps the legend starts expanded when it should be collapsed, or all filters are active when none should be. Filters that stack without clear logic or reset behavior create frustration, especially when users can't easily return to the starting state.

Each of these failures stems from under-specification. When you don't describe how disclosure should work, the system will guess, and those guesses accumulate into interfaces that feel arbitrary or overwhelming.

The solution is not to add more interactivity but to describe the interactivity you actually need, and to constrain what the map *will not* do. In the first edition of *Web Cartography*, I warned that legends either explain too little or dump so many irrelevant options on users that heads explode – metaphorically, of course. That warning still holds. You never want your legend to be so complex to navigate that it distracts from exploring the map itself.

Prompting Patterns

Effective disclosure prompts state what is desired and specify your synchronization rules explicitly. Consider the following examples:

```
Start with the legend collapsed. Allow one-tap expan-
  sion. Show only active categories. Update entries
  dynamically as filters change.
Default to no filters selected. Allow stacking within
  category groups but not across groups. Provide a
  prominent 'Reset to overview' button that clears all
  filters and returns the legend to its full state.
Limit the legend to categories currently visible in the
  map extent. When users zoom in, show only the subset
  of features within view. When they zoom out, expand
  the legend to reflect the broader context.
```

These prompts are directive without being prescriptive. They describe behavior – what happens when, under what conditions – without dictating implementation details. They also make dependencies clear: the legend responds to filters, filters respond to user actions, and reset controls return the map to a known state. Specificity here prevents drift and ensures that disclosure serves the map's purpose rather than overwhelming it.

Integration Reminder

Disclosure systems inherit constraints from earlier design decisions. Layout choices determine where legends and filters can appear without obscuring mapped space. Visual hierarchy governs which categories deserve prominence and which should be subdued. Semantic categories defined during data preparation shape what filter groupings make sense and how legends should organize information. Prompting for disclosure does not introduce these constraints – it enforces them. When disclosure is well-prompted, it feels less like an interface layer and more like a natural extension of the map's visual and conceptual logic.

Commitment Systems: Tooltips + Pop-ups (i.e., Information Windows)

If legends and filters govern what information is available, tooltips and pop-ups govern when that information is delivered. They are the primary instruments of pacing in interactive maps, controlling how quickly meaning accumulates and how much effort a user must invest to obtain it. The main difference between a tooltip and a pop-up isn't visual style – it's temporal commitment. Hovers are tentative; clicks are assertive. Think of the difference between smelling a freshly poured glass of wine and chugging it – or if you prefer, like I do, coffee. Tooltips respond to curiosity; pop-ups respond to intent. Try to reflect this difference clearly and consistently.

Temporal Commitment

Tooltips work best as glances. They are lightweight, reversible, and brief. An above-average tooltip answers several immediate questions – i.e., What the heck is this? Is this even worth clicking? – and then immediately gets out of the way, as shown in Figure 15.3.

Prompts that load tooltips with dense text, multiple metrics, or explanatory prose misunderstand the tooltip's role. If everything appears on hover, clicking becomes pointless. Worse, if everything appears on hover, someone on a mobile device may see nothing, because hover states don't reliably translate to touch interfaces.

Pop-ups mark a behavioral threshold. A click signals willingness to pause, read, and engage. This makes pop-ups an appropriate place for structured information, comparative context, or short narrative framing (see Figure 15.4 for examples). Prompting for pop-ups should almost always include guidance on visual hierarchy: what appears first, what is secondary, and what is intentionally omitted or purposefully subdued. More information is not automatically better information – in many cases, less information is better map design. In the first edition, I described pop-ups and info windows as callout labels on steroids, with a real drawback: they can cover too much mapped area, so they should stay compact and offer expansion only when needed. That principle still holds.

Pacing as Meaning

A new concept I didn't cover in the last book, but that requires attention now, is *pacing*. Pacing emerges from your design choices, rather than user interactivity decisions. If your map reveals too much too quickly, interpretation is flattened; if it hides everything behind numerous clicks, you raise the specter of user frustration or, worse, the user giving up.

It's important to confront this balance head-on. Instructions such as "show only the primary attribute on hover reveal secondary metrics on click or limit pop-up content to a single screen" are not inconsequential. They are explicit decisions about how meaning unfolds. Being specific helps your agents construct a map that suits your vision.

Timing matters as well. (And note, I'm highlighting this as someone who personally tends to have horrible timing; that's how important it is.)

FIGURE 15.3
Tooltips are not meant to mimic information windows. They should be concise, providing identifiable or crucial information in a flash and disappearing just as quickly.

Delays, fade-ins, and persistence subtly influence attention. A tooltip that appears instantly invites skimming; one that lingers too long can feel sticky or intrusive.

Fortunately, prompting allows these temporal qualities to be named. You can use phrases like "briefly, after a short delay," or "until focus moves away" to turn design intuition into specification. As with disclosure, pacing decisions do not exist in isolation. Visual hierarchy determines which attributes deserve immediate attention. Layout influences where pop-ups can appear without obscuring context. Data semantics shape what qualifies as primary versus secondary information.

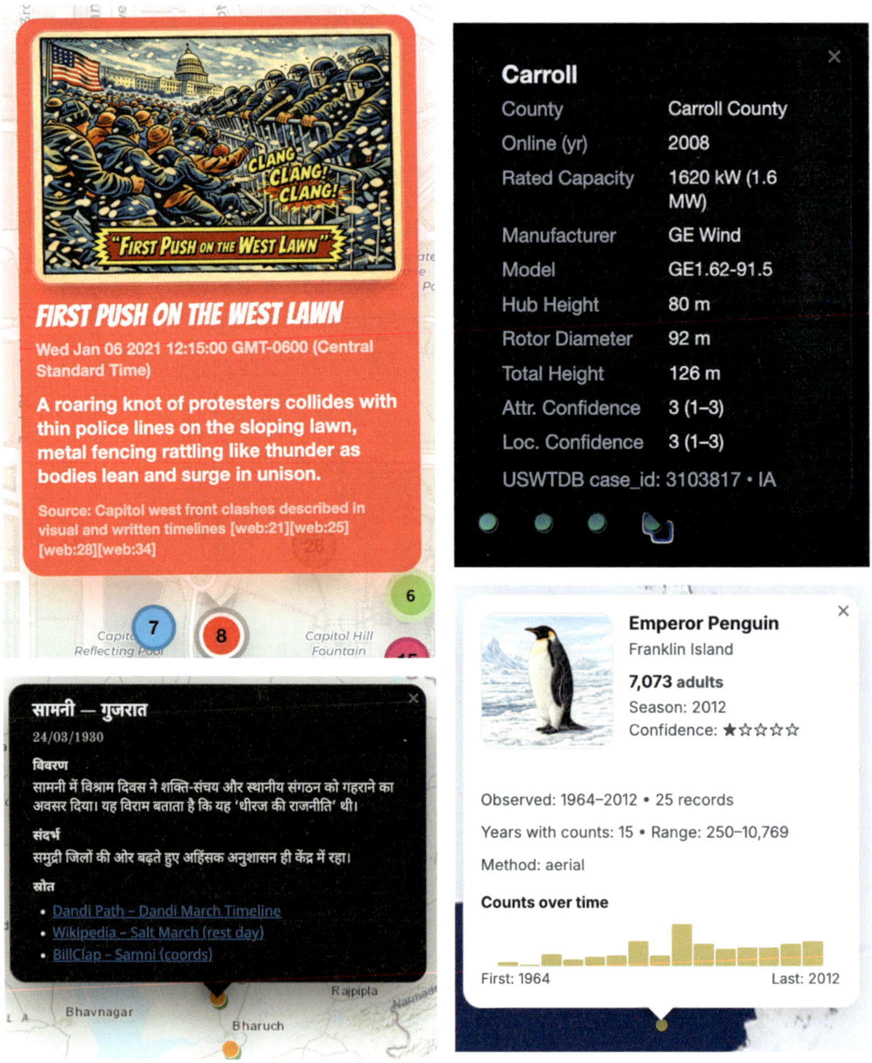

FIGURE 15.4
Information windows (i.e., pop-up windows) can take many forms. Some can include graphs to layer additional (and useful) information for your map users. Avoid including every field just because you can, as this stymies user experience. Remember you are narrating something. Guide your user to useful information – visually.

Your tooltip and pop-up design decisions should not introduce new meaning but rather sequence existing meaning that promotes your map's intent. When prompted thoughtfully, these ephemeral elements behave less like interruptions and more like conversational cues, inviting your map user to keep exploring the map one small commitment at a time.

Common Commitment Errors

Commitment systems fail in predictable ways. Overloaded hover tooltips with dense text or multiple metrics turn curiosity into cognitive work, discouraging exploration rather than encouraging it. Pop-ups that cover too much mapped area without compact defaults or controlled expansion obscure the very context they're meant to clarify. Inconsistent close behavior – especially on touch devices – frustrates users who can't easily dismiss information they no longer need. Another common mistake is over-disclosure by default, often rooted in bad habits developed using web map API defaults. Prompts that ask to *show details on hover* or *display feature attributes* often result in info windows showing six or seven fields, four of which are meaningless to the user. Remember to explicitly direct disclosure, and consider staging it: hover reveals name and one key metric; click provides name, metric, additional context, and an image.

Prompting Patterns

Effective commitment prompts specify content, timing, and behavior clearly. Consider these examples:

```
On hover, show only the feature name and one key metric.
  Add a short delay before display. Hide instantly on
  mouseout. On mobile, disable hover tooltips and rely
  on tap-to-open behavior.
Keep pop-ups compact. Show 2-3 primary fields first. Add
  a 'More details' expansion for secondary information.
  Allow click-outside-to-close. Make the close target
  large enough for touch.
Stage disclosure: hover reveals name and population;
  click provides name, population, year incorporated,
  median income, and a thumbnail image. Limit ini-
  tial pop-up height to avoid obscuring surrounding
  features.
```

These prompts are specific about what appears, when, and under what conditions. They acknowledge mobile constraints, specify timing, and define hierarchy within the information being disclosed. They also reflect an understanding that commitment is a threshold, not a binary – users should be able to glance, engage, or dive deeper depending on their interest and intent.

Memory and Persistence Systems: Panels + Dashboards + Persistent Context

Tooltips and pop-ups are largely ephemeral – hover or tap and they appear; do anything else and they're gone. Out of sight, out of mind. Panels and dashboards, on the other hand, do something else entirely: they remember. From a behavioral perspective, panels are not auxiliary displays or convenient places to put extra information. They are systems for maintaining context across interactions. As users select features, apply filters, or move through a sequence, panels preserve meaning that would otherwise be lost as the map redraws itself. They allow understanding to accumulate instead of resetting with every action.

Why Memory Matters

Without memory, maps force users to reorient themselves repeatedly. Each interaction becomes an isolated event rather than a step in a larger process of understanding. Panels provide continuity, allowing users to build on previous interactions rather than starting fresh each time. This is especially important in analytic and narrative maps, where meaning depends on accumulation – users need to see how their selections, comparisons, or exploration have shaped what they're viewing. Persistent interface elements serve as anchors, preventing the subtle disorientation that arises when interactions accumulate faster than understanding.

Prompting for panels shifts the design conversation away from visibility and placement and toward persistence and synchronization. What information should remain visible as the map changes? What should update automatically? What should stay fixed until the user explicitly clears or replaces it? These decisions matter. A panel that updates constantly can feel unstable, distracting users from focusing on what matters on the map and forcing them to reorient themselves repetitively. On the other hand, a panel that never updates can feel disconnected – like a grand opening sign on a store that never comes down – drifting out of relevance while occupying valuable screen real estate. By writing prompts that specify synchronization rules – *update on selection, persist across filters, reset only when cleared* – you turn these trade-offs into intentional design choices rather than accidental side effects.

Behavior Questions

Effective panel prompts answer specific questions about persistence and synchronization. What persists across interactions – selected features, applied filters, summary statistics? What updates automatically versus requiring explicit user action – does the panel refresh every time the map extent changes, or only when a user clicks a feature? What does "clear" mean, and when should it be offered – does clearing reset the entire interface or just the current selection? How should the panel behave during navigation – does it remain stable during pan and zoom, or does it collapse to save space?

These questions expose the logic of memory. When you can't answer them clearly, the panel's behavior becomes ambiguous, and ambiguous behavior erodes trust. Users start to wonder whether the panel is showing current information or stale data, whether their actions are being tracked or ignored, whether the interface is helping or just taking up space.

Prompting Patterns

Effective persistence prompts specify what stays, what changes, and what resets. Consider these examples:

```
Persist a running summary of user selections in the
  side panel. Update the summary only on click selec-
  tion, not hover. Keep the panel stable during pan and
  zoom. Include a 'Clear selection' button that resets
  the panel and removes all highlights from the map.
Display aggregated metrics for the current map extent
  in a dashboard panel. Update metrics automatically as
  the user pans or zooms. Provide a toggle to freeze the
  current values for comparison. Label frozen values
  clearly to avoid confusion.
Maintain a timeline of user interactions in a collaps-
  ible panel. Show the last five actions - filters
  applied, features selected, extent changes. Allow
  users to jump back to a previous state by clicking on
  a timeline entry.
```

These prompts make memory explicit. They define what accumulates, what triggers updates, and how users can reset or revisit earlier states. They also acknowledge that persistence requires clarity – users need to understand what the panel is showing and whether it reflects their current actions or a previous state.

Panels also operate under a different hierarchy than tooltips or pop-ups. Because they persist, panels are well-suited for contextual framing, aggregated metrics, or narrative cues that benefit from continuity. They are best used to summarize, not provide exhaustive information – dense attribute lists and long tables usually belong elsewhere. Your prompting can help enforce this discipline by stating intent directly: *summarize, compare, highlight change* rather than *show everything*. Reading order matters as well. Panels are often the only place where users encounter sustained text in an interactive map. Prompting for structure – *begin with a short summary, follow with key metrics, skip methodological detail* – shapes interpretation just as surely as typographic hierarchy. You wouldn't want to skip methodological detail if your target audience is comprised of scientists, but you also wouldn't want to include it for children, who may immediately tune out and find your map a bore. These choices echo principles discussed in earlier chapters, but when it comes to panels, they operate over time and through user interactivity with the map, rather than spatially. In summary, panels function as narrative anchors. When prompted carefully, they reduce cognitive load, reinforce continuity, and prevent disorientation.

Control Systems: Layer Toggles + Menus + Locator Maps

Turning layers on and off feels empowering to many users. It gives them a sense of control and invites engagement and exploration. GIS users tend to expect such behavior – it's baked into desktop software conventions and professional practice. However, in many circumstances, allowing this behavior defeats the purpose and intent of your map. Layer toggling can quickly spiral into what I'll call ADHD-style interactivity, pulling the user's attention away from your map's main purpose or distracting them from more important tasks they should be doing with the map. If there is one thing 20 years of social media has taught us, it's that people love clicking buttons and scrolling absent-mindedly. Our goal is to prevent your map from becoming a cartographic click-and-scroll time sink.

Architecting User Choice

Layer controls are choice architecture in action. Every toggle, every reordering option, every visibility rule shapes what users can do and what they're likely to discover. The question is not whether to offer control, but how much control serves your map's communicative purpose. A map designed for open-ended exploration may benefit from extensive layer controls, allowing users to build their own view of the data. A map designed to make a specific argument may need to lock most layers in place, offering toggles only for supplementary context. Large language models (LLMs) make it easy to add controls – perhaps too easy. An LLM agent might say, "Would you like me to add data export on each layer as well?" That sounds brilliant at first. Without thinking, you say yes. In development circles, they call this scope creep. In design circles, I call it chaos.

Your job as the director is to decide what each element should do, when it should do it, and – just as importantly – what it should not do. It's your responsibility to rein in your LLMs and the agents you're using. Prompting for control systems means specifying boundaries as much as affordances. Which layers are toggleable versus locked? Can layers be reordered, and if so, under what constraints? Are there scale-dependent visibility rules that prevent users from toggling layers that won't render meaningfully at certain zoom levels? Do layers get additional controls – zoom-to buttons, attribute tables, opacity sliders – and when should those controls be disabled?

Behavior Questions

Effective control prompts answer specific questions about what users can and cannot do. Which layers, if any, are allowed to be toggled visible and invisible? Can layers be reordered by the user, or is the layer stack fixed? Which layers are locked and always visible? Which layers are on by default, and which start hidden? Are there scale-dependent visibility rules – layers that only appear at certain zoom levels? Do layers benefit from added controls like zoom-to-extent buttons or attribute table access? If so, which controls, and under what conditions should they be enabled or disabled?

These questions expose the logic of choice architecture. When you can't answer them clearly, control systems become arbitrary, and arbitrary controls invite confusion. Users start toggling things on and off to see what happens, exploration devolves into clicking for its own sake, and the map's purpose gets lost in the noise.

Prompting Patterns

Effective control prompts specify what's toggleable, what's locked, and what additional affordances are available. Consider these examples:

```
Provide visibility toggles for the three thematic lay-
  ers: population density, median income, and land use.
  Ensure the basemap and boundary layers remain con-
  stantly visible. Default to showing all thematic lay-
  ers at the map's initial extent.
```

```
Add a 'Zoom to layer extent' button next to each toggle-
  able layer. When a layer is not visible, disable this
  button. Only enable the zoom-to button when the layer
  is active.
Lock the basemap layer and all annotation layers. Allow
  users to toggle the two data layers independently.
  Prevent users from reordering layers to maintain
  visual hierarchy.
```

These prompts define boundaries clearly. They specify what's flexible and what's fixed, what triggers enable or disable, and what defaults guide initial behavior. They also acknowledge that control is not unlimited – some layers need to stay put to preserve the map's integrity.

Menus: Hide Complexity Without Hiding Meaning

Menus exist because users want options, but options shouldn't bully the map. The challenge is organizing controls so they're discoverable without overwhelming the interface. Prompting for menus means deciding what lives in the main UI versus what gets tucked into an overflow menu or hamburger icon. Primary tasks – the actions most users will need most often – belong in the main toolbar or interface. Secondary tools – less frequently used controls, settings, or export options – belong in a menu, grouped logically and labeled plainly.

Avoid deep nesting. Menus that require users to drill down through multiple levels to find what they need create frustration and reduce discoverability. If a control is important enough to include, it should be findable within one or two clicks. Prompting for menu structure should specify grouping logic and labeling conventions:

```
Keep primary interaction tools - filter, legend toggle,
and reset - in the main toolbar. Move secondary tools -
data export, print view, and settings - into a single
'Tools' menu. Label menu groups with plain language.
Avoid nested menus deeper than one level.
```

This prompt makes hierarchy explicit. It distinguishes primary from secondary, specifies grouping, and constrains nesting to keep the interface navigable.

Locator Maps (Mini-Maps): Context + Navigation Alternative

Locator maps help users situate the main view and can double as navigation tools, but they should stay low in the visual hierarchy and be hideable when they aren't needed. A locator map that's too prominent competes with the main map for attention; one that's too faint gets ignored entirely. Prompting for locator maps means specifying visibility, interactivity, and visual hierarchy:

```
Show a small locator map in the lower-right corner.
Keep it visually subdued - use muted colors and minimal
detail. Allow users to hide and show the locator with
a toggle button. Allow dragging the extent rectangle
within the locator to update the main map view.
```

This prompt balances utility with restraint. The locator is available when users need context or quick navigation, but it doesn't demand attention or clutter the interface when it's not actively being used.

Scope Creep and the Director's Job

If you recall only one thing from this section, remember the old rule that still refuses to die: *design each map element with purpose, and be selective*. Map elements can absolutely make a map more usable and more memorable. LLMs make it to even easier to add interactivity that quietly decays communication. Your job as the director is to decide what each element should do, when it should do it, and – just as importantly – what it should not do. It's your responsibility to rein in your LLMs and agents. Now that we have a shared behavioral vocabulary, we can stop naming widgets and start examining how these elements operate together to form user experience systems that quietly choreograph attention and interpretation.

Motion and Continuity Systems

Transitions deserve the same scrutiny as all other interactive elements. A transition is not merely a visual bridge – it is an interpretive cue. Abrupt transitions imply spatial or temporal discontinuity, such as clicking on a bookmarked extent that instantly jumps the map to another place. Smooth easing implies relationship, like clicking "next" on a timeline and having the map glide to the next point of interest in the expedition you're mapping. If you don't explain to your LLM agents what type of transitions to use or their role in the overall narrative aesthetic of your map – see Chapter 8 again – the agent will decide for you. Smoothness, like motion, is not inherently meaningful and is not a neutral decision. You need to decide and explicitly state what you want.

Prompting Motion and Timing

Prompting for motion requires specificity. Vague instructions like *make it smooth* or *add transitions* will produce generic results that may or may not serve your purpose. Instead, describe the motion's role and specify timing, easing, and synchronization:

```
Animate the transition between time steps to show spatial continuity. Use a 500ms ease-in-out transition. Synchronize the map view animation with the timeline scrubber so users can see how the data evolves as they drag.
Fade in new features after filtering to signal change without distraction. Use a brief 200ms fade. Do not animate features that remain visible - only new arrivals.
When users click a location bookmark, zoom smoothly to the new extent over 800ms. Use ease-in easing to emphasize the shift in scale and context.
```

These prompts are directive about timing, easing functions, and what should and shouldn't move. They also specify when motion should stop – after one transition, after filtering, only during specific interactions – preventing motion from becoming ambient noise.

When Abrupt Is More Honest

If no meaningful relationship exists between states, abrupt changes can be clearer and more truthful than smooth continuity. Smoothness can mislead by implying connection where none exists. A map that switches between two unrelated datasets – for example, earthquake frequency and coffee shop density – should probably use an abrupt transition rather than a smooth one. The lack of continuity in theme is real and kind of the point.

Prompting for abruptness means naming it explicitly:

```
When switching between datasets, use an instant cut with no transition animation. Make the change immediate and obvious to signal that these are unrelated views.
```

```
Flash the map briefly to white when resetting filters, then return to the default view. Do not animate the transition - the reset should feel like a clean slate, not a continuation.
```

These prompts embrace discontinuity as a rhetorical tool. They acknowledge that sometimes the most honest design choice is to avoid smoothness altogether.

Motion and continuity are not afterthoughts – they are part of the map's argument. When prompted thoughtfully, transitions clarify relationships, emphasize changes, and guide attention. When prompted carelessly, they become distracting.

Behavioral Orchestration: Designing User Flows

Individual interactions rarely stand alone. A hover invites a click; a click triggers a filter; a filter reshapes the legend; the legend reframes the map – boom, Boom, BOOM! Interactivity presents cascading effects. Before long, the map is no longer responding to a single action but participating in a sequence, like a video game without the loss of player lives.

That sequence – what happens first, what follows, and what lingers – I've come to call user flow. In prior web mapping, user flows were often implicit. Designers assembled interface components and trusted that users would find their way through them – with varying success. Narrative mapping tools made this logic more explicit by introducing chapters, scroll positions, and scripted transitions, but the underlying structure still lived inside frameworks and event handlers.

User flows are often still implicit, but given that you need to describe behaviors in prompt cartography, sequences *no longer need to be* left to accidental user exploration. You can literally prompt for user flow and shepherd users quite efficiently, creating a type of path dependency that is very useful for communication, persuasion, and maintaining user focus – mind you, not necessarily in that order.

Prompting for user flow means articulating progression. Instructions such as *introduce the map at a regional scale, then reveal local detail only after filtering* or *guide the viewer through the dataset chronologically as they scroll* describe not just interaction but your directorial intention. They specify how the user's attention should move and when the map should ask for deeper commitment. Having explained to your agents what you want, they can then help devise ingenious schemes to engage and draw your users into the map.

This parallels what UX designers at social media companies do, though hopefully for better ends.

User flow also depends on transition conditions. What will trigger the next step? A scroll position? A selection? A threshold crossed?

You will need to state these conditions in your prompts. Vague instructions such as *guide the user through the data* tend to produce generic scrollytelling – formulaic, forgettable, and often annoying. Specific conditions – "after zooming below 1:50,000 scale, once a category is selected, when all filters are cleared" – produce legible structure.

Importantly, good user flows include moments of rest. Not every interaction should immediately lead to another. Prompts that specify pauses – "hold this view, allow exploration before continuing, do not advance automatically" – acknowledge that user comprehension takes time. These pauses are as much a part of flow as motion or change.

User flows are also where earlier design decisions converge. Blending interaction grammars – exploratory, narrative, and analytic – requires intentional orchestration. A map might begin exploratory, inviting users to roam and discover patterns. At a certain point, it might shift to narrative mode, guiding users through a specific sequence of insights. Finally, it might open up analytic controls, allowing users to drill down and test their own hypotheses. These shifts need to be signaled clearly in your prompts:

```
Begin in exploratory mode: allow free panning and zooming within the dataset extent. After the user has interacted with at least three features, reveal a guided tour prompt. Transition to narrative mode: walk the user through five key locations with accompanying text panels. After the tour completes, enable analytic filters and encourage independent exploration.
```

The above prompt choreographs attention across three grammars, specifying triggers, transitions, and shifts in agency. It also makes explicit when control shifts from user to designer and back again, preventing the disorientation that arises when those boundaries are unclear.

Common Behavior-Based Prompting Mistakes

Prompt-driven design lowers the technical barrier to adding interactivity for most cartographers, but it raises another major risk: the temptation to ask for everything. Perhaps this needs a name? I'll call it *interactivity compulsion*, the idea that web maps can never be interactive enough.

Many failed interactive maps are not broken in any obvious way. They simply try to do too much, too quickly, and with too little intent, or with datasets that don't afford as much exploration as the controls allow for. These web maps, of which I've made many of myself, remind me of the coffee maker in my kitchen that theoretically can steam milk, make iced coffee, hot coffee, tea, grind whole beans, accept coffee grounds, filter the water, and tell the time for me. I was sold, at first, by the sheer magnitude of its capability. It is an epic engineering feat that now sits largely neglected, because it's too much work to figure out how to use it compared to simply boiling water and pouring it over coffee grounds or tea. I feel duped. Don't dupe your map users like that coffeemaker company – redacted to avoid a lawsuit – duped me.

Under-Specification and Leaving Defaults Unspecified

One way to avoid falling into this trap is to double-check your prompts before clicking the go button. Requests such as "make it professional, add smooth animations, or improve the UI" sound reasonable, but they provide almost zero direction. They outsource design judgment entirely, encouraging generic results that may satisfy default aesthetics but will not directly serve your map's purpose.

Effective prompts name specific qualities, for example, "restrained typography, limited motion, clear hierarchy contrast," rather than vague aspirations. When state is left unspecified, agents will guess, and those guesses accumulate into interfaces that feel arbitrary or overwhelming. Perhaps your map opens with an expanded legend before the layer is even shown. Maybe your filter has ambiguous initial settings. Be as explicit as possible when prompting about states. If you do not describe them, the model will guess or forget to consider them altogether. This can lead to very messy refactoring and iteration. Prevention of bad design beats prescriptive prompting to fix bad design every time.

Over-Specification: Adding Too Many Affordances Without Purpose

Another frequent mistake is over-disclosure by default. Some of this, I think, can be blamed on bad habits developed using web map API defaults. Prompts that ask to "show details on hover" or "display feature attributes" will often result in info windows showing six or seven fields, four of which are meaningless to the map user. Remember to explicitly direct disclosure.

You can also stage disclosure: hover reveals name and population; click provides name, population, year incorporated, and image. In many maps, when everything appears at once, interaction stops adding value and maps become less engaging.

Scope creep happens when you keep saying yes to features that sound good in isolation but clutter the experience when combined.

Each new affordance – layer toggle, data export, comparison mode, timeline scrubber – adds cognitive load. The question is not whether each feature is useful, but whether the map needs all of them at once … or at all.

Inconsistent States Across Elements

A third pitfall is animation without justification – which we've mentioned already but cannot be emphasized enough. Asking for motion everywhere, including pulsing symbols (honestly, one of my favorite forms and one I do overuse), sliding panels (whiplash warning!), animated filters (subtleness adds latency), creates activity but may not improve understanding. A concise synonym for this: distraction. Your prompts need to specify why something moves and when it should stop.

Akin to the axiom that the quietest people in a room are often the smartest, maps with the right amount of visual quiet signal confident competence. Inconsistent states across elements create dissonance. Keep an eye out for a legend that shows all categories while the map displays only a filtered subset confuses users or a panel that doesn't update when filters change. Small synchronization rules (e.g., "update legend entries dynamically as filters change, reset panel when filters are cleared") prevent drift and keep the interface honest.

Gratuitous Animation and Novelty-First Design

Novelty can be seductive, especially when agents offer suggestions that sound impressive. But novelty wears off quickly, and what remains is whether the interaction actually helps users understand the data. Gratuitous animation – motion that doesn't explain change or reduce friction – becomes noise. Novelty-first design prioritizes what's new or flashy over what's clear or useful. The antidote is to ask, with every interactive element: does this clarify or clutter? Does this guide or confuse? If the answer isn't immediately clear, the element probably doesn't belong.

Validation Bias (Or: LLMs Are Paid to Praise You, Not Tear Your Hear Out)

Finally, beware of validation bias. LLMs are often eager to please (the companies making them want you to like them, after all), and when you ask them to critique your map, they will often default to superfluous praise rather than candid diagnosis.

If you want useful feedback, prompt for critique explicitly:

```
What aspects of this interaction design might confuse
users? Where does the interface create unnecessary fric-
tion? What happens if a user ignores the intended flow?
```

These prompts invite diagnosis rather than validation, and diagnosis is what you need during iteration.

All the above issues share a common root cause: treating prompts as shortcuts to get a job done rather than as design language. Your goal is not to ask for more, faster, but to ask for what is needed, precisely. When interactive behaviors are articulated clearly, restraint becomes almost automatic, and interaction becomes legible rather than loud. Now, will avoiding these common mistakes guarantee you a good map? No. However, avoiding them will prevent you from unleashing a lot of bad ones upon the world. Moreover, avoiding these mistakes will reduce the emotional torment of map review and critique, which is yet to come.

Testing and Refining Interactive Prompts

Testing interactive maps is different from traditional usability testing, though the goals remain similar. You still want to know whether users can accomplish their tasks, whether they understand what the map is showing, and whether the interface helps or hinders comprehension. The difference is that in prompt cartography, the design is encoded in language, which means iteration can happen quickly – but also means that small changes in prompts can produce unexpectedly large shifts in behavior. Testing becomes a way to validate not just the map but the prompts that generated it.

Test Communicatively: Does It Clarify or Clutter?

The first test is communicative: does this interaction clarify or clutter? Does it guide or confuse? These are subjective questions, but they're the right ones. A tooltip that clarifies identification is useful. A tooltip that dumps a paragraph of text clutters the experience. A filter that guides comparison is useful. A filter with ambiguous logic confuses users. Testing communicatively means watching how users respond to the map – not just whether they can complete tasks, but whether the tasks feel intuitive or frustrating, whether the map rewards exploration or punishes it.

Iterate by Subtraction: Constrain, Simplify, Remove

Iteration in prompt cartography often means subtraction rather than addition. When something isn't working, the instinct is to add more – more explanation, more controls, more motion. But often the solution is to remove: collapse the legend by default, disable hover tooltips on mobile, remove animation that doesn't explain change. Prompting makes subtraction explicit.

You can write "do not include a layer toggle for the basemap or disable tooltips on mobile devices," and the agent will comply. Subtraction was harder previously, where removing features often required deleting code and retesting dependencies. In prompt pipelines, subtraction (and debugging broadly) is just another instruction.

Variation Testing: Small Prompt Changes Reveal Sensitivity

Variation testing – making small changes to prompts and observing how outputs shift – reveals how sensitive the design is to language. If changing *collapsed* to *minimized* produces wildly different behavior, the prompt may be relying on unstable associations in the model's training data. If adding *update dynamically* significantly improves synchronization, that specificity was needed. Variation testing helps you understand which parts of your prompts are load-bearing and which are redundant, allowing you to refine language over time.

Over Time, Prompts Become Shorter Because Thinking Becomes Clearer

One encouraging pattern: over time, prompts tend to become shorter, not because you're cutting corners, but because your thinking becomes clearer. Early prompts are often verbose, hedging against ambiguity by specifying everything. As you develop fluency, you learn which details matter and which can be left implicit. You learn which phrases reliably produce the behavior you want and which invite misinterpretation. Shorter prompts are not lazier – they're more precise. They reflect an understanding of what the system needs to know and what it can infer.

Integrating Interactivity with Earlier Design Decisions

Interactivity should not feel bolted on – it should inherit earlier design constraints and extend them naturally. This is where prompt cartography's emphasis on structured pipelines pays off. Every interactive decision you make in this chapter builds on choices articulated in previous chapters: layout and visual hierarchy from Chapter 9, color from Chapter 20, typography from Chapter 12, and data semantics from Chapter 7. Interactivity is *not* a separate design layer. It is the behavior of prior decisions.

Layout choices constrain panel placement. If your layout reserves the right sidebar for contextual information, that's where persistent panels belong. If your layout emphasizes the mapped area and minimizes chrome, legends and filters should collapse by default. Prompting for interactivity within these constraints ensures coherence rather than conflict. Hierarchy governs emphasis, and emphasis carries through to interaction. If your visual hierarchy prioritizes one thematic layer over others, that layer should be on by default, its legend expanded, its filters more prominent. Secondary layers should start hidden or subdued. Prompting for interactive hierarchy means translating visual weight into behavioral priority.

Semantic categories from data preparation shape filter groupings and legend organization. If you spent time in Chapter 7 defining meaningful categories – grouping features by type, theme, or relevance – those categories should structure your filters and legends. Prompting should reference those categories explicitly: *group filters by the thematic categories defined in the data preparation phase* or *organize the legend to reflect the semantic hierarchy established earlier*. This prevents interactivity from imposing a new logic that conflicts with the data's structure.

Prompting makes dependencies visible and helps expose inconsistencies early. If your prompt asks for a legend that updates dynamically but a filter system that operates independently, you'll notice the friction when you test. If your tooltip shows attributes that don't exist in the dataset, the agent will flag the mismatch. This visibility is an advantage – it forces you to reconcile decisions across chapters rather than discovering conflicts late in the process.

Interactivity is the behavior of prior decisions, not a separate design layer. When integration is successful, interaction feels inevitable – like the map couldn't work any other way. When integration fails, interaction feels arbitrary, like controls and affordances were added after the fact to check boxes rather than serve purpose. Prompting for integration means continually asking: does this behavior extend the logic established earlier, or does it introduce a new, conflicting logic? If the answer is the latter, something needs to change – either the interaction or the earlier decision it conflicts with.

But Are Your Map Behaviors Working?

The question that matters most is not whether your interactive elements function technically, but whether they perform communicatively. Tooltips, filters, transitions – do they clarify or clutter? Do they guide or confuse? A tooltip that appears exactly as prompted but overwhelms users with information is technically correct and communicatively wrong. A filter that works perfectly but offers choices that don't align with the map's purpose is functional but purposeless. The distinction matters because prompt cartography makes it easy to generate technically correct outputs that fail communicatively.

This is where critique becomes essential. Changing interaction patterns required rewriting code, adjusting event handlers, and retesting dependencies – time-consuming work that discouraged experimentation. In prompt cartography, critique can happen earlier and more frequently because interaction is written in language. You can revise a prompt, regenerate the map, and test the new behavior in minutes rather than hours. This speed is both an advantage and a risk. The advantage is that iteration becomes cheaper, allowing you to test more variations and refine behavior incrementally. The risk is that speed can encourage superficial iteration – tweaking language without questioning whether the underlying design decision makes sense.

Recall that, because your interactions are written in language, they can be critiqued using language. You don't need to read code to understand what a tooltip is supposed to do – you can read your own prompt. This transparency invites critique from collaborators, stakeholders, and even users who wouldn't normally have access to the implementation layer. A prompt that says *show all feature attributes on hover* can be questioned immediately: do users really need all attributes on hover, or would a subset be clearer? A prompt that says *animate every state change* invites the question: does every change need animation, or only transitions that explain relationships? This linguistic transparency makes design decisions visible and debatable, which is exactly what critique requires.

You're now ready to walk through the entire workflow from start to finish. In Chapter 16 we will do so a bit fantastically, just to demonstrate that the workflow and individual pipelines are truly universal – both on Earth and Middle Earth.

Takeaways

- **Interactivity Is a Design Language**

  Timing, sequence, and restraint communicate meaning just as surely as color or typography.

- **Behavior Replaces Assembly in Prompt Cartography**

  You no longer build widgets; you describe how the map should respond.

- **Every Interaction Nudges Interpretation**

  Nothing is neutral – hovers, delays, filters, and transitions all shape how users understand the map.

- **Disclosure Systems Manage Cognitive Load**
  Legends and filters govern what is seen when, and defaults imply emphasis.
- **State Is a Design Decision**
  Collapsed versus expanded, active versus disabled, persistent versus transient – these are rhetorical choices, not cosmetic ones.
- **Tooltips and Pop-Ups Signal Different Levels of Commitment**
  Hover is curiosity; click is intent. Keep their roles consistent and mobile-friendly.
- **Pacing Shapes Understanding**
  Progressive disclosure reduces overload and sequences meaning deliberately.
- **Panels Preserve Narrative Memory**
  Without persistence, maps reset interpretation repeatedly, forcing users to reorient themselves with every interaction.
- **Motion Must Explain Change**
  Smoothness implies relationship and can mislead if no relationship exists. Animation earns its place by clarifying, not decorating.
- **User Flows Are Directed Sequences**
  Prompt them intentionally – specify progression, pauses, thresholds, and constraints.
- **Control Systems Are Choice Architecture**
  Decide what users can toggle, reorder, or access, and constrain what they cannot.
- **Common Mistakes Stem from Under-Specification and Over-Specification**
  Leave state undefined and agents will guess. Add affordances without purpose, and the map becomes cluttered.
- **Testing Turns Interaction into Hypothesis**
  Does it clarify or clutter? Does it guide or confuse? Revise explicitly based on answers.
- **Interactivity Must Inherit Earlier Decisions**
  Layout constrains placement, hierarchy governs emphasis, and semantics shape filters and legends.
- **Restraint Signals Competence and Builds Trust**
  The quietest interfaces are often the most confident.

Conceptual and Applied Exercises

Interactivity Archaeology

Find an interactive web map you use regularly – a news map, a municipal service locator, or a data exploration tool. Audit its behavioral systems:

- Which interactions feel like disclosure (legends, filters)?
- Which govern commitment (tooltips, pop-ups)?
- Which provide memory (persistent panels)? Now ask: which of these interactions clarify the map's purpose, and which feel like they were added because the designer could, not because the map needed them?

State Specification Practice

Write a structured prompt for an interactive map showing public library locations in your region. Your prompt should specify:

1. Legend state (collapsed or expanded by default, dynamic or static);
2. Filter defaults (preselected or cleared);
3. Tooltip behavior (what appears on hover, timing, mobile handling); and
4. Pop-up content hierarchy (what appears first vs. on expansion). Be explicit about every state – don't leave anything for the agent to guess.

LLM Behavioral Critique

Choose an interactive map – ideally one with multiple behavioral systems (legend, filters, tooltips, panels). Ask an LLM to critique its interaction design using diagnostic prompts, not validation prompts. Try:

```
What happens if a user clicks the legend before select-
ing a feature? Where might this interface create con-
fusion? What assumptions does this interaction design
make about user knowledge?
```

Compare the LLM's critique to your own. Did it identify failure modes you missed, or did it overlook problems you noticed immediately?

Flow Versus Chaos

Sketch or describe a simple user flow for a map showing changing wildfire risk over a 10-year period. Your flow should specify:

1. What users see first (default state);
2. What interaction invites them forward (hover, click, scroll);
3. What happens at each step (disclosure, animation, state change); and
4. Where users are allowed to pause or explore freely.

Now, reverse-engineer your flow:

- Where did you guide users, and where did you let them roam?
- Would a different interaction grammar (exploratory, narrative, analytic) serve the map's purpose better? How so?

For extended tutorials, hands-on labs, prompt templates, and updated examples related to this chapter, visit www.promptcartography.com.

16

From Prompt Cartographer to Interactive Web Map Director

The first time I saw a detailed map of Middle-Earth (as drawn by the author of *The Lord of the Rings* himself), I didn't read it the way I might read a novel. I read it like a cartographer.

I noticed what was named and what wasn't. Which regions on the map had more detail, and which were kind of vague and empty by comparison? His mixing of landmarks, place names, and mountain symbology worked together to suggest a story or a series of epic stories. It was world-building before the term became a video gaming and film cliché. Tolkien's map may be thought of as the original world-building moment – I'm sure some Jules Verne fan club (do these exist?) will call for my censure on that one.

That instinct – to read a map as a system of decisions rather than a neutral image – is exactly what Part II of this book has been cultivating in you… I hope. Similar to how film school "ruins" casual movie watching for graduates, it's my goal to turn you into someone who synthesizes everything we've gone through so far to view maps critically. It is this skill that will make you a better cartographer. The good news: you may go on enjoying watching bad movies because they're fun. Just not bad maps.

Up to now, each chapter in *Part II* has focused on a specific skill: framing data, naming visual styles, shaping layout, designing interaction, and learning how to critique maps responsibly. Those skills matter individually. But real-world mapmaking – especially prompt-directed mapmaking – doesn't happen one chapter at a time. There are two ways it can proceed: either as a workflow or haphazardly. We'll focus on the former here.

My goal is to help you see how all of these designed prompting and directing skills begin to operate together as a coordinated process – or beautiful mess. We're going to go through it from framing a dataset to directing visual meaning, shaping user behavior, listening and iterating through critique, and preparing a map for use beyond your own screen.

The good news? You won't be forced to learn anything new, per se. Instead, you'll see how what you have already learned works together as an entire workflow. You're more comfortable leaving command-line, IDE, and GUI-based software for natural language cartography when it's expedient and makes sense. (For the record: I'm not saying prompt cartography is a full-on replacement for other cartographic tools or processes. Just that it's a

 DOI: 10.1201/9781003797920-18

technological revolution that will inevitably envelop them, similar to how the text-based world wide web enveloped audio and video.)

The Prompt Cartography Workflow

By this point in the book, you are no longer learning isolated prompt techniques. You are learning how prompts work together.

Earlier we focused deliberately on individual skills: how to name a visual style precisely, how to interrogate and frame datasets, how to control layout, hierarchy, interaction, and critique. Each of those skills (and the pipeline tasks comprising them) matters. But professional mapmaking – especially prompt-directed mapmaking – does not happen one skill at a time. Your interactive web maps will be created via the convergence of numerous asynchronous pipeline tasks, large language model (LLM) assistant responses (and dead ends), directorial decisions, and feedback loops. Choices you make early on in one pipeline will have a cascading impact later on in the workflow for others. They may constrain, overindulge, or simply confuse your LLM and require you to go back to a previous pipeline and make changes.

This makes that workflow explicit.

At a high level, the prompt cartography workflow is nothing but a series of conversational pipelines:

1. Frame the dataset and its ethical boundaries
2. Establish a visual and narrative direction
3. Design symbology and hierarchy
4. Compose layout and spatial structure
5. Shape interaction and user behavior
6. Critique, revise, and refine (continuously!)
7. Prepare the map for publication (and reuse)

On any real project, you'll be hopping and poking around this list like one of the wild turkeys tearing up the ferns outside my office window right now. I'm not one for *always* following pipelines, even though I know they work.

What changes now is coordination. The goal is no longer to write this or that prompt well but to direct the sequence of prompts, diagnose breakdowns in the process, and keep the whole system from going off the proverbial through to final map.

This is where prompt cartography ups the stakes from avant-garde film class project to Hollywood blockbuster movie production.

The Prompt Cartography Stack

To make this coordination concrete, Figure 16.1 names the core pipelines that operate as a bundle. You can think of this workflow as a working prompt skill stack: you may move forward, loop backward, or pause at any stage depending on what your map demands. What matters is not rigid order, but intentional sequencing for each project.

This prompt stack is the real takeaway from Part II. Once you see it, you begin to recognize familiar professional behaviors: diagnosing problems without touching the data, adjusting rhetoric without changing geometry, and knowing when not to refine further.

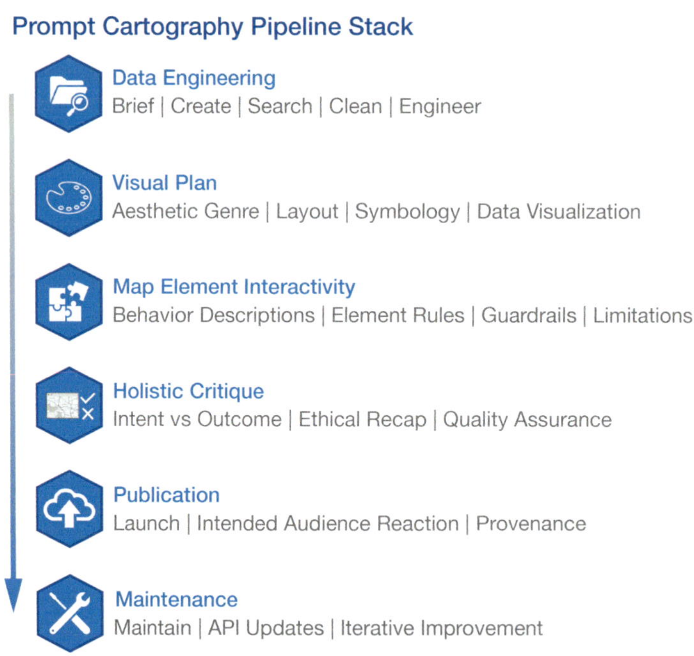

FIGURE 16.1
The prompt cartography workflow as a stack of pipelines. Each stage reflects a series of prompting skills explored and developed in earlier chapters, now combined into the pipelines comprising the overarching workflow.

What You Have Learned

At this point, it is worth pausing – not to recap content but to recognize capability.

By working through Part II of this book, you can now:

- Frame datasets with explicit scope, audience, and ethical boundaries
- Specify visual styles as named dialects rather than vague aesthetics
- Diagnose visual problems without defaulting to new data
- Design layouts as constraint systems, not decoration
- Shape user behavior through interaction design
- Critique maps continuously, not defensively
- Treat prompts as coordinated instruments rather than isolated commands

None of these skills depend on a particular dataset, software platform, or mapping genre. They are transferable precisely because they operate at the level of decision-making, not tooling. And no matter what the haters online may say about LLM-produced maps, it's true, they all require you, the human cartographer, to be built and work as they should. The LLM isn't making a map for you. You are making a map via LLM. And perhaps many, if not most, days you'll feel like you're totally winging it.

Don't be discouraged. If there's one thing I've learned in my cartographic career: when you're winging something, and yet, you're still optimistic about the eventual outcome, it's a sign that you've likely reached professional status.

A Training Arena, Not the Point

Before we go any further, let me say this out loud, because it matters:

> This hypothetical scenario is not the lesson. It's the training arena.

In this chapter, we'll work with a Middle-Earth place-based dataset – locations, regions, journeys, and landmarks drawn from Tolkien's world. The data is available online, and it's familiar to millions, richly named, and naturally narrative, which makes it a comfortable space to practice directing a full prompt cartography workflow without getting bogged down in domain complexity.

But the workflow you're about to see does not belong to fantasy maps!

It belongs just as much to real estate listings, genealogy records, invasive species sightings, public-facing government maps, or environmental risk dashboards. The reason we can move so fluidly here is because the skills you built up to now remain domain-agnostic. They operate at the level of framing, design judgment, and critique – not subject matter.

So if you're thinking, "I despise fantasy literature, and I don't care about mapping elves," relax. You're not learning to map Middle-Earth. You're learning how to direct an entire map, from start to finish. The point is the process, not the monsters.

Framing the Dataset Like a Director

The first move as a director in any visual medium is conceptual.

Before generating symbols, layouts, or interactions, you need to decide what the datasets you have are useful for. Not what they contain but what role they can play in your production. This is the data engineering pipeline, now operating in real time.

For this dataset, a producer-style framing prompt might establish:

- *The Intended Audience*: fans exploring places, not scholars analyzing lore
- *The Spatial Scope*: key regions and travel routes, not exhaustive geography
- *The Narrative Posture*: exploratory and atmospheric, not encyclopedic
- *Ethical Boundaries*: respectful naming, consistent language, no flattening of cultural distinctions

Notice what's happening here. You are not asking the model to just "make a map." You are setting data-based constraints – and it's actually those constraints that allow creativity to behave.

This is the same pipeline technique you would, or at least should, use when framing any mapping project or dataset, including, for example a:

- neighborhood housing map
- family migration map
- public health dashboard
- public awareness visualization
- map designed to go viral and get you advertising dollars

Different domains. Different goals. Same damn discipline! Cartography never dies, just evolves.

Style Emerges from Framing (Not the Other Way Around)

Once the dataset is framed, style stops being a guessing game.

Because you've already learned how to name and describe aesthetic dialects, you're no longer saying things like "make it cool" or "make it fantasy-looking." Instead, you're able to articulate something more akin to: "hand-drawn fantasy atlas," "muted parchment tones," "restrained ornamentation," "legible modern typography layered gently over historical texture."

That specificity matters – not because it sounds fancy but because it gives the model something to aim for. You're directing, not hoping.

And just as importantly, you now know when to pause and critique. If the map starts drifting toward a tourist brochure, or becomes visually loud, you don't fix that by regenerating everything. You diagnose the problem and intervene at the right level – style, hierarchy, or layout. Prompt cartography is not about selling your soul to LLM sorcery. It all comes down to mastering visual literacy.

WHEN WORDS FAIL

Assuming a picture is worth a thousand words, and life is one never-ending motion picture, there will often be times when you are unable to find the exact wording you need or want. Brain freezes are common – writer's block (in writing), tip-of-the-tongue (in speaking), and prompt panic (I just made that one up, take it or leave it). Fortunately, not all is lost – we have LLM agents that are experts at visual literacy.

```
Please provide a design description of the uploaded
image that I can use as part of a style-guide
prompt and schema. I will be creating a web map and
would like the map elements and web components used
in the map to mimic this image's aesthetic style,
tone, hierarchy, and visual elements.
```

Pop this into your LLM agent of choice, and you will be given a readable, interpretable schema. In fact, it will likely ask you if you would like its prompt written with a specific interactive web mapping API in mind. Your agents are your friends! They speak natural language better than we do. Ask them for help.

Ethical Dilemmas Can Show Up Early, Even in Fictional Places

We've arrived at a quiet but important moment in this project.

Middle-Earth place names are beautiful. They're also loaded with narrative weight, cultural meaning, and invented histories. When you decide which names appear, how they're spelled, and which places are emphasized, you are already making ethical and rhetorical choices – even in a fictional setting.

This is the ethics chapter's lesson arriving organically as part of the cartographic process. The difference now is that you're listening while the map is being built, not looking for ethical holes after it's finished. If something feels off – flattened cultures (e.g., don't amalgamate the Rohans with Gondors), inconsistent naming (e.g., dwarves incensed by too much Elven language), or an unintentional hierarchy (the Shire is too prominent) – don't shrug it off and move on. Now is the time to quickly adjust the framing prompt, tweak the symbology rules, or ask your agents to critique and rebalance the layout.

Ethics in this case is not about compliance; it's about empathy for Tolkien fans who adore different fictional species. (I've learned the hard way – *never* cross a fan of Hobbits!)

And that carries directly into real-world mapping, where names, boundaries, and visibility are never neutral.

Generalizing Beyond the Example

Let's zoom out for a moment. For people that haven't read Tolkien's books… I haven't either. I've learned everything I know vicariously from my daughter who is obsessed. The point is, everything you've just read also applies directly to:

- Real estate maps deciding whether to foreground *amenities* or *affordability*
- Genealogy maps balancing *accuracy* with *narrative clarity*
- Public planning maps communicating *options without overwhelming* residents
- Environmental risk maps where *tone, color, and hierarchy shape trust*

I cannot emphasize this enough – in cartography the datasets change but the overarching workflow persists. And this is the quiet power of what we've built working through Part II of this book. You're no longer dependent on a particular kind of data or, perhaps most excitingly, a favorite GIS and map-making tool. You're working with a transferable and portable way of map-making and harnessing cartographic expertise. It's *not* all on you anymore to learn and know everything. You no longer have to stick with certain APIs or coding languages or software packages, because you know them best.

This is what allows prompt cartography to scale and opens up repeatable and complex design possibilities for cartographers like never before.

Visual Direction: Style, Symbology, and Hierarchy

This is typically the moment when the room gets a little quieter.

Up until now, we've been talking about framing, intent, and workflow – important things, but mostly invisible. Visual direction is where prompt cartography starts to show its work. This is also where many maps either snap into focus or quietly fall apart.

The good news is that you already have the tools to handle this stage. Here I'm just asking you to use them together.

Style Logically

Let's start with style, because this is a stage where many projects still go off the rails. When you approach visual direction as a prompt cartographer, you are no longer asking the model to "make it look nice." You are naming a style dialect – a specific visual language with expectations, limits, and references.

In our training example, that might mean something like:

- restrained fantasy atlas aesthetics
- muted, parchment-inspired color palettes
- minimal decorative flourishes
- modern legibility layered over historical tone

Unless, that is, you are trying to create a universe crossover. Perhaps you are going for a post-apocalyptic fantasy vibe. What styles might you prompt for then?

What matters here is not fantasy per se. What matters is that the style choice emerges from the framing you already established. Because you defined audience, purpose, and tone earlier, style is no longer a guess – it's a consequence.

This is visual literacy in action! And here's the reassuring part: if the style doesn't land on the first attempt, that's not failure. That's feedback. You don't throw out the map; you refine the style prompt, clarify the references, or tighten the constraints.

That's directing!

Symbology Is About Meaning Before Beauty

Once style is set, symbology comes into play – and this is where hierarchy starts doing real work.

A common mistake at this stage is to treat symbols as decoration: icons, colors, shapes chosen because they look appealing in isolation. But effective symbology is about meaning first, aesthetics second. Every symbol on the map should answer a question for the reader: *What matters most here? Why should I be noticing this right now? Can I safely ignore this?*

This is what it means to think critically about your symbology.

Instead of asking for "distinct symbols," you're now able to specify:

- which features deserve emphasis
- which categories should recede
- how color, size, and contrast reinforce importance
- where simplicity beats completeness

If everything feels equally important, that's not a data problem – it's a hierarchy problem. And hierarchy problems are solvable with prompt-level adjustments, not wholesale redesigns.

Time for a Mid-Process Critique

This is a good moment to pause and do something that is easy to skip: critique before moving on.

At this stage, a short diagnostic prompt might ask:

- *Does any one symbol dominate unintentionally?*
- *Are categories visually distinct and semantically meaningful?*
- *Is the map readable at a glance, or only after study?*

Notice what you're not asking: you're not asking whether the map is finished. You're asking whether it is behaving the way you intended. You aren't asking the LLM what it thinks about the map – hint: it will almost always say it looks great, even when it doesn't.

That mindset – checking alignment rather than chasing polish (or praise) – is one of the biggest takeaways from what you have learned.

Your Hierarchy as Silent Hero

If there is one idea worth repeating at this stage, it's this: hierarchy design does more work for your map than color palettes ever will. Please, spend adequate time on hierarchy.

Strong hierarchy allows a map to remain legible even when your style preferences (including color palettes, fonts, etc.) change. When students tell me, "Something feels off, but I can't tell what," the answer is almost always found in their map's hierarchy.

And the fix is rarely dramatic. It's usually a matter of:

- culling emphasized categories or data layers
- increasing contrast where it matters and minimizing it elsewhere
- letting secondary information stay secondary

Fortunately, hierarchy is also where prompt cartography can really shine (see Figure 16.2). You can make these hierarchy adjustments deliberately, surgically, and iteratively without ever starting over.

Direction Before Composition

One final note before we move on. At this stage, you are still directing visual meaning, not spatial layout. You're deciding what deserves attention and how that attention should feel. Layout comes next – but layout only works when hierarchy is already doing its job.

Think of this as setting the volume levels before arranging the speakers. When style, symbology, and hierarchy are aligned, layout becomes a design problem instead of a rescue mission.

And that's exactly where we're headed next.

Layout and Composition as System Design

If visual direction decides what matters, layout decides how that meaning is experienced.

This is another dangerous point in our journey – not because the data is wrong, or the style is off, but because often you will find that there are too

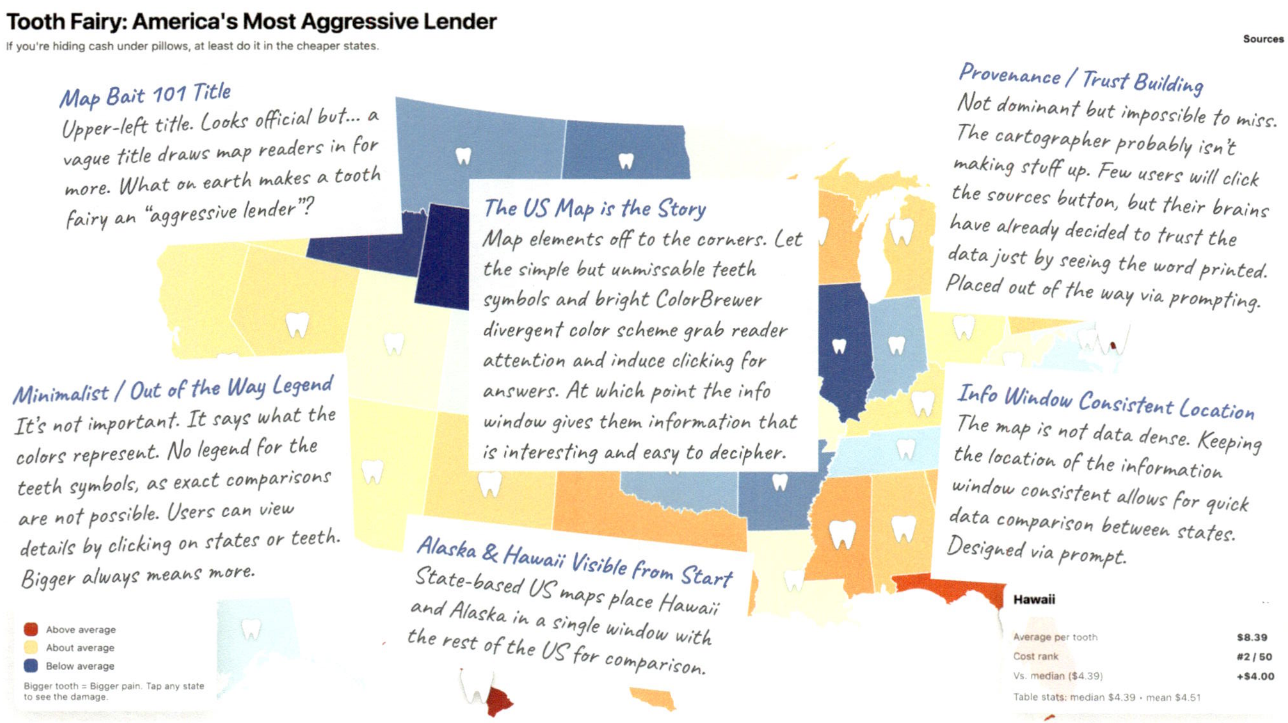

FIGURE 16.2
Visual hierarchy made explicit through symbol size, color contrast, and category grouping. Effective symbology communicates importance before detail.

many elements competing for attention. Legends fight with titles. Controls overcrowd the map. Important information ends up technically present but practically invisible. Or the worst sin of all, for some inexplicable reason, your map attribution winds up beneath the fold – i.e., it can't be viewed without scrolling the entire webpage down.

The key shift, introduced back in the map elements and layout chapter, is to stop thinking of layout as decoration and start thinking of it as system design. A good layout doesn't just arrange the elements via command. It enforces priorities via negotiation and contract.

Layout Is an Issue of Constraint

When you approach layout as a constraint system, the work becomes clearer – and calmer. Every map has limits:

- limited screen space
- limited reader attention
- limited tolerance for complexity

Your job is not to eliminate these constraints but to use them deliberately. Layout is how you decide what gets space, what gets proximity, and what gets pushed to the margins – literally and figuratively.

This is why layout prompts in any given pipeline are rarely poetic. They are precise. They specify regions, relationships, and relative importance. They describe structure, not mood.

Directing Layout with Intentional Prompts

At this stage, layout prompts typically describe:

- the dominant role of the map view
- the placement and visual weight of legends
- where filters or controls live – and where they don't
- how narrative text relates to spatial content

The important thing to notice is that these decisions are no longer happening implicitly. You are not hoping the model just "figures it out." You are directing the layout's composition, guiding it, bringing it to fruition with natural language (see Figure 16.3).

And when something feels off, you know where to look. If users miss key information, that's often a layout issue, not a data or style issue. If the map feels overwhelming, the fix is usually subtraction, not addition.

Layout as the Bridge Between Visuals and Behavior

One reason layout deserves its own moment in the sun is that it sits at the boundary between seeing and doing.

The map's layout shapes:

- what users notice first
- where they look next
- how confident they feel interacting with the map

In other words, layout quietly sets the stage for behavior. A clean, intentional composition invites exploration. A cluttered or ambiguous one discourages it, no matter how good the data is.

This is where the chapters on map layout and interactivity behaviors begin to overlap. Layout is not just spatial – it's behavioral.

Direction Locked In

By the time you leave this stage, three things should be true:

1. The map's visual hierarchy is clear
2. The layout reinforces that hierarchy
3. Nothing essential is fighting for attention

If you find yourself adding more panels, more widgets, or more explanatory text to "fix" confusion, that's usually a sign to step back and simplify the composition instead. If one of these is shaky, don't panic (no need to grab a towel), just go back one step and tighten the structure.

That instinct – to reduce, clarify, and reassert structure – is one of the clearest markers of growing design confidence.

With layout established, we're now ready to let the map move.

FIGURE 16.3
A compositional anatomy of a prompt-directed map layout, showing primary and secondary zones, supporting elements, and attention flow.

Interaction, Behavior, and User Experience

Up to this point, everything we've done has been about shaping what the map says. Interaction is where you begin shaping what the map does.

Everything you've learned about map elements has prepared you for this: interactive web maps are not passive images. They are media that behave and respond. And whether you intend it or not, they guide users toward certain actions while discouraging others.

The goal here is not to add interactivity for its own sake. The goal is to make the map's behavior feel natural, supportive, and purposeful.

Interaction Is Animated Rhetoric

Each interactive element – filters, hover states, animations, pop-ups – helps make an argument.

Each argues about what matters, what is optional, and what is worth a user's time. A cluttered control panel argues that everything is important. A carefully limited set of interactions argues restraint and clarity.

This is why interaction design straddles both style and layout. It is not a technical afterthought; it's integral to rhetorical design.

When you approach interaction through prompt cartography, you can be explicit about that rhetoric. You can ask the model to:

- encourage exploration without overwhelming
- prioritize discovery over precision (or vice versa)
- support casual browsing rather than analytical comparison

Those are not interface tweaks. They are behavioral decisions.

Designing for Confidence, Not Cleverness

A useful rule of thumb at this stage is simple: users should feel confident immediately.

If someone hesitates before interacting – wondering what a control does, or whether they might "break" the map – that hesitation is a design signal. It usually means the interaction is either too prominent, too ambiguous, or simply unnecessary.

This is where layout and interaction meet. Controls should live where users expect them, behave the way similar tools behave elsewhere, and disappear from attention when they are not needed.

Clever interactions are fun in demos. Confident interactions are what people actually use.

A Mid-Journey Behavioral Critique

Just as with symbology and layout, this is a moment to pause and listen. A short critique prompt here might ask:

- Do interactions clarify the map's purpose or distract from it?
- Is anything interactive that doesn't need to be?
- Can users explore without reading instructions?

Notice again that none of these questions are about aesthetics. They're about experience. If users struggle, the fix is often subtraction – removing options, simplifying controls, or slowing motion – rather than adding explanation.

That instinct to remove rather than add is another sign that you're now thinking like a director. (Hopefully the next Steven Spielberg, as opposed to say, George Lucas?)

Accessibility Is Interaction Design

One of the most useful realizations at this stage is that accessibility is not a separate checklist – it's a form of interaction design.

Contrast, font size, hover behavior, keyboard navigation, motion sensitivity all shape how different users experience the map's behavior. If an interaction only works for a narrow slice of users, it's not elegant – it's fragile.

The good news is that prompt cartography makes these considerations explicit. You can ask for reduced motion. You can specify contrast standards. You can request interaction alternatives for non-pointer input.

Letting the Map Breathe

By now, your map likely could do more. More filters. More animations. More explanatory text.

The question is whether it should. Strong interaction design often feels understated. The map responds when invited, stays quiet when it should, and never competes with its own message. When interaction is doing its job, users don't notice it – they simply feel oriented.

That sense of ease is not accidental. It's the result of deliberate choices layered throughout the workflow. With interaction in place, you now have a map that is alive (no worries: 100% metaphorically). It sees; it speaks; and it responds when it makes sense to. All that remains for you to do to the map is listen carefully, refine it responsibly (and ethically), and prep the map for release to be consumed by your target audience (i.e., for me, it will be my local gaming group).

Critique, Ethics, and Continuous Checking

At this point, it's tempting to think you're almost done. The map looks good. The layout holds together. Interaction feels smooth. This is often where earlier map production would stop – declare success, export, publish, move on.

But critique is not the final step in prompt cartography. It is just the through-line that keeps everything aligned.

The critique pipeline is not about asking "Is this map good?" It probably is good since you've put this much work into it. Instead, you're polishing and looking for an opportunity to refine it to make it even better. You want to be asking questions like:

```
What is the map emphasizing - intentionally or not?
What assumptions does it make about its audience?
Where might interpretation drift from intent?
```

Notice that these questions are literally diagnostic, not punitive. They help you hear what the map is saying on its own.

And importantly, critique doesn't require tearing anything down. Often it leads to small, precise adjustments: a renamed category, a softened color, a rebalanced legend. The power comes from knowing where to intervene.

Using the Critique Pipeline Throughout the Workflow

Fortunately, you have a pretty straight-forward pipeline at your disposal to quickly critique your map. Run it through the gamut now:

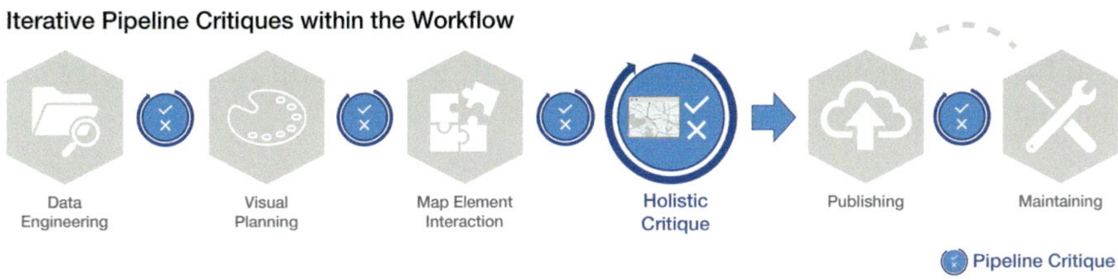

FIGURE 16.4
Critique embedded within the prompt cartography workflow, functioning as a continuous listening loop rather than a terminal evaluation step.

- ***Description***: What is present? What is visible?
- ***Analysis***: How are visual variables and interactions shaping meaning?
- ***Interpretation***: What story or argument is the map making?
- ***Evaluation***: Where does that align – or misalign – with intent?

We aren't done with the map, but you will quickly gain insight on whether your map or design decisions are starting to cause drift. This shift – from post-mortem map critique to continuous, expert critique and analysis every step of the way – is one of the most valuable things LLMs offer us (Figure 16.4). Bonus: it typically makes your final critique process a walk in the park as opposed to a nightmare scenario of … oh, yeah, I hadn't thought of that.

Ethics Lives in Ordinary Decisions

It's easy to think of ethics as something that only appears in obviously sensitive datasets. But ethics shows up much earlier – and much more quietly – than that. It shows up when you choose which names to include. It shows up when you decide which categories get visual priority. It shows up when some features are clickable and others are not.

In our training example, place names carry cultural and narrative weight. Choosing how they appear – or whether they appear at all – is not a neutral act, even in a fictional world. The same is true in real-world mapping, where names, boundaries, and visibility often reflect power as much as geography.

Ethical prompt cartography doesn't require moral grandstanding. (Personally, I despise moral grandstanding. In my experience, it's typically the moral grandstanders who are some of the shallowest people when it comes to making ethical decisions, but I digress.) No grandstanding required, but you do have to pay attention.

If something feels "beautiful but… wrong," that feeling is worth looking into further. It's often the first sign that your rhetoric and responsibility are starting to drift apart. Never ignore your gut instinct.

Critique Without Defensiveness

One of the most practical benefits of prompt cartography is that critique becomes easier to act on.

Because decisions are articulated in prompts, you don't have to defend them emotionally. You can revise them. Adjust them. Test alternatives. The map is not a reflection of your worth – it's a system you're tuning.

That mindset makes ethical reflection feel manageable rather than overwhelming. You are not trying to solve everything at once. You are making the best possible decision at each stage with the information you have. And then listening again.

Know When to Stop or You Will Never Finish

There is one final ethical decision embedded in critique: stop when something is good enough. Perfection is never the (real) goal. Clarity, honesty, and usability are. At some point, additional tweaks stop improving the map and start obscuring it. Or improving your map becomes so time-consuming that the costs of fixing it far outweigh any reward to gain from further iteration. Recognizing that moment is part of professional judgment.

If the map communicates what it intends, behaves as expected, and respects its audience, it is ready to move forward. If you still don't think so, it's not a problem that can be solved via map design tweaks; it's more likely a type of cognitive displacement on your part. (And this happens to every cartographer at some point, myself included, so I don't mean this pejoratively. Just in this scenario, it's not the map keeping you from releasing it to the masses anymore; it's your own brain.)

That readiness – not flawlessness – is what critique ultimately is meant to support. Not everyone is going to love your map. (If you've read this book, you know that's okay, because you would never design a map to appeal to everyone.) Some people will critique it harshly or meanly. (This is a reason to avoid social media, not a reason to avoid releasing your well-designed, imperfect map.) Once critique is complete, the map is no longer just a design exercise. It's an artifact prepared for use.

Publishing, QA, and Readiness

At some point, the map has to leave your proverbial desk, outbox, server…

Up until now, everything in the workflow has happened in a relatively safe space: prompts can be revised, layouts adjusted, interactions simplified. However, publishing changes the context. Once a map is released – shared with users, embedded on a website, circulated in a report – it will indubitably start encountering people you didn't design it for, in environments you didn't fully control.

This section is about preparing you for that moment. It pays to be ready and not be a people pleaser. (Difficult at times, I know.) Try to remember: you aren't chasing perfection. You design maps to make help clients, users, and give your life a sense of purpose (or pay the mortgage, or [fill in your preferred reason here]).

Publishing Is a Design Decision

It's tempting to think of publishing as a technical step: export the map, deploy it, move on. But publishing is actually a continuation of your design.

The moment you decide a map is "done," you're making a judgment about: who the map is for, how it will be encountered, and how much responsibility you're taking for the random batshit crazy interpretations that will result. (Go into this knowing there are a lot of bored dilettante trolls hiding behind the "critical cartographer" moniker. "Keystone cop cartographer" is probably a better name for them.)

That's why this chapter is *here* in the book and not at the very end. Publishing shouldn't just be an afterthought. A map that works beautifully on your screen but collapses on a phone, a projector, or when there is a slow connection is not exactly "finished"; more aptly, it's incomplete. And its limitations can kill your map before it ever gets to its intended audience.

REAL-LIFE QA EXAMPLE

My job has changed since I wrote the first edition. I am no longer teaching or a professor – though I still do academic research. Instead, I work as a product engineer on LLM and AI-based software for a large GIS company. The team I work on developed an insanely great LLM-based GIS tool. It was amazing! We were all thrilled. Until…

A member of my company demoed our tool to potential users. Our software worked well but… several users felt it was too slow. Like other

GIScientists the world over, my team and I had focused on how well the app worked from a certain perspective – functionality and accuracy as opposed to speed. We lost sight of the fact that some users value speed over capability and accuracy – shocking, I know, but true.

Lesson learned and shared here: quality assurance is about testing more than just answers and functionality. Test the app as your intended audience will use it to get the most robust feedback.

Quality Assurance Is About Fragility

Quality assurance (QA) often sounds intimidating, but at its core all it requires is asking a series of questions. (Followed by my favorite part of doing QA in my day job, trying to break the map or mapmaking software! If I'd only known that you can get paid for breaking stuff, I would have gotten into this role much sooner!)

The overarching question you should be asking yourself when testing and checking your map before publishing is quite basic: Where and how *will* this map break?

I include the word "will", because no map is perfect, and no software is unhackable, and those two things combined mean that your map can and will break. Your goal is to make sure it doesn't happen unexpectedly on a regular basis for your users.

Breakdown doesn't always mean error either. It might mean that your labels become unreadable on small screens; the interactions behave differently on touch devices or in one of those funky browsers that people much cooler than me insist on using. More commonly, certain color contrasts really do fail in bright light, meaning someone in sunny Algiers, Algeria, will think your map is unreadable but your cousin trapped in the fog banks of Grand Marais, Minnesota, won't have any problems.

The goal of QA is not to eliminate all such types of risks – again, definitely impossible – but to reduce your map's fragility. You want the map to remain intelligible, usable, and trustworthy across reasonable variations in context, particularly contexts your intended audience is most likely to be found in.

This is where many of the skills you've learned about quietly resurface. Systematic attention to visual hierarchy, interaction restraint, and clear framing aren't just useful for map building. They all make quality assessing your map easier.

Testing Without Overthinking

You don't need an elaborate testing lab to practice responsible QA.
Simple questions go a long way:

- *Does the map still make sense if someone skips the legend?*
- *Can a user figure out what to do without instructions?*
- *Does the map remain legible on a smaller screen?*
- *Are any interactions surprising in a bad way?*

If the answer to any of these is "not quite," that's useful information – not a failure. QA at this stage often leads to small, high-impact changes: simplifying labels, reducing motion, tightening interaction scope. Again, this is directing. You're choosing where to spend attention.

Readiness Is Not the Same as Perfection

One of the most important ideas to internalize here is that ready does not mean flawless. A ready map:

- communicates its purpose clearly,
- behaves consistently,
- respects its audience,
- and makes its limitations visible rather than hidden.

A perfect map, on the other hand, is often one revision away – and always will be. Knowing when to publish is a form of judgment. It reflects confidence in the process you've followed, not blind faith in the artifact itself.

Preparing for Reuse and Evolution

Finally, publishing is not always the end of the story. Prompt-directed maps are especially well suited for iteration. Due to the fact that decisions are encoded in prompts, revisiting a map later – updating data, changing emphasis, adapting to a new audience – becomes conversational, not code-heavy.

Thinking about reuse at the publishing stage will also help "future-proof" (always a loaded term) your work. Clear prompts, documented assumptions, and restrained design choices all make evolution possible without starting from scratch.

Peripheral Sparks: Seeing the Workflow Everywhere

Once you've directed a full prompt cartography workflow end to end, something interesting tends to happen. You start seeing it everywhere.

Not as a rigid checklist, but as a way of noticing how maps – and map-like systems – are constructed, framed, and guided. You recognize where a dataset was carefully scoped and where it clearly wasn't. You can tell when a layout is doing heavy lifting for a weak hierarchy, or when interaction has been added to compensate for unclear intent. Don't be alarmed. Take it as a sign of increased fluency.

Transfer Without Translation

One of the quiet takeaways from what you have learned thus far is that you *don't need to translate these skills to new domains.* They are already part of all cartography broadly.

No matter the map you're looking at – be it a pandemic dashboard or a city planning map – the same questions should naturally surface:

- *What is this map trying to help someone do?*
- *What decisions are being emphasized – or hidden?*
- *Where does clarity come from, and where does it break down?*

You're no longer impressed or disappointed by surface polish alone. You can read web maps (all maps, honestly) as systems of directorial choices. As already alluded to in the opening, that shift from consuming maps to interrogating them is one of the most valuable outcomes of learning how to be a cartographer of any kind.

You Have Directed Your First Feature Length Web Map

This chapter did not introduce new tools. It revealed how the processes you have learned about can be used together as pipelines within the prompt cartography workflow. By stacking prompts into staged pipelines, you've seen how data framing, visual planning, map element interaction, critique, and publishing reinforce and build on one another rather than compete. The most important takeaway is not a singular technique, but a shift in mindset. You are no longer asking, "What can this LLM tool generate?" but instead, "What do I want my map to do, and how can I direct my LLMs to help get me there?"

That question – asked repeatedly, at each step in process – is what turns prompt cartography from experimentation into direction. Now that you have reviewed how to produce an interactive web map from start to finish using this standardized approach, you're ready to learn about map maintenance…

Oh, wait. You didn't think your job ended when you hit publish, did you?

Takeaways

- **Prompt Cartography Is *Coordinated* Practice**

  Effective prompt-directed mapping emerges from how decisions work together. It requires coordination and planning across prompts (and agents when done in a multi-agent system).

- **Directing Is Key**

  The cartographer's role shifts from production to *guiding* sequences, *diagnosing* drift and *maintaining* alignment.

- **Your Pipelines Accumulate Meaning**

  Early framing will quietly constrain later choices. Management of the process is exercised through sequencing as opposed to having complete control (i.e., micromanagement).

- **Critique Travels with the Work**

  Evaluation is no longer a terminal step. Evaluation is continuous and can be used to reshape or fork away from previous decisions before they become hardened.

- **Ethics Lives in Ordinary Choices**

  Naming, emphasis, interaction, and omission carry ethical weight long before a map is published.

- **Restraint Is a Sign of Sound Judgment**

  Subtraction, simplification, and stopping at when it's ready are signs of maturity.
- **Publishing Is a Design Commitment**

  Releasing a map formalizes responsibility for how it will be encountered, misunderstood, and reused. Fortunately, in prompt cartography you will have a history of conversations showing that your will was good.
- **Prompt Cartography Enables Transfer**

  As your decisions are articulated in language, your pipelines should be adaptable across domains, datasets, and GIS or cartography tools with minimal-to-no translation.

Conceptual and Applied Exercises

From Maker to Director

This chapter reframed the cartographer as a director rather than a hands-on executor.

Reflect on a time in the past when letting go of direct control (over tools, steps, or details) improved one of a project's coherence or outcome. (It doesn't have to be a mapping project.) What kind of responsibility replaced that control?

Hierarchy Check

Identify a map you trust. Without focusing on its data or aesthetics, describe how its visual and interactive hierarchy guides attention and interpretation. What do you think the map's hierarchy is trying to protect the user from misunderstanding?

LLM Direction Probe

Ask an LLM:

```
What does it mean to "direct" a creative system rather
than operate it?
```

Assess whether the response emphasizes authority, orchestration, or accountability. Which emphasis best matches your role as the map director outlined in this chapter?

Knowing When to Stop

You've learned that readiness is not the same as perfection. Think of a map or project you have struggled to finish. What signals – conceptual, ethical, or experiential – might indicate that a map is ready to be released, even if it still feels incomplete?

For extended tutorials, hands-on labs, prompt templates, and updated examples related to this chapter, visit www.promptcartography.com.

17

Sustaining a Map That Has Left the Station

The Maintenance Pipeline

Interactive web maps don't move out of the house when you click *Publish*. They just start publicly broadcasting any and all shortcomings – whether they are the director's fault or not. Browsers update and cause new problems. APIs (application programming interfaces) deprecate and begin to fail. Color and interface styles shift – anyone remember the flat Metro style? Six months after production, a legend that felt modern and sleek now feels antiquated – like a fraying hem that is screaming to be re-stitched!

Shipping a map is a milestone, but it's rarely the finish line. It's more like a handoff of responsibility – from design to maintenance (see Figure 17.1).

From Prompt Spec to Live Deployment

Publishing a prompt-generated map is an act of translation. The model gives you polished specifications and HyperText Markup Language (HTML), JavaScript (JS), TypeScript (TS), cascading style sheet (CSS), versioning data, and spatial data files. But what then? Web servers don't read intention and gossipy conversations between you and your agents. Your job is to turn that structured output into something a host can serve reliably.

The basic pipeline is simple: Prompt → Map Spec → Repository → Live Map

Export the model's output. Save it as a versioned file. Commit it to a repository (e.g., git). Connect that repository to a hosting environment – GitHub Pages, hosting website, personal Linux Server running on an in-law's former Dell laptop … not that I'm speaking from experience or anything.

This is where prompt cartography meets web infrastructure. It can be scary, but it doesn't have to be. Particularly when you have large language model (LLM) agents that can help coach you through the process.

Most practitioners start small. Perhaps a single HTML map hosted on GitHub as proof of concept (Figure 17.2). From there, interaction, automation, and monitoring layer on naturally. (Note: the companion website has tutorials on using GitHub to host your first prompted web maps.)

DOI: 10.1201/9781003797920-19

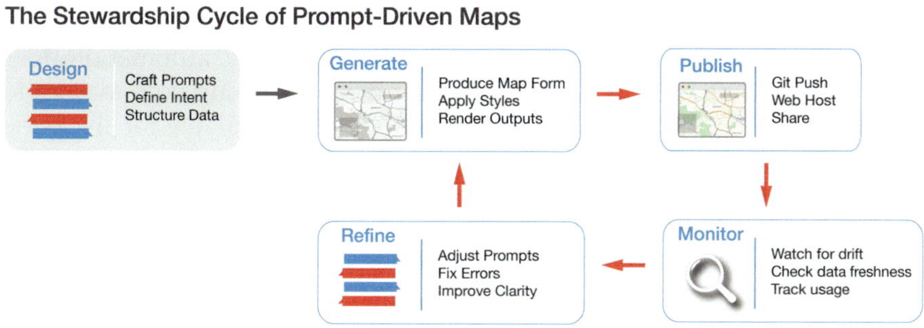

FIGURE 17.1
Your stewardship of any given map rarely ends when you publish it. In many ways, this is where it begins – anew.

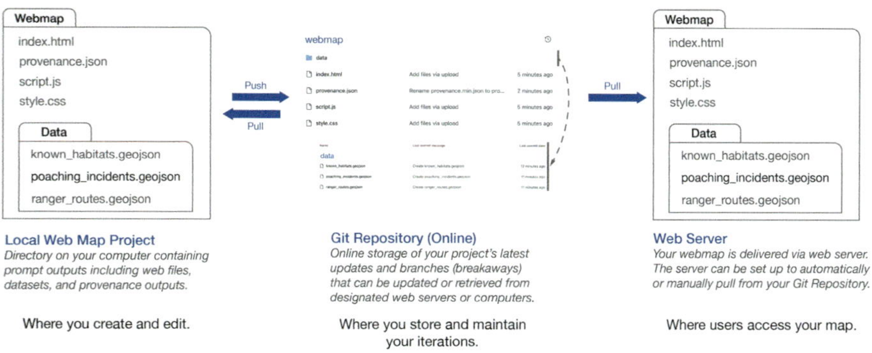

FIGURE 17.2
Example GitHub repository synced to your prompt cartography output files on your computer and linked to your web server (i.e., online website).

By this point in the book, you know how to design, critique, and audit your maps. This is where we wrap it all up. Shipping your product responsibly means assuming change – and planning for it.

Prompt-Driven Automation Pipelines

Once a map is live, maintenance begins. Data updates. Legends drift. A basemap style changes just enough to throw off contrast. None of this is dramatic. That's why it's dangerous. The slow failures are the ones that go unnoticed until someone else points them out.

Fortunately, automation exists to absorb this kind of change – not to replace judgment but to reduce fragility. In prompt cartography, automation does not mean handing control to a black box. It means teaching your system to repeat the boring parts reliably so your attention is available for the parts that actually require it.

At its simplest, a prompt-driven automation pipeline does three things:

- Regenerates a map when inputs change
- Checks the result against basic expectations
- Reports what happened, rather than assuming everything is fine

That's it. No mystique.

Automation as Continuity as Opposed to Convenience

Automation is often sold as speed. In practice, its real value is continuity. A manually updated map depends on memory: someone has to remember to rerun it, recheck it, and republish it. An automated map depends on routine. When new data arrives, the same sequence happens every time – generation, checks, and publication – whether or not anyone is watching.

This routine is often expressed in plain language:

```
When new data arrives, please regenerate the map speci-
fication, verify color contrast and legend complete-
ness, and publish the updated version of the map.
```

That instruction can be translated into many technical forms – a scheduled task, a hosted job – but the underlying idea stays the same. You are externalizing care to an agent.

Designing Small, Boring Pipelines

Effective automation pipelines are deliberately unambitious. They do not attempt to redesign the map. They do not make aesthetic decisions. They rerun known prompts – often keeper prompts – and verify that expected properties still hold. Common automated checks include:

- Does every layer still render?
- Is contrast above agreed baselines?
- Are credits and disclosures still present?

When one of these checks fails, the system won't fix the problem for you, but it will get your attention. It's a classic division of labor: automation detects problems, humans decide if it's worth fixing. Figure 17.3 shows a basic automat.

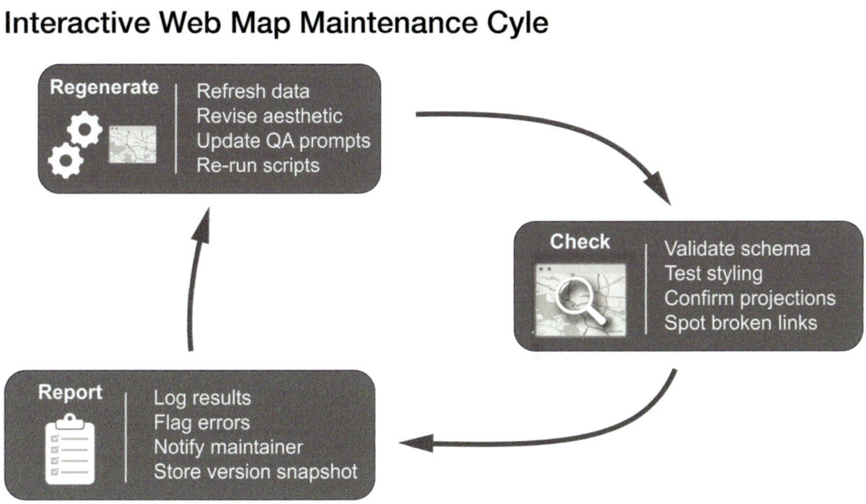

FIGURE 17.3
Annotated automation pipeline showing regenerate, check, report.

Automation Without Intimidation

Automation can sound intimidating, but rest assured, you aren't being asked to *master* automation tooling. The goal is to help you recognize when repetition should be handled by a system instead of a person. Even a weekly reminder that says "rerun the keeper prompt and glance at the output" might be described as a form of automation – automatic nagging. Start small; sophistication can come later. The goal is to develop the habit of automating things first.

Monitoring Drift and Version Integrity

Maps age like children: slowly at first … then seemingly all at once. A data provider renames a field, or an LLM update interprets a phrase a little more creatively than before. None of these changes are dramatic on their own, but they compound upon one another and can pull a map away from its original intent.

This is what I will call technical drift – different from aesthetic drift.

What Technical Drift Looks Like in Practice

Drift tends to show up in three familiar forms:

- Visual drift: colors warm or cool, fonts thicken, symbols subtly resize
- Data drift: fields disappear, units change, coordinate systems shift
- Semantic drift: the same prompt begins producing a different interpretation of the same request

The most frustrating thing about drift is that it often *looks intentional*. Nothing crashes. The map still loads and yet, one day, it *feels* all wrong. Monitoring exists to catch that feeling early, while it's still easy to correct.

Versioning for Map Integrity over Perfection

Monitoring drift does not mean freezing a map in time. Change and differences are expected, and honestly, sometimes change presents a welcome surprise.

What you're looking for in your versioning system are moments of *intentional change* as opposed to random or unintentional happenstances.

Map integrity means you can point to a specific moment in a pipeline and say, this is when we changed the classification scheme, this is when the palette shifted, or this is when a new model began interpreting the prompt differently. Without a versioning record, every change will look accidental, incidental – even when they weren't.

That's why monitoring pairs naturally with versioning. A map that drifts without a record loses credibility. A map that changes with documentation gains it.

Watching Entropy Happen Without Succumbing to Paranoia

Monitoring is the act of watching entropy run its course. Entropy, in a nutshell, is the law of physics that states all systems erode and become a proverbial mess. There is no escaping entropy in a closed system, so every map you create will suffer from entropy. Entropy isn't failure. You can expect it, if not still disdain it.

In other words, your prompts will produce different results at different times with the same exact datasets. Things won't always be exactly as you expect. I encourage you not to inspect every pixel after every update, or you will likely drive yourself mad. Instead, create a routine that tells you when something on your map may deserve a closer look. Perhaps create an agent that notifies you when it believes drift has crossed a certain threshold – visual, data, or semantic. Or set a personal nuisance threshold where you pause, investigate, and decide if entropy has gone too far. The goal is never control with LLMs and prompt cartography. It's contextual and cartographic awareness. You're the director, not the make-up artist. Let the make-up artist use a little more rouge on your proportional symbols when it needs to.

Human-in-the-Loop QA and Accessibility

Automation gets you consistency, but it won't give you judgment. Every automated pipeline needs a place where the human (i.e., you) re-enters – not to redo the work but to *interpret* it. At some point, someone (i.e., probably you) needs to look at the map and ask questions no script can answer:

- *Does this still make sense?*
- *Does it still feel clear?*
- *Would I trust this if I were seeing it for the first time?*

Automated checks are excellent at detecting violations: low contrast, missing layers, broken links. Humans are better at detecting mismatches – places where the map technically passes but conceptually stumbles.

Accessibility as Ongoing Care

Accessibility deserves special mention here because it is often treated as a one-time certification. In practice, it is ongoing maintenance.

As maps evolve, accessibility can quietly degrade. A new layer adds clutter. A color ramp becomes too subtle. Labels shrink to accommodate density. None of this is malicious. All of it accumulates.

This is where a human review matters most. Screen readers, keyboard navigation, and visual contrast checks can be automated – but interpretation cannot. Someone still needs to listen to the legend read aloud. Someone still needs to try navigating the map without a mouse. Someone still needs to ask whether the experience feels inclusive or merely compliant. Try not to treat accessibility like a box you need to check. Think of it as a condition of equilibrium your users want you to preserve.

The Rhythm of Review

Human-in-the-loop QA works best when it is routine rather than heroic. You are not swooping in to rescue a failing system. You are performing regular, light-touch reviews that keep small issues from becoming structurally debilitating ones.

Think in terms of rhythm:

- Automation runs frequently and quietly.
- Monitoring flags when something shifts.
- Humans review intentionally and briefly.

This rhythm respects attention. Your human time is limited and valuable. The whole point of having LLM agents design monitoring and run these things for you is so you can spend your time where it matters most – even if that means playing video games.

Maintenance as Creative Practice

Like college courses, prompt-generated maps are not finished objects. They are ongoing arrangements that evolve and shift based on new developments and different personalities in the classroom. In prompt cartography, changes to datasets, models, infrastructure, schemas, and audience will likely require that you maintain at least a few maps in your lifetime. Treating maintenance as an afterthought will not suffice, or if it does, it may make your life more difficult than it had to be.

Embracing the Change

Every update introduces a trade-off. A new dataset improves relevance but increases clutter. A refreshed basemap sharpens contrast but alters tone. An automated rebuild saves time but surfaces a design choice you didn't realize you were relying on.

None of these moments are failures. The creative work here is not invention. It is calibration – adjusting, pruning, and occasionally letting something go. You are not chasing perfection. You are maintaining coherence.

Shifting your mindset here matters. When maintenance is framed as design, your decisions will feel purposeful rather than reactive. You can stop asking *"Why is this breaking?"* and start asking *"What changed, and how should the map respond?"* I believe psychologists call it "conscious reframing." Whatever you decide to call it, asking questions from an alternative perspective can help you find better resolutions to your map's problems.

Stewardship over Control

Stewardship does not mean mastery. Again, I can't reiterate this enough; unlike previous types of cartography, you do not need to control every dependency, update, or aspect of your map. Micromanagement is sure to cause frustration. As director, you guide your team of agents (i.e., your own private agentic system) through change with attention and restraint.

This is why automation, monitoring, and human review can work well together. Automation handles repetitive tasks; monitoring provides alerts when something may be askew; your judgment decides when intervention is warranted.

Maintenance, in this sense, is not about preventing decay. It is about responding to it thoughtfully.

Maps age. Contexts shift. Audiences change. A steward accepts that entropy is part of the job.

Why This Matters Beyond the Map

The habits you build here extend beyond any single project – beyond prompt cartography. A cartographer who maintains maps responsibly leaves behind work that others can understand, inherit, and adapt from. Your prompts and conversations are archived like meeting notes. Decisions are documented and time stamped. Changes are linked to intention. The map becomes a documented contribution rather than an ephemeral piece of web litter.

I truly believe that prompt cartography presents us with an opportunity for professional maturity. Not because LLM tools are new – the internet was new 30 years ago, and we definitely aren't more mature now. Because cartography is and always has been about evolving with new technology. And LLMs offer us capable agents that we can direct and manage to produce a map rather than do all the grunt work ourselves. That includes *publishing and maintaining* our maps on the internet.

Takeaways

- **Publishing Begins a Map's Second Life**

  Once released, a map enters an environment of changing data, platforms, and expectations, which is why your design responsibility continues beyond launch.

- **Stewardship Depends on Continuity**

  Prompt-driven workflows benefit from pipeline routines that regenerate, review, and document changes so intent remains legible over time.

- **Automation Delegates Maintenance**

  When routine checks and updates are handled consistently, human attention stays available for interpretation, judgment, and decision-making (and outside hobbies).

- **Monitoring Preserves Alignment**

  Observing visual, data, and semantic shifts allows designers to respond thoughtfully as systems evolve, because … drift happens.
- **Human Review Sustains Meaning**

  Automated checks identify changes, but (at this point) only human judgment can evaluate whether a map still communicates clearly, fairly, and accessibly.
- **Maintenance Is a Creative Practice**

  Ongoing calibration, map pruning, and adjustment keep maps coherent as contexts shift, extending the life and relevance of design work.

Conceptual and Applied Exercises

A Map's Second Life

Here we focused on what happens after a map is released. Reflect on a map you have seen reused, repurposed, or reinterpreted beyond its original context. What aspects of its design enabled – or failed to prevent – that reuse?

Drift in the Wild

Design drift often becomes visible only after deployment. Identify one way a map's meaning could change over time without any alteration to its code or data. What early design decisions might have reduced that risk?

LLM Maintenance Probe

Ask an LLM:

```
Why is maintenance an important part of responsible
system design?
```

Evaluate whether its answer treats maintenance as technical upkeep or as ongoing stewardship. Which interpretation aligns most with what you've learned here?

Knowing What to Preserve

Versioning and keeper prompts privilege some decisions while letting others fade. Think about a design principle you would actively preserve across future versions of a map – and one you would deliberately allow to evolve. What distinguishes the two?

For extended tutorials, hands-on labs, prompt templates, and updated examples related to this chapter, visit www.promptcartography.com.

Part III

The Prompt Cartography Seminar

By now, the work of prompt cartography should feel slower – and more deliberate – than it did at the beginning of this book. Up to this point, I confess (and you've probably sensed) that this book has been a bit methodical, almost like it was building toward something. This was done by design, and now… Here we are.

You've learned to speak a new design language. Not just individual words like "hue" or "tooltip," but full sentences that coordinate color systems, typographic hierarchies, interactive behaviors, and ethical constraints simultaneously. You've practiced translating cartographic knowledge that once lived in your hands and eyes into explicit, promptable instructions.

This is much harder than it sounds. It requires a lot of thinking, learning, and (well) reading. In an era of instant gratification, I often wonder if this is one reason so many people haven't started trying prompt cartography earnestly yet (as of the time of this writing), even though the payoffs for mapmakers promise to be huge!

Traditional cartographers could rely on intuition, muscle memory, and iterative tinkering. Prompt cartographers must articulate intent clearly enough that a language model – which has no spatial intuition, no aesthetic judgment, and no ethical compass – can act as a capable collaborator. You've essentially been learning to direct and manage the expertise of a team rather than execute tasks yourself, which requires a fundamentally different way of thinking and results in a new type of cognitive load.

DOI: 10.1201/9781003797920-20

If Part II felt overwhelming at times, that's appropriate. You weren't just learning about how to prompt an interactive web map into existence; you were learning about the entire prompt cartography workflow from the ground up. Data discovery through semantic search, visual form through Dondis' framework, interactivity as behavioral vocabulary, critique as continuous conversation – each chapter added another layer to your directorial responsibilities.

But here's what matters: you've *completed* the foundation. Before moving forward, take a moment to acknowledge what you've built. You can now *frame datasets* like a producer, *declare style specifications* that resist drift, *design interactive behaviors* that align with user intent, and *critique outputs systematically* rather than defensively. These aren't small accomplishments. (They're huge!)

Unfortunately, mapmaking techniques can be learned via reading but applied learning typically doesn't happen in tidy chapters. Cartography is messy. Everything, everywhere, all at once… not only a *great* movie but a perfect metaphor for how the cartographic process often unfolds. In this part, we highlight that messiness and walk through how the pipeline and systems you've been introduced to can help you persevere through the decision-making maelstrom that is web map design.

This book concludes with a variety of end-to-end, seminar-like prompt cartography scenarios. The goal of these chapters is not to demonstrate a "right" way to build a map with prompts, but to provide you with the experience of seeing how and where the techniques and processes you've learned come into play, contextually. Each case begins with a simple natural-language intent and ends with a finished, published map – which can be found online on the companion website. In between, you'll see decisions, revisions, missteps, and course corrections – the actual moments that define cartographic practice.

In this book, you will find representative prompt structures and key decision points. Full code listings and periodically refreshed implementations live on the companion site. This is intentional on my part. I would like you to follow along conceptually and visit the website for updated codes and schemas as you begin doing prompt cartography yourself.

If we printed full prompt transcripts, setup instructions, and deployment scripts, this book would be hundreds of pages long – and outdated before it ever reached your desk. More importantly, it would obscure the most valuable takeaways from each chapter:

- how to approach problems,
- what to prioritize, and
- when to intervene as prompt director

One final note before we begin.

You are no longer learning *what* prompt cartography is. You understand and can explain this already. The next step is application – you are about to start practicing prompt cartography, graduate-seminar style.

The following case studies are meant to be read like you're in a class with other students. Internally, you should be questioning, expanding upon, and eventually surpassing the thought processes found therein. Whenever you find yourself wanting to change things – different datasets, different map tone, different design priorities – or even disagreeing… that's not a sign that you're wrong or don't understand something. It's a sign that you're thinking critically and pushing yourself. Like pain during exercise, it's a sign of progress; that you're nearing the peak of the Bloom's Taxonomy!

This is the moment we've been building up to… Grab your proverbial director's chair and a favorite beverage. It's time to go prompt some web maps into being!

18

Mapping Community Data with Accountability

Maps made for civic engagement are not my favorite. What makes them difficult isn't because they are technically more complex. It's the audience. Maps made for constituents or a diverse group of people are bound to irritate a certain percentage of your audience, because the audience will always look to see how they are represented in the map. They don't look at the dataset as a collection of fallible data points but rather look to find their neighborhoods, their homes, their way of life, daily routines, etc. When they don't like how they perceive their world being represented, or when your map presents a reality that threatens geographies they hold dear (e.g., proposed new condominium development in place of their beloved, though decrepit, primary school), they tend to get mad.

If you ever want proof that maps are never neutral artifacts but geographic arguments, go to a local planning meeting. Questions come quickly, and not all of them are about the data, current geography, or even local politics. Some are about fairness. Others are about intent. Some are about what the map seems to be saying, whether or not you meant it to. Often, members of the audience will accuse you of sinister negligence based on economic theories from the nineteenth and twentieth centuries (e.g., I've accused of being both a socialist and neoliberal ideologue in the same meeting by different audience members). There's a reason I ditched my career as a planner and opted for graduate school after six months. Since government and civic maps are some of the most difficult, we will begin here.

Civic Maps

The goal of what I will call civic maps – e.g., maps made by government, non-profit, or consultant agencies for constituents or planning purposes – is rarely to win arguments or settle debates. Typically, these maps are to inform public understanding while minimizing harm, misinterpretation, and accidental accusation. That is a harder design problem than it sounds. Civic datasets are often plentiful, but they are rarely neutral. Enforcement data, safety

DOI: 10.1201/9781003797920-21

records, service requests, infrastructure layers, new development proposals… all carry implicit judgments simply by being mapped.

Prompt cartography does not remove or mitigate any of these risks. In fact, by lowering the barrier to producing maps at a steady clip, it may even increase them. So in this first case study, I figure it makes sense to focus on accountability. Before style, before narrative, before orchestration, there is the simple question every map director must answer – explicitly or not:

Who is this for, and what happens when it is misunderstood? (Because it will be!)

Intent Under Scrutiny

Every map has intent, even when it claims not to. In civic contexts, that intent is rarely invisible to the audience. Residents, journalists, planners, and advocates are accustomed to reading between the lines. They know that data choices, boundaries, and classifications often stand in for values and priorities.

In this case study, the map's intent is slim. The map does not attempt to explain *why* patterns of enforcement and public risk vary across neighborhoods. It does not claim causality. It does not rank communities or identify "problem areas." Instead, it asks a more modest – but still consequential – question:

> Where do patterns of enforcement and public risk diverge across neighborhoods, and how can those differences be visualized responsibly?

That phrasing matters. It limits what the map is allowed to say. It also limits what it should *not* be asked to say. Prompt cartography is particularly useful here, because intent can be written down, revised, and tested before any symbols appear on the screen. When intent remains implicit, it tends to drift. When it is explicit, it can be defended.

In civic mapping, intent is rarely a single sentence. It is a set of constraints. In this civic scenario, those constraints include:

- Comparing neighborhoods without ranking them
- Normalizing data without pretending normalization removes bias
- Showing differences without assigning blame
- Designing for public interpretation, not expert analysis

These constraints shape every downstream decision. They determine which datasets are included, how they are aggregated, how they are symbolized, and – just as importantly – how the final map is framed in text.

One of the easiest mistakes in civic cartography is to assume that careful data preparation guarantees careful interpretation. Never! People do not perceive your maps as the workflow they came from. They encounter them as images and interfaces with their own biased personal social contexts. That is why intent must be designed into the map itself, not just found in your cartographic analysis behind the product.

On the plus side, prompt cartography makes your intent visible. By writing intent in natural language and revisiting it throughout the design process, you are forced to confront uncomfortable questions early. What might a hostile (i.e., someone who doesn't want their home bulldozed for a streetcar line) reader see here? What false assumptions might a well-meaning reader make (e.g., that the red line represents danger, not a new bike lane)? Where might the map be doing rhetorical work it was never meant to do (e.g., is it underselling the amount of green space in the community)?

None of these questions, or similar ones, has tidy answers. But asking them – explicitly, and before publishing – and trying to resolve as many as possible goes a long way toward differentiating your civic maps from a hodge-podge collection of civic datasets.

Choosing and Shaping Civic Data

Civic data has a way of feeling authoritative simply because it exists. Parking citations are issued by a city. Collision records are logged by police departments or transportation agencies. Neighborhood boundaries appear official enough, even when they are journalistic or historically contingent. None of that guarantees that these datasets deserve to be placed side by side on a map.

In the upcoming scenario, I'm consciously limiting the data choices you have available. The goal is not to capture everything that might matter in a city like Los Angeles, but to work with a small set of datasets that are legible, interpretable, and likely to provoke (hypothetical) questions rather than obvious conclusions. That constraint is part of the scenario.

The first layer – parking citations – represents enforcement. It is an imperfect proxy, but an instructive one. Enforcement data reflects policy priorities, patrol patterns, reporting practices, and human discretion. It is tempting to read it as a direct measure of behavior. It is not. Still, it is a dataset that residents recognize immediately, often viscerally, which makes it powerful and dangerous in equal measure.

The second layer – traffic collisions – represents public risk. Like enforcement data, it is shaped by reporting practices and access to infrastructure. Collisions occur everywhere, but they are recorded unevenly. Severity thresholds, reporting requirements, and human judgment all influence what ends up in the dataset. That uncertainty is not a flaw to be corrected; it is a condition to be acknowledged.

The third element is normalization. Raw counts are almost always misleading in civic contexts, particularly when working across neighborhoods of different sizes and densities. Normalization does not fix that problem, but it does change the question being asked. A per-capita or per-household rate invites comparison, *but* it also introduces assumptions. We need to be careful, because rates invite comparisons, and unchecked (and unintended) comparisons in civic cartography can lead you down a rabbit hole to hell. Remember: people frequently look at areas of impact on their lives and compare from there. Thus, guiding them to compare what you want them to in your map is not as easy as when creating a map with comparisons of far-off places.

If there is one recurring theme in civic mapping it is this: restraint is often more honest than completeness.

So at this state, with our datasets determined, we can begin to critique what and how we want to show this data. Instead of rushing toward symbolization, which is easy to do, we can articulate to our agents – in plain language – what we want each dataset to represent, and what we want to avoid. This articulation, or clear directive, will act as a brake. It slows the cartographic process just enough for us to ensure that wayward assumptions don't accidentally slip through.

Design Decisions with Consequences

Once civic data is aggregated and normalized, the temptation is to make it legible at all costs. Clear colors with sharp contrasts, perhaps some rankings, and a legend that makes it all feel official – i.e., decisive. Sure, many such choices would be appropriate in analytic settings, if you're exploring the data in an attempt to create new knowledge. However, given your target audience and the situation, they may not all be appropriate in a civic setting.

In this case study, the design goal is clarity without emphasis. Yes, our map should be readable and discernible, but it shouldn't pejorative or segregate. It should invite areal comparison across the city but not suggest a judgment of those areas – apples to oranges, not first, second, third. That is a narrow design space, and it requires saying no to many familiar cartographic techniques.

Ranking neighborhoods, for example, is to be *deliberately* avoided. Ordered lists and extreme values are rhetorically powerful, but they also flatten context and invite stigma. Similarly, it's wise to use strong sequential color ramps sparingly – if at all. Color implies intensity, and intensity implies urgency, and from urgency it is a small step to emergency (at least as far as rooms in hospitals go). In civic mapping, urgency should be real, not a default aspect of your preferred symbolization technique.

Given this… perhaps side-by-side views should be favored over layer overlays. This keeps "enforcement" and "risk" visually and causally distinct, while they retain their conceptual relationship. The reader may notice divergence or connections, but won't be misguided into interpreting it as statistically significant. These little decisions of distinction matter a lot. They can shift any map, not just civic ones, from an argument to an invitation – i.e., toward reconciliation.

Titles and captions can do more work here than they might in other contexts. Use them well and they will explicitly limit what the map claims to show – otherwise known as covering one's ____. They will remind the reader things like these are "rates," not "totals," or "observations," not "significant findings." This is exactly what I meant earlier in the book when discussing "responsible framing."

One of the more uncomfortable realizations in civic cartography is that good spatial information design does not prevent misuse. A map can be carefully designed by you, or anyone, and still be misread, excerpted, or weaponized by an online influencer with two-million-plus followers that become incensed. The goal, then, is not to eliminate risk – which is impossible, as every map is a risk – but to reduce the probability of risk – i.e., make harmful interpretations less likely and easier to challenge.

Ironically, prompt cartography kind of supports this more than any other type of cartography because it literally treats language as part of the design system, rather than an afterthought. Titles, legends, and explanatory text are drafted alongside symbol choices, not appended at the end. When those elements drift out of alignment, the map becomes rhetorically unstable, and you have the skills now to notice this and iterate.

At this point in the process, the map is not finished, and that is intentional. Civic maps, like book manuscripts I've discovered, benefit from pauses. (Though, perhaps not one-year pauses, as I took while drafting this revision. Sorry, editor!) They benefit from being shown to colleagues, stakeholders, bosses, and even skeptical readers from the other side of the proverbial political aisle before production. Prompt-based critique – asking a language model to describe what the map appears to be saying from a certain roleplayed perspective, or to identify possible misinterpretations from constituents – may raise issues that traditional cartographic checks will not. (Isn't nice having innumerable agents to direct and ask for help?!)

Just remember: design decisions in public (much less civic) contexts are never neutral. The best one can do is make them visible, defensible, and revisable. Also, don't forget to run your map through a gauntlet of roleplaying agents first – and ask it to come up with strong counter arguments to any issues it finds. You'll be prepared and thankful you did.

Setting Up Interpretation without Accusation

In my experience, at some point every civic map reaches an uncomfortable threshold. For example, geographic patterns become visible enough to invite interpretation by the public, but not necessarily robust enough to justify explanation. This is the moment where many civic maps accidentally overstep – not because what the datasets are showing is wrong, but because your audience is human. They have feelings.

In this case study, let's assume this threshold arrives as soon as enforcement and public risk are viewed side by side. It will be difficult not to ask *why* certain neighborhoods appear one way and not another. It will be even more difficult to prevent those questions from drifting toward blame, whether aimed at residents, policymakers, or institutions.

Like the ending of a good art-house drama, you won't find any answer to those questions here. That choice may feel unsatisfying, especially to readers trained to treat maps as analytical tools. I get it but… civic cartography often requires a different kind of discipline. The ethical risk here is not ignorance; it is overconfidence. When a map implies causality it cannot support, it does more than misinform – it reshapes public perception in ways that are hard to undo.

One useful framing is to think in terms of interpretive distance. How far does the map allow the reader to travel from what is observed to what is inferred? In this design, that distance is *not* accidentally short. The map allows comparison and the user to explore where their curiosity takes them. It does not allow for diagnosis, but honestly, contrary to what amateur online true-crime sleuths believe, the goal of your map is rarely to let members of the public diagnose society's most complex problems. If it were, you may end up with flat-earthers in charge or… people who want Greenland because it looks big on a map.

Prompt cartography makes this ethical pause easier to formalize. By explicitly asking what the map should *not* be used for, the cartographer introduces a powerful constraint. Language becomes a way to draw boundaries – not spatial ones, which your map product will do well enough, but *rhetorical* boundaries.

This is also where omission becomes a legitimate design choice. (Omission is *not* a sin in cartography!) Variables that might strengthen an explanatory argument are surreptitiously excluded, not because they are unimportant, but because their inclusion would suggest a level of causal certainty the map cannot defend. Leaving them out is not a failure of analysis; it is an act of restraint.

Ethical civic mapping is rarely about finding the perfect dataset. More often, it is about knowing when to stop adding layers.

Publishing and Knowing When Not To

Publishing a civic map is not the end of the design process. In many ways, it is the beginning of a different one.

Once a map enters the public sphere, it acquires a life of its own. It may be shared without context, excerpted without captions, or interpreted by audiences the cartographer never anticipated. Those realities do not invalidate civic mapping, but they do change the responsibility attached to it.

In this case study, publication is treated as a governance decision rather than a technical milestone. The question is not simply *is the map finished?* but *is the map ready to be read without us in the room?* That distinction matters.

Prompt cartography supports this shift by encouraging designers to simulate public interpretation before publishing. Asking a language model to summarize what the map appears to show, or to identify possible misreadings, can surface issues that internal review might miss. These exercises do not replace human judgment, but they do provide a low-cost way to stress-test framing choices.

Sometimes, the outcome of that process is revision. Other times, it is deferral. Civic cartography includes the option to not publish – or to publish in a more limited context – without treating that decision as a failure. A map that generates more heat than light may be better served as an internal exploratory tool than as a public artifact.

There is no single, "correct" publication strategy to present here. Rather, the moral of this is that publication is part of design, not something that happens after design is complete. Decisions about audience, access, and longevity are inseparable from symbolization and data choice. Knowing when *not* to publish is, in many cases, just as important as knowing how – even though the decision can be difficult.

Revisiting What You Just Applied

What was the goal of this chapter, you might ask? It wasn't a walkthrough or a step-by-step guide to anything. I want to learn how to do prompt cartography. I want specifics!

I would like my investments to return 30% annually on a consistent basis too, but no financial advisor has given me "the specifics" to achieve this feat year-after-year yet either. My point being: a step-by-step tutorial is not going to help you at this point. You have all the knowledge to explore on your own and make your own design decisions. The way you will grow is through practice, experience, and regularly thinking of hypothetical scenarios.

Case in point: you just practiced designing for accountability over optimization. You worked through a scenario using data that invites interpretation while deliberately limiting what the map claims to explain. You treated boundaries, normalization, and aggregation as rhetorical decisions rather than neutral steps. And you encountered the reality that careful design does not eliminate ethical risk – it manages it.

Perhaps most importantly, you observed how to make intent explicit (CRAFT). By articulating what the map is for, *and what it is not for*, you created a framework that could be defended, revised, and communicated to others. That framework and acquired knowledge matter more than any particular prompting for symbol choice in today's large language models (LLMs).

This civic prompt cartography exercise represents an uncomfortable, complex mapping scenario. It requires that designers slow down, anticipate disagreement, and accept that some questions should remain open – unanswerable. Prompt cartography does not make that work easier, but it does make it more visible – and therefore more accountable. You learned how to cover your proverbial…

The scenarios that follow will relax some of these constraints and tighten others. Each offers a scenario with different demands. But the habit you are honing here – of treating language as a core design tool rather than a finishing touch – is timeless.

19

Mapping Presence, Absence, and Uncertainty

The maps in the previous situation dealt directly with human geography – i.e., enforcement, infrastructure, public risk. Those systems are imperfect, political, and uneven, but they are also designed, administered, controlled, and documented by institutions. Someone is responsible for them, even when that responsibility is contested.

Ecological or environmental maps present us with a different set of challenges. Here, the thing being mapped does not belong to anyone. There is no issuing agency for wildlife presence, no comprehensive reporting system for ecological absence, and no agreed-upon threshold for what counts as "enough data." What we observe is shaped as much by where people go, what they notice, and what they choose to record as by the underlying ecological processes themselves.

Chapter 5 reviewed how to design maps responsibly when data carries human consequence; here we ask how to design maps honestly when uncertainty is *the* defining feature. The challenge here is no longer avoiding accusation or stigma but avoiding false confidence. In ecological cartography, the greatest risk is not misinterpretation – it's *overinterpretation*.

At first glance, observational ecological data can feel refreshingly straightforward. Points appear where sightings occur. Patterns seem to emerge. But *obvious clarity* is often *false clarity*. Every ecological or environmental observation reflects a human moment: someone was there, at that time, with the ability and motivation to notice and report something; or if not a human moment, a socially constructed moment via remotely sensed data. (Human-produced tools are merely extensions of our own biases and belief systems.) The datasets are placed on a map that records data from a particular moment, not a monitored organism itself or reality or truth or anything of the sort. In the forthcoming scenario, we are going to focus on the fallibility of data and the everyday danger of maps becoming substitutes for reality in the minds of map users – and, just as dangerously, map creators.

As we proceed, it's important that you remind yourself of the objective. Rather than attempting to infer or argue for an ecological truth with the observational data, your goal is to design a map that communicates *what was observed*, *where*, and *under what conditions*. This is tougher than it sounds. Can you design a map about ecology in a manner that averts your audience, and

DOI: 10.1201/9781003797920-22

yourself, from collapsing the dataset into claims about presence, abundance, or habitat suitability? This is a much more difficult goal than that of most ecological maps – which are typically designed to make an explicit rhetorical argument that conflates the data with reality. You can do this!

What Observations Can and Cannot Say

In outlier scenarios like this, it's often best to begin with constraints – what we don't want to do with this map. Outlining limitations right away will filter our options moving forward, providing direction like solid stripes on a highway in a no-passing zone.

This map *does not* aim to model species distribution. It *will not* attempt to predict future presence. It *does not* claim that areas without observations are ecologically empty, degraded, or irrelevant. Instead, you will ask a simpler, but more honest, question:

> *What does observational data reveal about where humans have ecological encounters – and what might be obscured or missed by focusing on the encounters themselves?*

That framing shifts the role of the map. It is no longer a proxy for ecological reality but instead a scientific investigation of interaction between people and ecosystem. It's also an important framing, because it changes how uncertainty should be treated here. Uncertainty here is *not* something to be reduced through better analysis, the layering of multiple datasets over time, or statistical analysis. Rather, *uncertainty is the message*. Our goal is to communicate this message through design.

Overtly Highlighting Dataset Bias

Our primary dataset will consist of wildlife observations reported through a citizen science platform. Such records are rich in detail and broad in coverage, but they are also dishearteningly uneven. Observations might cluster near roads, trails, parks, and (always) more populated or visited areas. Observations tend to thin out where human access is limited, even when ecological presence may be high. Or counterintuitively, they may cluster in the limited natural areas that humans frequently visit and be totally missed in natural areas that humans don't visit in masses. Such patterns are *not* errors; they *are* the signal in the noise.

The intent of the map, then, is not to correct or even account for this known bias but to make it inherently visible within the map's meaning. Prompt cartography supports this by encouraging the cartographer to articulate, in advance, what the map *will not* attempt to resolve. By naming those boundaries early, the design process resists the familiar slide toward explanation.

This is where ecological cartography diverges sharply from civic mapping. In the last scenario, the ethical risk was misattribution – assigning blame or responsibility where none could be supported. The risk for us, and our hypothetical map users, is projection – reading ecological significance into patterns that reflect human behavior more than ecological systems.

Designing against that risk requires a different posture. It requires comfort with ambiguity, a willingness to leave questions unanswered, and a design language that avoids visual cues associated with intensity, danger, or dominance. The map should feel observational, not diagnostic.

As with the civic case study, intent is not a single sentence. It is a set of guardrails that shape every downstream choice. In ecological cartography, those guardrails are often about subtraction rather than addition: fewer claims, fewer colors, fewer analytical moves. What remains is not less informative; it is simply more careful.

Choosing Observational Data

Observational ecological data is appealing precisely because it feels direct. Someone saw something. They recorded it. The record exists. Compared to civic datasets that pass through layers of institutional mediation, to me at least, observational data feels refreshingly close to the world it describes. (But it's not!)

In this case study, the primary dataset consists of wildlife observations reported through a citizen science platform. (Please refer to the companion website for ideas and links to dataset sources.) I love animal sighting datasets, because they are so rich and personal. Not only do many share a lot of contextual data – species names, dates, locations, and sometimes even really cool, amateur photographs of the animals themselves – but each sighting is tied to a real human experience: "Hiking with my son when an 18-point buck suddenly stepped onto the trail in front of us." Each feature in a sightings dataset reflects a unique, though specific, kind of human interaction with the environment – one shaped by access and attention.

Attention matters a great deal. How many potential ecological encounters were observed by animals but completely missed by humans – likely staring at their phones? Personally, I'm horribly inobservant. Last year, I literally walked into a deer in my yard – i.e., okay, it was about two meters away. I was looking up at the pine trees, assessing their health, not observing where I was going. Between the deer's loud grunt and my ear-piercing shriek, I'll never be sure who was traumatized more. What is observed is not simply what is present; it is what is cognitively *encountered*.

That distinction matters. Observations cluster along trails, near roads, around parks, and in places where people feel safe and welcome spending time. Observations are likely to thin out in private lands, rugged terrain, and areas that are ecologically significant but socially inaccessible. I'm going to speculate here, but it's likely one will find more wildlife observations where cellphone reception is nonexistent, not just because the place must be remote but because humans won't be distracted.

Choosing to work with this kind of dataset is, therefore, a commitment. It means accepting that the map will reflect human movement and attention as much as the reality of ecological systems. Rather than compensating for that bias through modeling or correction, it's often better to treat it as part of the story the map tells.

Prompt cartography supports this approach by making the data limits explicit before design decisions are made. Instead of asking how to improve coverage, the cartographer asks how to *signal* uneven coverage. Instead of smoothing variation, the cartographer asks what variation might actually represent.

It's a subtle shift but essential. Your data selection is *always* a design choice; it's never a prerequisite. You can make maps without standardized or official datasets – you just make up your own data from memory or … make it up. (Some famous cartographers have done this throughout history, in fact.) As discussed earlier, datasets are never just there – you create them or find them, you curate and clean them, and then *you* – and *only* you, as director of the map production – decide to visualize that dataset. Thus, no dataset is a neutral input. It is a design decision leading to the next step in your pipeline.

Designing for Uncertainty

Designing ecological maps often involves a quiet struggle against visual habits. Many cartographic techniques are optimized for clarity, emphasis, and comparison. (Again, if you recall Chapter 8, this emphasis is why many cartographers come up short when it comes to overall aesthetic form.) These optimizations work well when the goal is to highlight patterns decisively. They are less suited in situations where decisiveness itself can be misleading. (The old: it's not black and white, it's gray, issue.)

In this case study, uncertainty is not treated as a secondary annotation or a footnote. It is a primary design constraint. The map must communicate that observations are partial, uneven, and contingent without turning uncertainty into visual noise.

Symbolizing Less to Show More

One of the first decisions, then, is what *not* to use. Density surfaces, heat maps, and interpolated fields should probably be avoided. Why? Even though technically defensible, these techniques imply continuity and completeness. Animal ranges are continuous but human observations – these are discrete and incomplete. Using a continuous visualization technique will encourage your readers to see animal hot spots and dearth zones that your dataset simply can't support.

An alternative approach may be to have your map favor sparsity over density. Individual observations remain visible as points, with opacity and symbol sizes used conservatively to avoid encoding magnitude as relevant. What if you try minimizing the magnitude of sightings in the visual hierarchy and instead let the gaps in the dataset speak more loudly. I don't have your dataset in front of me, but I imagine the result will be a map that feels quieter and raises a different set of questions – why are there no sightings in those areas? What might be going unnoticed in these spaces?

Contextual and Base Map Layers

Next, let's look at contextual layers. Does your base map selection include land cover or protected areas? What questions do, or might, the inclusion of these elements raise for the map user? Do they adequately provide environmental and location references without becoming subliminal explanatory devices for the observations themselves? The goal of the map is not to suggest that certain land cover types *cause* these observations to occur. Is it adequately situating these observations within a broader landscape without pushing for causation?

Map Elements and Interactivity

Legends and captions can do a lot of useful work here – i.e., remind your user that absence of evidence is not evidence of absence. You can design your legend to emphasize observations rather than presence, scarcity over abundance. You can use language in the title, subtitle, labels, and pop-ups that purposefully disengages the dataset from showing true coverage, completeness, or intensity. In fact, you can design all the elements to support and guide map users into alignment – align their expectations with what the data can support and reinforce what the map shows.

The design posture here is noticeably lighter than in the last scenario. There is less emphasis on comparison and more emphasis on pattern recognition without resolution. It's rare in life that things truly get resolved – at least until death, I suppose, though I have no way of confirming that yet. Remember: *your maps don't have to resolve anything*. I can't think of a good map that ever has. The best maps, like the best narrative art forms, tend to whet the appetite of viewers and purposefully leave them hanging (which is why so many people are *still* upset that the *Firefly* television series was canceled). Here, your map's goal is to invite curiosity, but it is under no obligation to provide certainty.

Prompt cartography helps you maintain that posture, fortunately, by keeping design language in view throughout the process. When you are creating this map and you see your symbol choices drifting toward emphasis, it's a good time to revisit your written intent prompt or have a large language model (LLM) agent revisit it and report back about where the map is drifting. You have agents that can help provide correctives to other agents. Discussions with your agents and the use of language can actually be used as a stabilizing force, preventing the map from gradually becoming more assertive than the data allows.

It's crucial I note, however, that designing for uncertainty doesn't mean you intentionally design vague or less-detailed maps. No, no, no! It means you design maps that are openly specific about what they *do not, or cannot, claim*. I'm not encouraging you to make a map that shows nothing. (What's the point of that?!) Rather, you are practicing designs that specify what your map doesn't show. That specificity is one of the hardest skills to master in cartography, and perhaps in the sciences broadly – and one easily overlooked in a world full of big GIS datasets. With practice, it becomes second nature.

Temporal Thinking Without Prediction

When observational data spans multiple years, it invites questions about trends, change, and movement. Looking at temporal-spatial data makes it easy to start asking questions about whether a species is expanding, retreating, or responding to environmental pressures, for example. Those questions are not unreasonable, but because data collection methods, sampling techniques, sample sizes, and reliability change dramatically over time, they are typically not as easy to answer as looking at some datasets may imply.

In the case study here, temporal information is treated descriptively rather than analytically. Observation dates are used to provide context, but you don't want to suggest trajectories. Your map can show when observations were recorded, but you don't want to attempt to explain why they occurred at those times, much less imply the timeliness of future sightings.

Temporal patterns in observational data often reflect changes in reporting behavior as much as changes in ecological systems. New apps, increased participation, seasonal recreation, and cultural events can all influence when and where observations are recorded. Without careful framing, a map can easily suggest ecological dynamics that are more likely social ones. Case in point, I grew up in northern Minnesota, with a continental climate comparable to Siberia's. I observed lots of wildlife in the woods ten months of the year. However, when it dipped to around −30 degree Celsius in January and February, I didn't observe much. I was inside next to the space heater, fighting with my siblings for a round of *Pole Position* on the family Commodore 64.

Design choices reinforce this restraint. Time is represented in broad bins rather than continuous scales. Animations, when used at all, emphasize recurrence rather than progression. The map does not move toward a climax. It simply accumulates.

Prompt cartography supports this approach by making predictive language easier to identify and remove. When captions or annotations begin to drift toward implication – *increasing, declining, shifting* – they can be revised back toward observation. The result is not a weaker map but a more honest one – one where temporal thinking is contextual, not predictive.

The Ethics of Mapping Uncertainty

Ecological and environmental cartography often carries with it, or at least implies, an implicit (and I would argue often false) promise: with enough data, uncertainty will eventually disappear and clarity will become self-evident about [insert your preferred physical geography belief here]. That assumption was purposefully challenged here to instead embrace and endorse the uncertainty found in many ecological and environmental datasets. Remotely sensed and human-observed data is not real – it is a biased and always tainted sample of reality. (Note: this is obviously true of all data, not just environmental and ecological datasets. However, I feel that this is most frequently overlooked when it comes to environmental and sustainability datasets. Understandably, I suppose. The data is often scary and does suggest inconvenient truths. But just because environmental datasets are from "authoritative" corporate, nonprofit, and government organizations, or collected using the latest and greatest remote-sensing technologies, doesn't make them immune to bias and uncertainty.)

The ethical risk in such contexts is subtle. It lies not in overt claims but in the quiet authority of a well-designed map using a "scientific" dataset. Even without explanatory text, visual patterns invite interpretation. Clusters feel meaningful. Gaps feel significant. The human mind is remarkably good at constructing stories from spatial cues. You can use maps to induce such story construction – great for arguing. But if you are trying to present evidence and data without rage baiting, hopefully, this scenario has taught you about the power of underselling your dataset by taking a moment to ask a simple question:

> *What stories might this map encourage, and am I prepared to stand behind and support them?*

This is not a call for paralysis. Ecological and environmental maps can and should be made. I've been arguing for years that they should be more rhetorical, not less. But maps should always be made with an awareness of how easily observation can be mistaken for explanation. (Cartographers are not immune to this either, as I noted. Perhaps we are more susceptible to this, as we become engrossed with our datasets and the map production itself.) Prompt cartography can help here. It won't ever resolve uncertainty; indeed, it raises new types, but it can help us build guardrails early in the process and keep us on task by embedding the goal of the map in language and outlining our intent early on.

What You Just Practiced

If you feel a little whiplashed after going from a civic mapping situation to this one, it's okay. That was the goal. This scenario required a different type of direction from the first one – like going from directing a science fiction action film to a family musical. No two maps are alike, though, so diverse exposure is good for learning. (At least, that was the thought when I started outlining these final situations to work through.)

Instead of accountability, you practiced humility. Instead of explanation, you practiced description. Instead of completeness, you practiced restraint. You worked with data that reflects human attention as much as ecological systems, and you designed a map that does not pretend otherwise.

You practiced treating uncertainty as a design constraint rather than a problem to be solved. You learned to recognize when familiar cartographic techniques – smoothing, interpolation, emphasis – might overstate what the data can support. And you saw how language, when used deliberately, can help keep a map aligned with its intent.

Most importantly, you practiced being *precise about limits*. You defined what the map could say, and you resisted the urge to let it say more. (You've learned your lesson now. You're officially unleashed from this uncomfortable position again.) The scenarios that follow will largely relax this constraint, but they will tighten others. As you will see, the cultural cartography scenario will reintroduce subjectivity. The speculative cartography scenario (a personal favorite of mine) will raise the stakes of interpretation. The last scenario is an integrated amalgamation of all the others. There, we will see if you can maintain multiple cartographic postures at once.

But the habit developed here – the willingness to sit with uncertainty and design honestly within it – will hopefully carry forward. You may need to channel this skill at different points in your cartographic career. Hopefully, now you have a better appreciation of why and how.

20

The Main Feature Presentation

Visualizing Complex Connections

If there is one lesson that keeps arising, I hope it is this: maps do not merely show information – they *perform* it. Every map is a production. Every map invites inference and leaves an imprint on viewers' and users' belief systems. Your maps suggest relationships, whether or not you intended them to, and here's the kicker: you will never control how they are interpreted, merely (when designed well) guide map users toward your preferred interpretations. Here, in the final scenario, we are going to highlight how dataset integration can raise the difficulty of guiding user interpretation dramatically.

The moment you combine any 2, 3, 4, 50 datasets, you imply that they belong together. As the director, you automatically imply compatibility and comparability. You are implying the datasets are relevant not just on their own but together. An analogy: actors sharing a scene in a movie are associated in the viewer's mind. Datasets together are linked in your map user's mind – regardless of how carefully you explain that they arrived there by different paths.

In this scenario, we focus on this issue. Not the technical challenge of connecting data sources, but the cartographic challenge of deciding which connections are defensible. Orchestration among datasets is not about showing everything; it's about deciding what belongs together. That is the professional discipline we have been building toward across all of these scenarios. Civic mapping addressed dataset consequence; ecological mapping embracing uncertainty; cultural mapping framing of limitations; speculative mapping the concept of persuasion sans explanation. Now we combine all those lessons in a more complex scenario – and we direct (the heck) out of it.

From Prompting Maps to Managing Systems

Throughout cartographic history, maps have been looked at as artifacts. They were composed, printed, published, and – at least conceptually – finished. Even interactive maps often retained this logic. They might allow panning,

 DOI: 10.1201/9781003797920-23

zooming, or filtering, but the underlying design assumptions have largely been perceived as fixed.

But dominant model is starting to wither. Many contemporary maps function less like finished compositions and more like complicated production systems – e.g., film production. They are assembled from live data sources. They support multiple views at once. They require the input of innumerable experts (some human, some large language model [LLM] agents, some Generative AI tools). They invite ongoing and even collaborative interaction across platforms, datasets, and LLMs.

As already reviewed in this book, this shift is changing the nature of cartographic responsibility. When a map becomes a system, design decisions no longer stop at symbolization. They extend to direction and production decisions: which datasets are allowed into the system, how are they allowed to interact, and under what conditions are they presented together? Such decisions are never trivial or neutral. They shape interpretation long before a user sees, clicks, or prompts anything.

Production Does Not Equal Optimization

It is tempting to think of map production as an engineering problem – particularly given how the last 20 years or so have seen cartography become as much about software engineering and database management as map design. How do we connect more data sources, for example? Or how do we make our data sources update automatically? Do we let users toggle layers, filter attributes, and explore combinations, and if so… how?

Production, in a cartographic sense, is closer to directing than optimizing. (A-ha! The *Prompt Director* comes storming back in the final act of the book! Naturally.) A Hollywood director does not simply put all actors on screen at once and let the audience sort it out. They decide who appears together, in what order, under what lighting, and to what effect. They understand that set proximity creates meaning, even without dialogue.

Here, the map is your medium. It is your set. You must direct the myriad set pieces and actors (the system) the same way.

For example, if civic infrastructure data appears alongside environmental indicators on a map, it implies a relationship by default. (The proverbial first law of geography – close things are related.) If you have demographic data and it is placed side by side or overlapping a public risk dataset, it pleads for map readers to interpret the two as intrinsically linked. Fine if that's what you want to narrate or shepherd your use into seeing, but it's on you to ensure the dataset allows for this. On the other hand, leaving out speculative data automatically communicates sound and ethical judgment just as clearly as inclusion does.

Directing a cartographic system means embracing this reality rather than pretending it does not exist. It means acknowledging that design is not just about enabling exploration, but about shaping the map space in which this exploration occurs.

Deciding What Belongs Together

In this scenario, I'm asking you (and this is weird, because I don't necessarily know who you are, but if you've read this far, I presume I'd enjoy hanging out and talking maps with you) to find a variety of datasets from different realms to create a more complex map than we have so far. The goal of this map scenario is to produce a map that will prove useful for decision-making. I leave it up to you to find a context or scenario requiring decision-making – government, university, family-trip planning…? If you can't come up with anything, you can find some ideas on the prompt cartography companion site under labs.

Requirements

Instead of one or two datasets, please create a project with at least three different datasets, preferably one ecological or environmental, one social- or civic-based, and as well as one or two others. Hopefully, you have a project you've wanted to work on for some time or dream map you've been pining to create. If not, I'll have some ideas posted on the prompt cartography companion site.

This is your prompt cartography capstone, so to speak. No online fees. Once you finish this, you can explore, expand upon, and evolve from these concepts along with LLM and Generative AI technologies for the rest of your career. As I mentioned, this book is a fledgling start – not a destination. I encourage you to share your capstone results on social media with the hashtag #promptcartography or similar. I'm excited to see what you and others produce in the near and distant future.

No worries, I'm not about to put you through the wringer with everything you've learned in the previous scenarios. I will ask, though that you exclude cinematic or fiction-based datasets – not because I think they lack value, obviously, but because their narrative and data fuzziness typically doesn't crossover well with real-world datasets for orchestration purposes. Even more firmly, no offense to my uncle, I encourage you to exclude speculative datasets as well, because their inclusion implies decision-making that can't be supported. (Oh look: there are 42 UFO sightings near dolphin pods. Maybe Douglas Adams was onto something?)

Again, such exclusions are not technical limitations. They are sound cartographic judgments. What belongs together is not determined by compatibility alone. It is determined by consequence. A system that allows certain datasets to coexist implicitly endorses their relationship. Once that endorsement is visible, it is difficult to retract. This is why orchestration cannot be left to default settings, user choice alone, or technical convenience. You need to direct.

Designing for Separation Without Isolation

One of the central challenges of this project for you is to avoid producing false coherence *and* avoid creating a map with dataset silos. Dataset silos are produced when a map includes numerous datasets, but they are so compartmentalized and unintuitively integrated that they might as well exist on their own. They have virtually no crossover impact or meaning with the other datasets being shown on a map.

To achieve this, you'll want to think about how your datasets can be visualized in coordination with one another, rather than fused, amalgamated, or completely isolated. Perhaps you present related datasets in parallel views rather than as a stack of layers. Or maybe you can tweak some other aspects of visual alignment to support comparison without forcing an artificial synthesis of information.

Use your agents to get ideas and help. Prompt cartography assists here by making these structural needs explicit. You can create a schema for your agents to follow. But don't be afraid to ask your agents for help and ideas. You are *not* on your lonesome. Use your prompting skills to articulate ground rules, limitations, generate map components, but at first, if you are stymied or hit a creative wall, use them to ask for ideas. Prompt cartography works best when you collaborate with your agents. Direct, don't command.

Responsible Map Design over Accurate Map Design

How do you know if you passed? Does the map have to look spectacular? Should it be bug and error free? Should the color system be both aesthetically appropriate and accessible?

Honestly, I don't care about any of that. You should, if you want to become a better cartographer, of course. But presumably that's why you read this book and you are actually doing the exercises and labs – to get better at web cartography via prompting broadly.

The goal of this final capstone scenario is not to produce an optimized and accurate web map. (Though, I hope you have.) It is whether the system directs user interpretation responsibly. *Does your map make its limits visible? Does it prevent inappropriate dataset combinations? Does it resist the temptation to imply completeness where it's not warranted?*

Directing a web map means accepting that neutrality is not an option. Perfection is a mirage. Every possible allowed interactive and dataset combination is a directorial choice – or failure if unintended. Excluded datasets are also statements – based on intent, audience, and editorial decision-making. Your role as prompt cartographer shifts from map author and creator to collaborative steward – or, more accurately, to map director.

I Finished the Capstone... What Now?

This scenario, which required you to complete a capstone lab to get the most use from, was a practice in synthesis. You brought together multiple cartographic postures without flattening them into a single voice. You treated integration as an ethical act rather than a technical achievement. You practiced deciding what belongs together – and what does not.

You learned that directing a map into a being via a team of agents is different from designing a map in the traditional sense. It's not easy. It's not one-off prompting. It requires attention to language, discernment about intent, judgment about datasets, restraint in claims, and a willingness to say "no," even when all the agents at your disposal make saying "yes" and "give me more, faster" effortless.

Taken together, the previous scenarios and this capstone ride-along scenario have shown that prompt cartography is *not a shortcut* to certainty. It's not a half-assed and fast method to mapmaking. It is a framework for making intent explicit, managing consequence, and directing meaning responsibly in a world where maps are no longer static artifacts. It's a new technique for web cartography that depends on how humans first conceptualize space, geography, and their environments – via natural language.

That work does not end with this book. If anything, it is just beginning. It's a great time to be a cartographer.

Conclusion

We are in the very infantile days of an unpredictable but huge technological revolution.

The other day, I was reminded of my first experience using a modem. For Christmas, I received a 1200 baud modem that I could use on my Commodore 64C. There were two phone numbers in my entire city that I could call with the modem. (The two lines were busy a lot! It took patience.) Once I was logged into the bulletin board system (BBS), I could type messages for my one other friend in the entire city that had a modem. Then I'd log off, play some Frogger, and call the BBS later to see if he or anyone else had responded. This was absolutely revolutionary back in 1988! And at the time, I would not have fathomed how networked computers were going to evolve.

If there's one thing I hope has come through in this book, it's that prompt cartography is very much in its early days. That's not a disclaimer. It's an observation – and a familiar one.

Anyone who has been around long enough to remember early web cartography, early internet mapping, or any of the myriad geovisualization endeavors that have come and gone will recognize the pattern.

The terminology starts off awkward. In hindsight, the tools that feel cutting-edge at the time are always clunky. The interfaces feel magical one moment and ridiculous months later. And yet, embedded in this messiness are ideas, concepts, and technologies that continually evolve from their origins and never go away.

Large language models (LLMs), AI tools, and agent-based systems as they exist today will look as comical and archaic as AOL, Gopher, and MySpace do today as I write this book – in much less time. Or if you want to go even further back, as antiquated as Windows 3.1, the Commodore 64, Atari console, and punch-card programming.

This book will show its age more quickly than I would like, given how much effort I've put into it. I started writing this well aware of this fact. The terminology will change. The models will improve in ways that are impossible to fathom right now. The prompts that feel elegant today will look like PASCAL in the future.

Indeed, I personally imagine that maps themselves – what we mean by map and how they are used – will change dramatically. Perhaps they will become augmentations layered on top of reality, spatially embedded in our environments, or something that no longer fits comfortably under the word "map" at all. LLM brain implants anyone? (It may sound ridiculous today, but so did home PCs in the early 1980s.)

DOI: 10.1201/9781003797920-24

Mapmaking evolves; and that's totally okay by me. I personally just missed the darkroom and hand-drawn map production eras, and honestly, I'm grateful for it. What tends to persist through all mapmaking technological revolutions, though, even as the techniques and paradigms shift, are the underlying practices.

In the case of prompt cartography, the practice is not simply "using AI" or "writing prompts," but learning how to communicate map design clearly using natural language – explicitly, purposefully, and early in the process. This is something central to cartography throughout history. While writing this book, I began to realize that in many ways, technology has taken mapmaking from its roots. Mapping wasn't originally about accuracy and data provenance. It was about visualizing human descriptions of the spatial environment – word of mouth about our world in pictures.

Cartographers have always used language to design maps. Prompt cartography just allows those of us that are writers, speakers, and oral thinkers, rather than coders, graphic designers, engineers, or mathematicians, a chance to create interactive web-based maps without all the prerequisite scientific or technical skills – i.e., in a manner that is more comfortable for us.

This shift matters – not just in cartography but across all scientific disciplines, I believe. Science has never been about the ivory towers. It's been about disproving what we know so that we may know more by asking questions and testing them. LLMs and AI technologies allow anyone, from layperson in a field to expert, to start framing and asking powerful questions in a manner that may result in increasingly insightful answers.

On the other hand, natural language interfaces don't remove the need for judgment. Hopefully, you've learned that if anything, they increase it. When it becomes easier to generate maps quickly, it also becomes easier to generate maps carelessly. The danger is not that machines make maps. The danger is that humans stop thinking about the maps they are making with those machines. This is why explicit language matters. It forces decisions into the open. It reveals assumptions. It makes uncertainty harder to hide and easier to acknowledge.

Finally, I argue (vehemently, in fact) that prompt cartography allows *all* mapmakers (trained or not) to be more creative, not less. Writing code is *not* creativity; it's grunt work. Clicking buttons in an esoteric graphical user interface is not creativity; it's production-line work and a waste of energy if you don't have to do it. Working with LLMs that have agency and can implement what you ask, directing your team of agents and assistants to bring your spatial visualization to life… in my mind, this is liberating.

I'll stop ruminating now. But before I leave, I want to be clear about what this book is *not* trying to do.

It does *not* make anyone an expert in prompt cartography. It provides grounding to become an expert with practice, critical thinking, and lifelong learning.

This book does *not* offer a concrete workflow, universal schemas, or future-proof recipes. It does not suggest that other methods of mapmaking are inferior, less efficient, or that other cartographers don't think about the same things (workflows, pipelines, ethics) that I recommend prompt cartographers do in this book.

What this book does offer you is a way of thinking – a vocabulary for intent, a set of habits for critique, and a framework for making LLM-based design decisions explicit rather than accidental. It highlights a new technique for making interactive web maps using language and agentic experts. It is a method I work on implementing in my day job as a product engineer, and it is a method I've been honing on my own for fun as a hobbyist at home. I believe it is a method that will only become more popular in the future, but popularity does not equal superiority. (Thankfully! If that were the case, reality television would win more awards.)

If this book helped you transition more smoothly into this emerging mode of mapmaking – if it helped you ask better questions, slow down at the right moments, and articulate what your maps are trying to do and why – then it has done its job. The rest will change. It always does.

So yes, these are early days. That's not something to be embarrassed about. It's something to be transparent about. The techniques will evolve. The interfaces will improve. The maps will surprise us. But the ability to think clearly about spatial meaning, and to communicate that thinking through language, will remain important for anyone creating maps or map-like visualizations – whatever they end up looking like.

Thanks for reading. Thanks for being. I hope you are as excited about the future as I am.

Happy mapping!

Glossary of Terms

Accessibility as a design constraint: The idea that maps and interfaces should be usable by people with different abilities, including those with visual, motor, or cognitive impairments. It means planning colors, text, and interactions so that more people can successfully read and use the map. It treats accessibility as a core design requirement, not a special add-on.

Aesthetic defaults (in LLMs): The typical visual styles that language models fall back on when you do not give clear design instructions, such as generic color ramps or glossy gradients. These defaults come from patterns in the model, training data, not from cartographic best practice. If you do not override them, they quietly shape how your maps will look.

Aesthetic drift: A slow change in the look and feel of a map or map series over time, even when no one consciously decides to change the style. In LLM workflows this can happen as prompts are copied and lightly edited, allowing small shifts in color, typography, or symbol choices to accumulate. The result is that your maps no longer match the original design intent, even though each individual step felt minor.

Aesthetic style (cartography): A recognizable look or, feel, of a map, shaped by choices in color, typography, texture, and composition. Examples include minimalist, sketch-like, or atlas-inspired styles. Aesthetic styles help communicate tone and context, not just decoration.

Agent: A software helper that can act somewhat independently once it receives instructions, such as searching for data, generating code, or drafting map specifications. In prompt cartography, agents are often language-model based and handle specific tasks in a larger workflow, like miniature teammates. They still follow your intent and constraints rather than making their own high-level decisions.

Agentic system: A setup where multiple agents work together in a coordinated way to complete a complex task, such as going from idea to live interactive map. Instead of a single tool doing everything, responsibilities are divided across several specialized agents that pass results between each other. You remain the director who defines goals, guardrails, and when to accept or revise their output.

Archiving prompts: Saving important prompts, specifications, and conversations so you can show how a map was produced and revisit decisions later. This record helps with trust, teaching, and accountability because others can see not just the final map but how you instructed the system along the way. It also lets you reuse and refine successful approaches over time.

Basemap: A background map layer that provides geographic context, such as coastlines, roads, or terrain, onto which thematic data is placed. Choosing an appropriate basemap affects how prominently the thematic information stands out and what assumptions readers make about the, normal, landscape. A busy basemap can easily overpower your main message.

Behavioral system (map): A way of thinking about a map in terms of how it behaves over time, what moves, what responds to user input, and how states change, rather than only how it looks at one moment. Seeing the map as a behavioral system helps designers plan interactions, transitions, and feedback loops more deliberately.

Chunking (RAG): The practice of breaking source material (like this book or your style guide) into smaller pieces so a system can retrieve only what it needs for a given task. Good chunking keeps together ideas that belong together while separating those that would cause confusion if mixed. It is an editorial act because it decides how knowledge will be remembered and recombined.

Civic atlas: A collection of maps focused on public life and community issues, such as housing, infrastructure, and services, gathered into a coherent whole. A civic atlas is intended to support public debate and decision-making, not just private exploration. The case study uses it to show how accountability and ethics enter prompt-based workflows.

Coherence as trust: The idea that when all elements of a map, colors, symbols, labels, and interactions, work together consistently, viewers are more likely to trust what they see. Coherence makes the map feel deliberate rather than accidental. Inconsistent or clashing design choices can quietly erode confidence even if the data is sound.

Color palette: A chosen set of colors used across a map to represent data and structure the visual hierarchy. Thoughtful palettes can highlight patterns, support accessibility, and set tone, while careless ones can distort meaning or exclude some readers. Palettes in LLM workflows need to be specified clearly so defaults do not take over.

Conceptual prompt-to-map pipeline: A high-level sequence of stages that describes how a map moves from idea to published artifact in a prompt-based workflow. Typical stages include a data brief, visual plan, interface plan, specification, review, and publication. The goal is to make the order of key decisions visible so they can be repeated and critiqued instead of staying in your head.

Conversation pipeline: A way of structuring a series of prompts and responses so that an LLM knows, which step of the workflow it is helping with. Rather than treating each query as isolated, a conversation pipeline labels stages like data selection, styling, or critique, and moves through them deliberately. This helps prevent the system from jumping ahead or mixing goals.

CRAFT framework: A simple checklist for writing better prompts that stands for Context, Role, Audience, Format, and Tone. It reminds you to clarify the situation, what kind of assistant you want, who the output is for, the form of the answer, and how it should sound. Using CRAFT makes prompts less vague and more likely to produce usable results.

Cultural cartography: A branch of mapping that focuses on stories, media, and cultural phenomena rather than only on physical or administrative features. Examples include maps of fictional worlds, film locations, or fan communities. In this book, cultural cartography is used to show how prompt workflows handle messy, narrative-rich data.

Data brief: A short document or prompt that describes what a map is really about, whom it is for, what area and time it covers, and what claims the data can and cannot support. Instead of jumping straight into software, you pause to clarify purpose and limits. In an LLM pipeline, the data brief anchors later prompts so the system does not drift into topics or interpretations you never intended.

Data drift: A gradual change in the datasets or fields a map relies on, without an equally careful update of the design or explanation. New sources or time periods may creep in, shifting what the map actually represents while it still appears the same to readers. In automated pipelines this can happen silently unless someone actively checks alignment between data and message.

Dataset silo: A situation in which multiple datasets appear in the same map but remain functionally isolated. The layers coexist visually yet fail to interact, contextualize, or meaningfully inform one another, resulting in false coherence rather than integrated insight.

Design drift: The overall tendency for a map or map series to become something different from what was originally intended, often through many small, reasonable changes. It can involve style, data, or ethics, and is especially easy to miss when using fast, generative tools. Naming design drift makes it easier to notice and correct before trust is damaged.

Design QA toolkit: A set of focused checks and prompts you use to review whether a map still matches its intent, communicates clearly, and treats data responsibly. It includes questions about hierarchy, readability, scope, and experience rather than just whether the code runs. In prompt cartography, QA steps are often turned into reusable critique prompts.

Design rule: A guideline about how to use visual variables, data, and layout to produce clear and responsible maps, such as, do not use rainbow color ramps for ordered data. Rules can be followed, bent, or broken, but they give a starting point for critique and teaching. In prompt cartography, many design rules are translated into reusable text snippets.

Design-by-prompt: An approach to mapmaking where most design decisions are expressed in natural language instructions rather than direct manipulation in a GUI or code. You describe intent, style, behavior, and ethics in prompts, and the system generates specifications or renderings that you then refine. It keeps human judgment in charge but changes how that judgment is written down.

Disclosure (UI): The controlled revealing of information or controls over time, such as expanding panels, tooltips, or progressive details. Good disclosure prevents users from being overwhelmed while still allowing deeper exploration when they are ready. Prompting for disclosure patterns helps structure complex map interfaces.

Dondis framework: A visual literacy framework from Donis A. Dondis that breaks down visual communication into basic principles such as balance, contrast, and rhythm. The book uses it to connect traditional design education with prompt-era cartography. Referencing the framework helps students see maps as part of a broader visual language.

Ecological cartography: Mapping focused on ecological phenomena such as species presence, absence, and environmental uncertainty. It pays particular attention to gaps, uneven sampling, and the limits of observation. In the book's case studies, ecological cartography highlights how prompt workflows can communicate uncertainty without overstating what is known.

Embedding (text): A numerical representation of words, sentences, or documents that captures their meaning in a high-dimensional vector space. Two pieces of text with similar meaning end up with embeddings that are close to each other. Embeddings allow systems to perform semantic search rather than simple keyword matching.

Ethical drift: A slow shift in what a map implies or permits ethically, such as normalizing sensitive classifications or revealing more details than is safe or fair, often without explicit discussion. In automated or LLM-assisted workflows this can creep in as defaults or quick fixes accumulate. Regular ethical pauses help catch such shifts before they become standard practice.

Ethical pause: A deliberate stop in the workflow where you ask whether a map should be made, released, or changed at all, given its impacts on people and places. Instead of focusing on what is technically possible, you consider harm, misinterpretation, and consent. Ethical pauses are built-in checkpoints rather than last-minute worries.

Experiential claim: An implied promise about what a user will feel or be able to do with a map, such as explore your neighborhood in detail or to see how risk changes over time. Even when not stated directly, design choices set experiential expectations. The QA toolkit asks whether the map actually fulfills these claims.

Experiential clarity: The degree to which a user can understand what a map is offering and how to use it without confusion or guesswork. It covers issues like overload, pacing, and interaction design, not just legible labels. Experiential clarity is checked in the QA toolkit as a sign of whether the map respects the reader's time and attention.

FAISS: A widely used software library that helps systems quickly search through large collections of text or data that have been turned into numerical vectors. In RAG workflows, it acts like an index or card catalogue, deciding which chunks of your stored material to pull up for a given prompt. It does not make value judgments itself.

Form variable: A basic visual building block of map design, such as size, shape, color, texture, or orientation. By adjusting form variables, cartographers control what stands out, what recedes, and how relationships are perceived. In prompt workflows, form variables must be described in language so the model can use them consistently.

Framing (map): The way a map defines what the topic is, who is involved, and what questions matter, often through title, extent, data selection, and surrounding text. Framing can subtly steer readers toward certain interpretations while hiding others. Prompt-based workflows make framing choices more visible because they are spelled out in language.

Guardrails (prompting): Explicit limits and rules built into prompts that tell an AI system what it must not do, in addition to what it should do. Guardrails narrow the space of possible outputs, reducing harmful or off-target results. They are especially important when mapping sensitive topics or working with noisy data.

Hallucination (LLM): When a language model produces confident but false or unfounded statements, such as inventing a dataset, citation, or map feature that does not exist. In cartography this might show up as made-up locations, incorrect data sources, or fictitious statistics. Poor or vague prompts often increase the risk of hallucination because the system must guess what you want.

High-dimensional space: An abstract mathematical space with many numeric dimensions used to store embeddings of text or other data. Items that are similar in meaning end up close together in this space, even if they share no exact words. LLM tools use distance in high-dimensional space to perform semantic search.

Information scent: Subtle cues, like labels, button text, or preview states, that help users guess what will happen if they click or interact with something. Strong information scent makes exploration feel safe and purposeful

Intent alignment: The degree to which the generated map, narrative, or interface still matches the purpose and audience described in the data brief or initial prompt. Good alignment means the final artifact supports the claims you meant to make and avoids extra claims you did not authorize. Misalignment signals that prompts, data, or defaults have drifted.

Interactivity as design language: The idea that interactive features, hover states, filters, animation, and transitions, are not just utilities but a vocabulary that can support or undermine the map's message. How and when information is revealed, and with how much motion, affects interpretation. Treating interactivity as language encourages more careful, story-aware interaction design.

Interface element: A visible control or component, such as buttons, sliders, legends, or panels, that lets users interact with a map. The book treats these not as neutral widgets but as part of the map's language, since they suggest what is important and what kinds of actions are expected. Poorly chosen interface elements can nudge people toward the wrong questions.

Interface planning: A stage where you decide how people will encounter and manipulate the map before you worry about specific widgets or code. Questions include who the audience is, what tasks they have, and how much time and attention they can give. Clear interface planning keeps interactivity aligned with purpose rather than becoming a collection of random features.

Label density: How crowded or sparse text labels are in a given area of the map. Too many labels create clutter and fatigue, while too few can leave readers disoriented. In prompt workflows, you can describe target label density so the system knows when to simplify or group names.

Language audit: A focused review of the words used in titles, legends, annotations, and surrounding text to see what they imply or promise. The goal is to spot loaded, speculative, or over-confident phrasing that could mislead readers. In speculative or sensitive maps, language audits help keep narrative tone honest about uncertainty and limits.

Large language model (LLM): A type of AI system trained on vast amounts of text so it can generate and interpret language in flexible ways. In prompt cartography, LLMs act as assistants that can help write code, brainstorm designs, search semantically, or critique maps, but they still need clear intent and constraints from you. They are powerful pattern matchers, not human experts.

Layout system: An underlying structure for arranging map frames, legends, titles, and other elements on screen or page. Layout systems define columns, grids, or regions that keep designs consistent and readable across scales and devices. In LLM workflows you can describe layout systems in prompts so the model places elements with purpose.

Legend tone: The emotional and rhetorical feel of the legend text, such as neutral, urgent, playful, or cautious. Legend tone influences how readers interpret the seriousness or certainty of the data. In prompt cartography, specifying legend tone helps keep narrative voice aligned with the map, Äôs ethical and scientific stance.

Map element: A fundamental component of a map, such as the title, legend, scale bar, north arrow, inset, or main data frame. Each element plays a role in guiding understanding and setting expectations. Even in automated workflows, deciding which elements appear and how is a core design responsibility.

Map form: The overall structural type of a map, such as choropleth, dot map, flow map, or isarithmic map, combined with its layout and presentation. Map form is separate from topic.

Map function: What a map is meant to do in practice, for example, persuade, explain, warn, help navigate, or invite exploration. The same data can be turned into very different maps depending on function. Alignment between form and function is a recurring theme in the book, Äôs design guidance.

Mapped area (extent): The geographic region shown in the main map frame, defined by its boundaries and scale. Extent decisions strongly affect what patterns are visible and what appears cropped or missing. The myth of infinite zoom reminds us that even digital maps always show some places and hide others.

Memory design: The process of deciding what a system should remember, how that knowledge is organized, and what should be allowed to fade over time. Instead of treating memory as a technical afterthought, you treat it as a design problem that shapes consistency, voice, and bias in LLM-assisted mapmaking. Good memory design supports reuse without locking in outdated assumptions.

Messy metadata: Descriptive information about data that is inconsistent, incomplete, or expressed in many different ways, such as varying place names or vague locations. Rather than trying to erase all this messiness, the book argues for understanding and clearly framing it. Prompt tools can help organize messy metadata, but humans must decide how much uncertainty to expose.

(Sane) Metrics: An approach to quantitative evaluation of maps where metrics like load time are used as helpful signals, not as inflexible grades. The goal is to inform judgment rather than replace it. This framing encourages students to see metrics as conversation starters in critique.

Multi-agent workflow: A workflow where several specialized agents each handle a different part of the cartographic process, such as data search, styling, interaction design, and critique. These agents may call one another and pass structured outputs forward, but the human prompt director still oversees the pipeline. This setup scales complex projects while keeping responsibilities visible.

Narrative containment: Designing text and visual elements so that speculative or imaginative content cannot easily be mistaken for factual claims. It involves careful wording, legends, and visual cues that keep the story within safe bounds. Narrative containment is especially important in speculative geographies.

Narrative memory: The way a map interface remembers what a user has already seen or done, such as visited views, filters applied, or previous steps in a story. Good narrative memory helps users feel oriented and acknowledged instead of starting over each time. Prompting for narrative memory can shape bookmarking, highlight states, or progress indicators.

Natural Language GIS (NLGIS): An emerging way of doing GIS and cartography where many tasks are carried out through natural language instructions rather than traditional tools alone. Instead of clicking through menus or writing code, you describe the map you want in everyday language and let the system translate this into operations. NLGIS shifts the main skill from technical manipulation to clear articulation.

Natural-language cartography: Mapmaking workflows in which the main communication with tools happens through everyday language rather than specialized code or menu sequences. This includes both high-level design discussions and low-level instructions expressed as prompts. The term emphasizes a shift in interface, not in the seriousness of the work.

Neatline: A border drawn around a map frame or page to visually separate it from surrounding content. Neatlines can provide structure and focus, but on screens they may also feel ornamental or unnecessary. The book questions when neatlines still help and when they are just inherited habits.

Opening state (UI): The initial visual and interactive condition a map loads in, what is visible, what is hidden, and what appears, ready to use. Opening states strongly influence which patterns are noticed and which paths users take. They should be designed deliberately, not left as mere defaults.

Opening studio moment: A short narrative scene at the start of a case-study chapter where students are asked to imagine a concrete mapping situation. It is used to shift mindset from abstract theory to practical, studio-style problem solving. These moments act as gentle entry points into more complex prompt workflows.

Orchestrated cartographies: Map systems where many components, datasets, views, interfaces, and agents, are coordinated as a whole rather than treated as isolated single maps. The focus moves from one-off designs to ongoing systems whose behaviors and relationships are directed over time. This makes questions of responsibility, consistency, and authority more complex and more important.

Performance (map): How quickly and smoothly a map loads, responds, and redraws across devices and network conditions. Poor performance can undermine even excellent design by frustrating users or making features feel unreliable. In prompt workflows, performance considerations can be encoded as constraints on data size, symbol complexity, or interaction patterns.

Pipeline (design): A structured sequence of steps that takes a project from idea to finished artifact, with clear handoffs and checkpoints. In this book, pipelines help make prompt-heavy workflows repeatable and inspectable. Far from being, dirty, pipelines are framed as supportive scaffolding for creativity and ethics.

Prompt: A written instruction or set of instructions given to an AI system that describes what you want it to do. In this book, prompts are treated as design specifications in natural language, not casual questions. Their clarity, structure, and constraints strongly influence the quality and reliability of the outputs you receive.

Prompt-based workflow: The ordered set of stages and feedback loops that take a map from idea through data, design, critique, and publication in an LLM-assisted environment. A good makes room for briefings, ethical pauses, and iterative critique instead of only rapid generation. It turns fast tools into a deliberate, teachable practice.

Prompt cartographer: A mapmaker who uses natural language prompts and AI agents as primary tools for designing, critiquing, and producing maps. This role still requires cartographic judgment about purpose, ethics, and form, but expresses that judgment mostly through language instead of direct drawing or coding. Prompt cartographers learn to think and speak design fluently to guide their tools.

Prompt cartography: A mode of mapmaking where the entire cartographic process, from data discovery to design, interaction, and critique, is carried out largely through prompts to LLMs and related agents. It keeps human intent central but changes how that intent is recorded, shared, and executed. The book presents prompt cartography as the next major iteration after GUI- and code-based workflows.

Prompt director: The person who coordinates an AI-assisted mapping workflow by specifying intent, constraints, and priorities rather than doing every technical step by hand. Like a film director, a prompt director does not operate all the tools but is responsible for coherence and meaning across the whole production. This role makes technical judgment more visible instead of less.

Prompt literacy: The ability to express design intent, constraints, and expectations clearly enough in language that an AI system can act on them without distorting your goals. It includes knowing how to structure prompts, anticipate misreadings, and revise wording when outputs drift. Prompt literacy is presented as a durable professional skill, not a passing trick.

Prompt pipeline: A structured sequence of prompts that move a project from initial idea to finished map, with each stage having a clear purpose and handoff. Instead of improvising each time, you reuse and adjust pipeline steps such as briefing, visual planning, specification, and critique. This helps stabilize quality and makes the process easier to teach and audit.

Prompt specification: A prompt or set of prompts that is meant to be reused as a stable template, rather than as a one-off experiment. It spells out roles, constraints, outputs, and formatting in enough detail that collaborators or agents can rely on it across projects. Prompt specifications function more like contracts than casual requests.

Prompt type (system, user, tool): A way of classifying prompts according to their role in a conversation with an AI system. A system prompt sets boundaries and values, a user prompt states the immediate task, and a tool prompt defines what an auxiliary tool or agent should do. Thinking in terms of roles helps keep responsibilities clear.

Provenance (cartographic): The documented history of where data, prompts, and design decisions came from in a mapping project. Clear provenance lets others trace how a map was made and judge whether its sources and methods are trustworthy. In AI-assisted workflows, provenance includes saved prompts and retrieved context, not just datasets and code.

Retrieval (in RAG): The step in a retrieval-augmented workflow where the system, looks up, relevant pieces of stored information, such as your style guide, examples, or notes, before generating a response. Retrieval narrows what the model is allowed to consider right now, guiding its behavior toward your curated knowledge instead of only its general training. The choice of what is retrievable is an editorial decision.

Retrieval-augmented generation (RAG): A method where a language model first searches a curated library of documents or notes you provide, then uses both that material and its general training to craft a response. RAG does not retrain the model.

Rhetoric (first impressions): The combined effect of title, opening view, legend, and early interactions in shaping what readers think the map is about. First impressions are hard to undo and can anchor later interpretation. Designers are encouraged to script these impressions intentionally rather than letting them happen by accident.

Semantic fingerprint: An informal way of describing the combined meaning of all the phrases in a prompt as seen by a semantic model. This fingerprint determines which datasets or documents feel closest in vector space. Changing adjectives or constraints changes the fingerprint and thus the search results.

Semantic search: A search method that looks for results based on meaning rather than exact keyword matches. Both your query and candidate documents are turned into numerical vectors, and items whose meanings sit closest to your intent are returned. For cartographers, this allows more natural requests like recent flood risk data for low-income neighborhoods, instead of rigid keyword strings.

Specification (LLM-era): A structured description of what a map, dataset, or interface should be like, written so that both humans and AI agents can act on it. Unlike strict code commands, specifications in this book are treated as agreements about scope, hierarchy, tone, and behavior that still leave room for interpretation. They are central artifacts in prompt cartography workflows.

Speculative geographies: Maps that explore what if scenarios or unseen aspects of the world without claiming to be literal truth. They might show possible futures, imagined spaces, or ambiguous patterns. The book emphasizes that such maps must clearly signal their speculative nature so they do not get mistaken for factual evidence.

Stochastic parrot (critique): A critical nickname for large language models that suggests they only remix existing text patterns without understanding. The book acknowledges this critique but argues that, in practice, LLMs can still be useful partners if cartographers express intent and constraints clearly. The metaphor is a reminder not to confuse fluent output with deep knowledge.

Style drift (LLM): A slow change in a map visual style caused by default design choices in language models and small prompt edits over time. If you do not keep your style language and specifications consistent, newer outputs may no longer match earlier ones in tone, palette, or hierarchy. Watching for style drift is important for long-lived projects and institutional branding.

Symbology: The system of visual symbols, points, lines, polygons, and icons, used to represent real-world features on a map. Good symbology balances clarity, aesthetic appeal, and cultural sensitivity. In prompt workflows, you describe symbology rules so the model can generate or critique them consistently.

Tacit knowledge (in cartography): Know-how that experienced cartographers carry in their heads or habits but rarely write down, such as sensing when a layout feels off or a color choice feels misleading. Tacit knowledge is powerful but hard for AI systems to use unless you externalize it into prompts, guidelines, or examples. RAG workflows push this tacit knowledge into explicit form.

Temporal thinking without prediction: Reasoning about how phenomena unfold over time without claiming to forecast the exact future, such as showing past patterns or multiple plausible scenarios. This approach respects uncertainty while still helping users think beyond a single moment. It is used in ecological and speculative mapping case studies.

Threshold effect (visual): A sudden change in how a map is perceived when a design choice crosses a certain limit, for example, when label density goes from manageable to overwhelming. Small adjustments may feel harmless, but once a threshold is crossed, understanding collapses. Recognizing such thresholds helps in critique and iteration.

Title rhetoric: The way a map's title frames what the map is about, and how strongly it makes claims. Titles can invite curiosity, signal caution, or push a particular interpretation. In prompt cartography, you can specify title tone and scope, so AI-generated titles do not overpromise or oversimplify.

Visual hierarchy: The ordering of visual importance among map elements so that some are noticed first and others later. It is built with contrast, placement, size, and color choices. Strong hierarchy helps readers follow the intended story instead of wandering aimlessly across the page or screen.

Visual literacy: The ability to read, interpret, and create visual messages, including maps, charts, and diagrams. Visual literacy involves understanding how form, color, and layout influence perception and meaning. The book treats prompt literacy as a close cousin of visual literacy.

Visual planning: A stage in the workflow where you decide what should be visually emphasized, what should recede, and what overall mood the map should convey before any final drawing is done. It is about clarifying visual intent, not picking exact colors or fonts. In prompt workflows, visual planning turns gut feelings into explicit language that models can follow.

Widget (UI): A small interactive control in an interface, such as a button, slider, toggle, or dropdown. Widgets are often treated as neutral building blocks, but the book argues they also carry messages about what is important and what actions are expected. Prompting for widgets means describing their purpose, not just their appearance.

References and Further Readings

Andrienko, Natalia V., Gennady L. Andrienko, and Peter Gatalsky. 2005. "Impact of Data and Task Characteristics on Design of Spatio-Temporal Data Visualization Tools." In *Exploring Geovisualization*, edited by Jason Dykes, Alan M. MacEachren, and Menno-Jan Kraak, 201–222. Oxford: Elsevier.

Anthropic. 2024. "Contextual Retrieval." *Anthropic Engineering Blog*, September 19, 2024. https://www.anthropic.com/engineering/contextual-retrieval

Bender, Emily M., Timnit Gebru, Angelina McMillan-Major, and Shmargaret Shmitchell. 2021. "On the Dangers of Stochastic Parrots: Can Language Models Be Too Big?" In *Proceedings of the 2021 ACM Conference on Fairness, Accountability, and Transparency (FAccT '21)*, 610–623. New York: Association for Computing Machinery.

Bertin, Jacques, and William J. Berg. 1983. *Semiology of Graphics*. Madison: University of Wisconsin Press.

Brauen, Gwen. 2006. "Designing Interactive Sound Maps Using Scalable Vector Graphics." *Cartographica* 41 (1): 59–72. https://doi.org/10.3138/5512-628G-2H57-H675

Brown, Stuart L., and Christopher C. Vaughan. 2009. *Play: How It Shapes the Brain, Opens the Imagination, and Invigorates the Soul*. New York: Avery.

Carter, Rob. 2002. *Digital Color and Type*. Crans-Prés-Céligny, Switzerland: RotoVision.

Cartwright, William. 2007. "Development of Multimedia." In *Multimedia Cartography*, 2nd ed., edited by William Cartwright, Michael P. Peterson, and Georg Gartner, 11–34. Berlin: Springer.

Cartwright, William, and Michael P. Peterson. 2007. "Multimedia Cartography." In *Multimedia Cartography*, 2nd ed., edited by William Cartwright, Michael P. Peterson, and Georg Gartner, 1–10. Berlin: Springer.

Case, N. 2007. "Taking Apart Cartography: Our Field as a Graphic Tradition." Paper presented at the *North American Cartographic Information Society Annual Conference*, St. Louis, MO. http://nat.case.home.mindspring.com/nacis07G.pdf

Chernoff, Herman. 1973. "The Use of Faces to Represent Points in k-Dimensional Space Graphically." *Journal of the American Statistical Association* 68 (342): 361–368.

Christophe, Sébastien, Christophe Hoarau, Anouchka Kasbarian, and Aurélie Audusseau. 2012. "A Framework for Pop Art Maps Design." Paper presented at the *GIScience Conference*, Columbus, OH.

Dent, Borden D., Jeffrey S. Torguson, and Thomas W. Hodler. 2008. *Cartography: Thematic Map Design*. 6th ed. New York: McGraw-Hill.

DiBiase, David, Alan M. MacEachren, John B. Krygier, and Catherine Reeves. 1992. "Animation and the Role of Map Design in Scientific Visualization." *Cartography and Geographic Information Systems* 19 (4): 201–214, 265–266.

Dondis, Donis A. 1973. *A Primer of Visual Literacy*. Cambridge, MA: MIT Press.

Dorling, Danny. 1992. "Stretching Space and Splicing Time: From Cartographic Animation to Interactive Visualization." *Cartography and Geographic Information Systems* 19 (4): 215–227, 267–270.

Ehlschlaeger, Charles R., Andrew M. Shortridge, and Michael F. Goodchild. 1997. "Visualizing Spatial Data Uncertainty Using Animation." *Computers & Geosciences* 23 (4): 387–395.

Field, Kenneth. 2018. *Cartography*. Redlands, CA: Esri Press.

Field, Kenneth. 2022. *Thematic Mapping: 101 Inspiring Ways to Visualise Empirical Data*. Redlands, CA: Esri Press.

Field, Kenneth, and Ian Muehlenhaus. 2024. "On the Ethics of Cartographic Ethics." *The Cartographic Journal* 61 (4): 288–305. https://doi.org/10.1080/00087041.2025.2479647

Fisher, Peter. 1994. "Animation and Sound for the Visualization of Uncertain Spatial Information." In *Visualization in Geographical Information Systems*, edited by H. M. Hearnshaw and David J. Unwin, 181–185. Chichester: John Wiley & Sons.

Gigerenzer, Gerd. 2007. *Gut Feelings: The Intelligence of the Unconscious*. New York: Penguin.

Gilbert, Jason. 2013. "HAPIfork: Buzzing Fork Offers Ultimate First-World Solution to Overeating." *Huffington Post*, January 8, 2013. http://www.huffingtonpost.com/2013/01/08/hapifork-buzzing-fork-solution-overeating_n_2433222.html

Gladwell, Malcolm. 2005. *Blink: The Power of Thinking Without Thinking*. New York: Back Bay Books.

Google Cloud. 2026. "Introduction to Prompt Design." *Vertex AI Documentation*. Updated January 8, 2026. https://docs.cloud.google.com/vertex-ai/generative-ai/docs/learn/prompts/introduction-prompt-design

Gray, Dave, Sunni Brown, and James Macanufo. 2010. *Gamestorming: A Playbook for Innovators, Rulebreakers, and Changemakers*. Sebastopol, CA: O'Reilly Media.

Gruget, Maïeul, Guillaume Touya, and Ian Muehlenhaus. 2023. "Missing the City for Buildings? A Critical Review of Pan-Scalar Map Generalization and Design in Contemporary Zoomable Maps." *International Journal of Cartography* 9 (2): 255–285. https://doi.org/10.1080/23729333.2022.2153467

Harrower, Mark. 2004. "A Look at the History and Future of Animated Maps." *Cartographica* 39 (3): 33–42.

Harrower, Mark. 2007. "The Cognitive Limits of Animated Maps." *Cartographica* 42 (4): 349–357.

Harrower, Mark, and Sara Fabrikant. 2008. "The Role of Map Animation in Geographic Visualization." In *Geographic Visualization: Concepts, Tools and Applications*, edited by Martin Dodge, Mary McDerby, and Martin Turner. Chichester, UK: Wiley.

Hewing, Michael, and Vincent Leinhos. 2024. *The Prompt Canvas: A Literature-Based Practitioner Guide for Creating Effective Prompts in LLMs*. arXiv. https://arxiv.org/abs/2412.05127

Hoffman, Donald D. 1998. *Visual Intelligence: How We Create What We See*. New York: W. W. Norton.

Jenny, Bernhard, Hannes Jenny, and Stefan Raeber. 2008. "Map Design for the Internet." In *International Perspectives on Maps and the Internet*, edited by Michael P. Peterson, 31–48. Berlin: Springer.

Jiang, Bin. 1996. "Cartographic Visualization: Analytical and Communication Tools." *Cartography* 25 (2): 1–11.

Kelion, Leo. 2012. "Here's How Nokia Creates Its Maps." *BBC News*, November 2012. https://www.bbc.co.uk/news/technology-20497719

Koch, Tom. 2004. "The Map as Intent: Variations on the Theme of John Snow." *Cartographica* 39 (4): 1–14.

Kraak, Menno-Jan. 2007. "Cartography and the Use of Animation." In *Multimedia Cartography*, 2nd ed., edited by William Cartwright, Michael P. Peterson, and Georg Gartner, 317–326. Berlin: Springer.

Kraak, Menno-Jan, and Ferjan Ormeling. 2010. *Cartography: Visualization of Spatial Data*. 3rd ed. Harlow, UK: Pearson Education.

Krug, Steve. 2005. *Don't Make Me Think: A Common Sense Approach to Web Usability*. 2nd ed. Berkeley, CA: New Riders.

Krygier, John B. 1994. "Sound and Geographic Visualization." In *Visualization in Modern Cartography*, edited by Alan M. MacEachren and D. R. Fraser Taylor, 149–166. Oxford: Pergamon.

LangChain. n.d. "Build a RAG Agent with LangChain." *LangChain Documentation*. https://docs.langchain.com/oss/python/langchain/rag

Lewis, Patrick, Ethan Perez, Aleksandra Piktus, Fabio Petroni, Vladimir Karpukhin, Naman Goyal, Heinrich Küttler, et al. 2020. "Retrieval-Augmented Generation for Knowledge-Intensive NLP Tasks." arXiv. https://arxiv.org/abs/2005.11401

Li, B. 2025. "Study on the Long-Term Cognitive Effects of Excessive LLM Use Among Students and General Users." *Journal of Contemporary Educational Research* 9 (10): 203–214. https://doi.org/10.26689/JCER.V9I10.12112

MacEachren, Alan M. 1994a. *Some Truth with Maps: A Primer on Symbolization and Design*. Washington, DC: Association of American Geographers.

MacEachren, Alan M. 1994b. "Time as a Cartographic Variable." In *Visualization in Geographical Information Systems*, edited by H. M. Hearnshaw and David J. Unwin, 115–130. Chichester: John Wiley & Sons.

MacEachren, Alan M. 1995. *How Maps Work: Representation, Visualization, and Design*. New York: Guilford Press.

MacEachren, Alan M., and David DiBiase. 1991. "Animated Maps of Aggregate Data: Conceptual and Practical Problems." *Cartography and Geographic Information Systems* 18 (4): 221–229.

Markillie, Paul. 2011. "The Greening of Blue." *The Economist*. https://www.economist.com/node/21537977

McMaster, Robert B., and Susanna McMaster. 2002. "A History of Twentieth-Century American Academic Cartography." *Cartography and Geographic Information Science* 29 (3): 305–321.

Monmonier, Mark S. 1990. "Strategies for the Visualization of Geographic Time-Series Data." *Cartographica* 27 (1): 30–45.

Monmonier, Mark S. 1993. *Mapping It Out: Expository Cartography for the Humanities and Social Sciences*. Chicago: University of Chicago Press.

Muehlenhaus, Ian. 2012. "If Looks Could Kill: The Impact of Rhetorical Style on Persuasive Geocommunication." *The Cartographic Journal* 1–16. https://doi.org/10.1179/1743277412Y.0000000032

Muehlenhaus, Ian. 2013a. "The Design and Composition of Persuasive Maps." Cartography and Geograhpic Information Science 40 (forthcoming): http://doi.org/10.1080/15230406.2013.783450

Muehlenhaus, Ian. 2013b. *Web Cartography: Map Design for Interactive and Mobile Devices*. CRC Press. http://www.ian.muehlenhaus.com/webcartography

Muehlenhaus, Ian. 2014. "Looking at the Big Picture: Adapting Film Theory to Examine Map Form, Meaning, and Aesthetic." *Cartographic Perspectives*, no. 77. https://doi.org/10.14714/CP77.1239

OpenAI. n.d.-a "Best Practices for Prompt Engineering with the OpenAI API." *OpenAI Help Center*. https://help.openai.com/en/articles/6654000-best-practices-for-prompt-engineering-with-the-openai-api

OpenAI. n.d.-b "Prompt Engineering." *OpenAI Platform Documentation*. https://platform.openai.com/docs/guides/prompt-engineering

OpenAI. n.d.-c "Retrieval." *OpenAI Platform Documentation*. https://platform.openai.com/docs/guides/retrieval

OpenAI. n.d.-d "Retrieval-Augmented Generation (RAG) and Semantic Search for GPTs." *OpenAI Help Center*. https://help.openai.com/en/articles/8868588-retrieval-augmented-generation-rag-and-semantic-search-for-gpts

Peterson, Michael P. 2007. "Elements of Multimedia Cartography." In *Multimedia Cartography*, 2nd ed., edited by William Cartwright, Michael P. Peterson, and Georg Gartner, 63–73. Berlin: Springer.

Peuquet, Donna J. 1994. "It's About Time: A Conceptual Framework for the Representation of Temporal Dynamics in Geographic Information Systems." *Annals of the Association of American Geographers* 84 (3): 441–461.

Peuquet, Donna J. 2002. *Representations of Space and Time*. New York: Guilford Press.

Pickles, John. 2004. *A History of Spaces: Cartographic Reason, Mapping, and the Geo-Coded World*. London: Routledge.

Pinecone. 2025. "Chunking Strategies for LLM Applications." *Pinecone Learn*, June 28, 2025. https://www.pinecone.io/learn/chunking-strategies/

Robinson, Arthur Howard. 1952. *The Nature of Maps: An Examination of Cartographic Design*. Madison: University of Wisconsin Press.

Saravia, Elvis. 2023. *Prompt Engineering Guide*. DAIR.AI. https://www.promptingguide.ai/

Slocum, Terry A., Robert B. McMaster, Fritz C. Kessler, and Hugh H. Howard. 2008. *Thematic Cartography and Geographic Visualization*. 3rd ed. Upper Saddle River, NJ: Pearson Prentice Hall.

Smith, Neil. 1991. *Uneven Development: Nature, Capital, and the Production of Space*. Oxford: Blackwell.

Sutherland, Stuart. 2013. "Bright Screens Could Delay Bedtime." *Scientific American Mind* 13.

Swaine, Jon. 2010. "Google Maps Error Sparks Invasion of Costa Rica by Nicaragua." *The Telegraph*, November 2010. https://www.telegraph.co.uk/news/worldnews/centralamericaandthecaribbean/nicaragua/8117902/Google-maps-error-sparks-invasion-of-Costa-Rica-by-Nicaragua.html

Theja, Ravi. 2023. "Evaluate RAG with LlamaIndex." *OpenAI Cookbook*, November 6, 2023. https://cookbook.openai.com/examples/evaluation/evaluate_rag_with_llamaindex

Tobler, Waldo R. 1970. "A Computer Movie Simulating Urban Growth in the Detroit Region." *Economic Geography* 46: 234–240.

Touya, Guillaume, Maïeul Gruget, and Ian Muehlenhaus. 2023. "Where Am I Now? Modelling Disorientation in Pan-Scalar Maps." *ISPRS IJGI* 12 (2).

Touya, Guillaume et al. 2021. "Please, Help Me! I Am Lost in Zoom." In *Proceedings of the ICA*, Florence, Italy. December 14-18, 2021.
Tufte, Edward R. 1983. *The Visual Display of Quantitative Information*. Cheshire, CT: Graphics Press.
Tufte, Edward R. 1991. *Envisioning Information*. Cheshire, CT: Graphics Press.
Ware, Colin. 2008. *Visual Thinking for Design*. Burlington, MA: Morgan Kaufmann.
Weinschenk, Susan M. 2011. *100 Things Every Designer Needs to Know about People*. Berkeley, CA: New Riders.
Yau, Nathan. 2011. *Visualize This: The FlowingData Guide to Design, Visualization, and Statistics*. Indianapolis: Wiley.

Index

Pages in *italics* refer to figures, pages in ***bold-italics*** refer to boxes.

A

absence and uncertainty, *see* mapping presence, absence, and uncertainty
accessibility, 242, 244, 247, 295, 311, 314; *see also* color-vision deficiency; human-in-the-loop QA
  automated checks and human interpretation, 311
  ongoing care, 311, 314
  reduced-motion preferences, 242, 244, 247
accessibility in palette design, 175, 179–182
  color-vision deficiency, 179–182
  WCAG AA contrast, 175
accurate map design, *see* responsible map design
aesthetic defaults (in LLMs), 132, 134–135
aesthetic drift, 177
aesthetic styles (cartography), 117, 123–131, 134–136
  data visualization style, 125, *126*
  experiential or artistic styles, 127, *129*, 130
  historical or retro styles, 125, 127, *128*
  minimalist modern styles, 123–125, 130
  realistic or photographic styles, 127, 130
  as visual dialects, 117, 123, 130, 134
agentic systems, 8–9, 21–22
agents, 62–63, 65, 70–71, 249, 265, 267–268, 271–272, 274, 276, 278
  LLM, 249, 265, 267–268, 271–272, 274, 276, 278
aggregation, 65, *66–67*, 69–71
AI agents, 339, 341–345
AI assistants and agents, 2–3
AI tools, *see* generative AI
AI-generated maps, *see* large language models (LLMs), color associations and palette generation
animation, 239–248; *see also* accessibility; motion and continuity systems; visual variables
  duration and simplification, 245, 247
  explanatory vs. decorative motion, 239–240, 246–247
  motion constraints, 243–244, 246–248
  review questions, 245–246
  types of, 243, 247
  user controls, 241, 244, 247–248
artificial intelligence (AI), 343–344
  cartographic use, 343–344
attribution, 60–61, 69–70; *see also* transparency
  data sources and credits, 60–61, 69–70
  style attribution, 60, 70
authorship, 28, 31, 33, 38–39, 41, 43, 52–54, 56
  in LLM-assisted mapping, 43, 52–54, 56
automated checks, 308, 311, 313–314
automation, 308
  as continuity, 308
automation in cartography, 2
automation pipelines, 307–309, 312–313
  prompt-driven, 307–308
auto-play animation, 242, 244, 247

B

base map, *see* basemaps
basemap context, 201, *203*
basemaps, 60, 114, 127, 130–133, 135, 265–266, 274

locked layers, 265–266, 274
permission and attribution, 60
Bertin, Jacques, 119, 122
visual variables, 119, 122
Bertin's variables, *see* visual variables
bias in maps, 62–63, *64*, 69–70, 72
explainable style decisions, 62–63, 70
explicit and implicit bias, 62, *64*
binning, 222–223, 237–238
Brewer, Cynthia, 179

C

cartograms, 209, 228–230, 235, 237
contiguous and non-contiguous, 229–230
distortion and rhetorical emphasis, 228, 230, 235, 237
reference maps and orientation, 230, 237
cartographer as director, 138, 164–166
cartographic decisions, *see* cartographic judgment
cartographic design, 2
decision-making and automation, 2
cartographic design for uncertainty, 331–336
contextual and base map layers, 333
legends and captions, 333
sparse point symbolization, 333
techniques to avoid, 333, 336
visual hierarchy and visual noise, 333
cartographic intent, 74
cartographic judgment, 5, 9–10, 14, 31, 37, 39, 41, 340–341; *see also* cartographic responsibility
cartographic orchestration, 338, 340–341
cartographic pipelines, 3
cartographic process, 318
messiness of, 318
cartographic responsibility, 5, 10–12, 14–15, 73
delegation and accountability, 5, 10–12, 14–15
cartographic systems, 339–340, 342
cartographic techniques, *see* cartographic design for uncertainty, techniques to avoid
cartography fundamentals, 14
cartography process, *see* cartography process as knowledge recall
cartography process as knowledge, *see* cartography process as knowledge recall
cartography process as knowledge recall, 28–42
cartography workflow, *see* prompt cartography, workflow
case studies, 318–319
prompt cartography, 318–319
chart maps, 233
choice architecture, 264–265, 277
choropleth maps, 209, 211–215, *219*, 231, 235–238; *see also* thematic mapping
classification schemes, 211, 213–214, 235, 237–238
equal-area projections, 211, 214, 237
internal uniformity, 211, 214, 236
raw data distortion, 211, 214, 238
chunking (RAG), 32–33, *35*, 40
chunk size and context, 33
as editorial decision, 32–33, *35*, 40
civic cartography, 321–328; *see also* map misinterpretation
audience sensitivity, 321–322, 326–327
data selection and restraint, 323–324, 326, 328
publication decisions, 327
civic data, 321, 323–324
civic mapping, *see* civic cartography
civic maps, *see* civic cartography
clarity, 74
accountability and subjectivity, 74
classification logic, *see* legends
classification schemes, 211–215, 235, 237–238
equal intervals, 213–214, 235, 237–238
natural breaks, 213–215, 235, 237–238
quantiles, 213–214, 235, 237–238
standard deviations, 213, 237
unclassed, *212*, 213–215, 235
cognitive load, 231, 233–234, 237, 239, 243, 245–246, 252, 254, 256–257, 272, 277, 317

animation and, 239, 243, 245–246
in prompt cartography, 317
collision rules, 183, 191–193
labels, 183, 191–193
color as rhetoric, 169, 171, 174–175, 179–182
color choices, *see* color as rhetoric
color harmony, 177, 179
color palettes, 114–115, 123, 127, 131, 135
color properties, 169, 171–172, 174–175, 177, 180–182
hue, 169, 171–172, 175, 180–182
saturation, 171–172, 174, 177, 180–181
value, 169, 171–172, 174, 180–182
color schemes, 174–177, *178*, 180–181, 211, 213–214, 220–221, 223, 225, 237–238
categorical schemes, 174, 177, 180–181
diverging schemes, 174, 177, *178*, 180–181
rainbow schemes, 220–221, 223
sequential palettes, 214, 221, 223, 225, 237–238
sequential schemes, 174–175, 177, *178*, 181
color theory, 169, 171–175, 177, 179–181
color value, *see* color properties, value
ColorBrewer, 179; *see also* color schemes
commitment systems, 257–261, 277–278
pop-ups, 257–258, 260–261, 277–278
tooltips, 257–259, 261, 277–278
community data, mapping with accountability, 321, 323, 325, 327–328
community data with accountability, *see* community data, mapping with accountability
companion website, 12–13, 16, 318
code listings and implementations, 318
consistency, 36, 39, 41
and stylistic drift, 36, 39, 41
constant critique pipeline, *see* critique pipeline
contrast, 169, 171, 173–175, 179–181
color contrast, 169, 171, 173–175, 179–181
control systems, 264–267, 272, 274, 277–278
layer toggles, 264–266, 272, 274, 277
locator maps, 264, 267
menus, 264, 266
CRAFT framework, 24–26; *see also* prompt framing
elements of, 24–25
as framing checklist, 24–26
critique, 45, 52–53, 56, 318
as continuous conversation, 318
as design stage, 45, 52–53, 56
prompts, 53
critique maps, *see* critique pipeline
critique pipeline, 75–79, 81, 83, 85, 87–91, 288, 295–298, 303
continuous critique, 76, 89–91
from conversations to pipelines, 90
description, analysis, interpretation, evaluation, 78–79, 91, 297
critique prompts, 76, 79, 81–84, 89, 91
command prompts contrasted with, 82–83
as conversation, 82–83, 91
interaction behavior and state, 82–83
CSS, 184, 188, 191–192
font stacks and typography, 184, 188, 191–192
cultural interpretation of color, 169, 172, 177, 179–180; *see also* color as rhetoric
curse of LLM, *see* LLM defaults
curse of LLM defaults, *see* LLM defaults

D

data and phenomena, 209–210, 236
data brief, 46–48, 52, 56, 93, 106, 111–113
elements, 47
as lodestar for LLMs, 47
purpose and editorial role, 46–48, 52, 56
data cleaning, 104–106
data discovery, 94–99, *110*, 111, 318; *see also* semantic search
semantic search, 318
data drift, *see* drift, data
data engineering, 45–48, 56

in prompt cartography, 45–48, 56
data engineering pipeline, 93–99, 101–111, 113; *see also* data engineering
agent orchestration, 108–111
data cleaning, 104–106, *110*, 111
data discovery, 94–99, *110*, 111
validation and critique, 102–103, 108–109, 111
data ethics, *see* datasets, ethics of
data export, 265–266, 272
data fidelity and communicative clarity, 101, 111
data logic, 174, 177, 181–182
data quality, *see* dataset integrity
data selection, 332
data sources, 60, 69–70
credits and licenses, 60, 69–70
data type, 177, 180, 182, 209–210, 236, 238
method appropriateness, 209–210, 236, 238
palette choice, 177, 180, 182
data visualization, 119, 122, 125, *126*, 134
as aesthetic style, 125, *126*
data with accountability, *see* community data, mapping with accountability
dataset bias, 330–332, 335
dataset framing, 280–281, 283–286, 303
ethical boundaries, 281, 284, 286
dataset integration, 338, 340–342; *see also* cartographic orchestration
compatibility and comparability, 338, 341
ethical implications, 340–342
exclusion of datasets, 340–342
false coherence and silos, 341
parallel views and visual alignment, 341
dataset integrity, 99, 102–104, 111
authority, 103, 111
bias and coverage, 103–104, 111
completeness and recency, 102–103, 111
granularity, 103
dataset silos, 341; *see also* false coherence
datasets, 93–96, 98–104, 106–113
ethics of, 102–104, 106–108, 111–112
evaluation criteria, 99, 102–103, 111
fields and attributes, 96, 98, 102, 106, 112–113
metadata, 102, 104, 106–107
selection as design decision, 93, 99–101, 109, 111
visual accent and tone, 99–101, 111–112
decision-making maps, 340–341
decorative color creep, 206
defaults, 78, 81, 85, 88, 91
in map critique, 78, 81, 85, 88, 91
delegation to LLMs, 28, 31, 40–41
derivative works, 60
design artifacts, *see* prompts
design choices, *see* design intent
design decisions, 184, 188, 191–192, 324–325, 327
clarity without emphasis, 324–325
typography as explicit, 184, 188, 191–192
design drift, 79–81, 88–89, 91–92; *see also* critique prompts
aesthetic drift, 80, 91–92
data drift, 81, 91–92
diagnosis rather than blame, 80–81, 91
ethical drift, 81, 89, 91–92
design intent, 21, 25–27, 76–77, 82, 88, 90, 92, 114, 131, 134–135
design judgment, 76, 83–84, 90–91, 144, 165; *see also* metrics
design language, *see* prompting
design quality assurance process, 88–89, 91
credibility and care, 89
experiential clarity, 89, 91
intent alignment, 88
differential privacy, 69
disclosure, 59, 61, 70, 72
disclosure systems, 252, 254–257, 271–272, 275, 277–278
common failures, 256

filters, 254–257, 271–272, 275, 277–278
legends, 254–257, 271–272, 275, 277–278
Dondis, Donis A., 115–116, 118–119, 122, 134
visual elements, 118–119, 122
Dondis' framework, 318
dot maps, 209, 218–220, 231–232, 236
default lie, 218, 220, 236
overlap management, 218, 220
drift, 309–310, 314; *see also* monitoring; versioning
aesthetic, 309
data, 310, 314
design, 314
semantic, 310, 314
technical, 309–310
visual, 310, 314

E

ecological cartography, 329, 331, 333, 335–337; *see also* uncertainty as message
ethical risk of overinterpretation, 329, 331, 335–336
uncertainty and restraint, 331, 333, 335–337
effective symbology, *see* symbology, meaning before beauty
embedding (text), 30
enforcement data, 323, 326
engineering pipeline, *see* data engineering pipeline
entropy, 310, 313
ethical considerations, 47, 53
in data briefs and critique, 47, 53
ethical constraints, 317
in promptable instructions, 317
ethical drift, *see* design drift, ethical drift
ethical judgment, 73
ethical pause, 326
ethics, 58–61, 63, 65, 69–72; *see also* privacy
everyday design choices, 58, 70–72
prompt pipelines, 69–70
ethics and privacy, *see* ethics
ethics in cartography, 284, 286, 297–298, 303, 305
explanatory legend, *see* legends, explanatory

F

FAISS, 36
fallback font stacks, 187–189, 191, 193; *see also* graceful degradation
false coherence, 341
Federal Geographic Data Committee (FGDC), 104
filter confetti, 149
flow maps, 209, 225–228, 235–237; *see also* thematic mapping
animation, 228
bundling and filtering, 226, 228, 237
default lie, 228, 235
heat flow maps, 228
fly-through animations, 243
font, 184
typeface distinction, 184
font families, 185, 192
limit to two, 185, 192
font properties, 186, 192, 194
hierarchy encoded by, 186, 192, 194
font selection, 183, 188–193
map tone and rhetoric, 183, 188–193
framing (maps), 5, 9–10, 15, 62, 69–70, 72, 140–143, 325–327
map titles and first impressions, 140–143
maps and prompts, 5, 9–10, 15
responsible framing, 325
rhetorical boundaries, 326
full prompt, *see* prompts, structures and transcripts

G

generative AI, 339–340
geometric abstraction, 199, 206–208
in symbols, 199, 206–208
GIS software, 1

GitHub Pages, 306
graceful degradation, 187, 189, 193
typography, 187, 189, 193
graduated symbology, 200
graduated symbols, *66*
privacy-preserving aggregation, *66*
grammar layout, *see* layout systems, spatial grammar
graphical user interfaces (GUI), 17, *18*
gratuitous animation, 272
guardrails, 8–10, 32, 45, 53, 96, 107, 112, 137, *154*, 234–238
prompting, 8–10, 96, 107, 112, 137, *154*

H

hallucination, 8, 106, 111
LLM confabulation and prompt error, 8
heat maps, 209, 220–223, 228, 235–238
binning alternatives, 222–223, 237–238
kernel density and smoothing, 220, 223, 235–237
rainbow color schemes, 220–221, 223
hexadecimal color values, 173, 175, 179–181
hexbins, *see* binning
high-dimensional space, *see* semantic search, embeddings and vectors
hosting environments, 306
human cartographer, 61, 63
human judgment, 344
in AI-assisted mapmaking, 344
human-in-the-loop QA, 311–312, 314

I

improvisation, 43–44, 57
limits in mapmaking, 43–44, 57
information windows, *see* pop-ups
inset maps, 137, 160–163, 165, 167
interaction behavior, 77–78, 82–83, 88, 90
interaction design, 283, 294–296, 300, 303–304; *see also* accessibility
animated rhetoric, 294
confidence, designing for, 294–295
restraint, 295–296
interaction grammars, 251–253, 270, 279
analytic, 251–253, 270, 279
exploratory, 251–253, 270, 279
narrative, 251–253, 270, 279
interactive behaviors, 317–318
interactive maps, 138, 144, 150, 168; *see also* web maps
interactive web, *see* web maps
interactive web map director, *see* map director
interactive web maps, 1–2, 21–22, 27, 281, 285, 294, 299–303, 344–345; *see also* prompt cartography
publication readiness, 299–301
from start to finish, 303
interactivity, 249, 251–254, 267, 269–277
versus animation, 253–254, 277
common prompting mistakes, 270–273
integration with earlier design decisions, 274–277
interactivity compulsion, 270–271
interface choices, 49–50, 56
as rhetorical choices, 49–50, 56
interface elements, 147–151, 164, 166; *see also* visual hierarchy
accessibility, 147–150
as affordances, 148–150
excess, 149–150, 164, 166
interface planning, 49–50, 56; *see also* interface choices, as rhetorical choices
anticipated use, 49–50, 56
questions, 49
interpretive distance, 326
isarithmic maps, 209, 223–225, 235–237
interpolation uncertainty, 224–225, 235, 237
isometric and isoplethic, 223

J

JSON, 32, ***33***
schemas, ***33***
JSON schemas, *see* JSON
judgment visibility, 43–44, 55–57

K

keeper prompts, 84–85, 91, 308–309, 315
criteria for, 84–85
knowledge documents, 3

L

label density, 183, 191, 193; *see also* collision rules
label placement, 183–184, 186–187
area labels, 187
line labels, 187
point labels, 187
language models, 43, 46–48, 317, 325, 327–328
as cartographic collaborators, 317
and explicit workflows, 43, 46–48
language models LLMs, *see* large language models (LLMs)
language of map form, *see* map form
language of map form and style, *see* map form
large language models (LLMs), 1–3, 7–8, 10–12, 17, 20–24, 27, 29–32, ***33***, 35–42, 76, 86, 92, 114, 116, 119, 122–123, 130, 132–136, 169, 172–175, 177, 179–184, 186, 188–189, 193, 197–198, 200, 204–206, 208, 213, 216, 234, 238, 242–243, 265, 272, 278, 281, 283, 285, 288, 299, 303–304, 306, 309–314, 339–340, 343–345; *see also* critique prompts; LLM agents; prompting, typographic intent; prompts
agents, 2–3, 306, 312–313
animation defaults, 242–243
assistants and agents, 20–22
color associations and palette generation, 169, 172–175, 177, 179–182
as critique partners, 76, 86, 92
emergence in cartography, 1
general knowledge versus retrieved knowledge, 29–31, 35–37, 39–40
lack of perception, 119, 122, 134
limitations of, 20
outputs and user responsibility, 8, 10–12
specificity in prompting, 122, 134
symbol generation, 197–198, 200, 204–206, 208
thematic map prompting, 213, 216, 234, 238
training data and stylistic associations, 119, 122, 130
translation between humans and computers, 7–8
typographic prompting and reasoning, 183–184, 186, 188–189, 193
large language models LLMs, *see* large language models (LLMs)
layout and composition, 280–281, 285, 289, 291–292, *293*, 295; *see also* interaction design
bridge between visuals and behavior, 292
constraint system, 291–292
layout systems, 153–159, 164–165, 167; *see also* visual hierarchy
balance, 155, 158–159, 164
compartmentalized layouts, 155, *157*, 158
fluid layouts, 155, *156*, 158
spatial grammar, 153, 155, 158–159, 165, 167
learning to speak via prompt, *see* prompts
legends, 137–138, 142–143, 149–151, 159–160, 162–163, 165, 167–168; *see also* disclosure systems; inset maps
classification logic, 160, 162–163, 165

explanatory, 160, 162–163, 165, 168
inventory, 160, 162
placement and prominence, 151, 159, 162–163
linked views, 231, 233, 237
LLM agents, 2–3, 10, 29–30, **33**, 36–38, 41, 45, 51–54, 56, 74, 114, 119, 123, 133, 135, 137–138, 141, 164, 166, 285–256, 303, 306, 312–313, 334
autonomous software agents distinguished from, 3
book-specific usage, 3
reviewing map intent, 334
as smart assistants, 51, 53
in workflow coordination, 45, 51–54, 56
LLM assistants, 12
LLM assistants and agents, 93–95, 104, 106–113
specialist roles, 108–109, *110*
user responsibility for, 106–108, 111–112
LLM defaults, 50, 137–139, 142–145, 147, 151, 155, 159, 163–164, 166–168
curse of, 137–139, 151, 163, 166–167
omission as design decision, 137–138, 142, 144, 147, 151, 163, 166
LLM maintenance probe, 314
LLMs, 58, 60, 62–63, 69, 72–73
memory and design, 73
LLMs as palette critics, 175, 180–181
locator maps, 243, 247
with path or fly-through animation, 243, 247
looping animation, 240–241, 244–247

M

maintenance, 312–314; *see also* stewardship
as creative practice, 312–314
as stewardship, 312–314
maintenance pipelines, 306–309, 312–313
from prompt spec to live deployment, 306
regenerate, check, report, 308–309
map communicates, *see* map communication
map communication, 10–12, 14–16; *see also* maps, persuasion and rhetoric
map design, *18–19*, 20–21, 26, 344–345
code-based, *19*
communicating through natural language, 344–345
GUI-based, *18*
prompt-based, 20–21
map design notes, 32, *34–35*
map director, 280–281, 283, 303–304, 338–339, 341–342
collaborative stewardship, 342
dataset choices, 338–339, 341–342
from maker to director, 303–304
map elements, 137–140, 143, *145*, 147, 149, 151, 153, 159, 163, 165–168, 285, 291, 294, 303
as conceptual intentions, 138, 140
defaults and specification, 137–138, *139*, 166
map extent, 252, 257–258, 261, 263, 265, 267–268, 275
map form, 114–115, 117–119, 122, 130–131, 133–136; *see also* aesthetic styles (cartography)
building blocks of, 118–119, 122
function of, 130–131, 134–135
language for describing, 114–115, 117–118, 134–135
map framing, 330, 335
map integrity, 310
map intent, 322–323, 328
map interactivity, 20, 27
map interpretation, 329, 331, 333, 336
projection by map users, 329, 331, 333, 336
map layout, *see* layout and composition
map misinterpretation, 321–323, 325, 327
map navigation tools, 147–148
map production, *see* maps as production systems
map purpose, 93, 99–103, 106, 109, 111–112
map rhetoric, 282, 294, 298

map specifications, 59–60
map style, *see* aesthetic styles (cartography)
map symbol libraries, 204
map symbology, 195–201, 203–208
  default web map symbols, 195–197
  effective symbology, 197, 207
  symbols as visual arguments, 195, 207
map symbols, *see* map symbology
map users, *see* map interpretation, projection by map users
mapmaking, 343–345
  technological change and future evolution, 343–345
mapped area, 143–146, 165–168; *see also* zooming, infinite zoom
  extent, 143–146, 165–168
  as narrative boundary, 144, 146, 165, 167
mapping presence, *see* mapping presence, absence, and uncertainty
mapping presence absence, *see* mapping presence, absence, and uncertainty
mapping presence, absence, and uncertainty, 329–337
  absence of evidence, 333
  avoiding false confidence, 329–330
  observations versus reality, 329–332
maps, 11–16, 73, 321, 325
  audience, 13–16
  communication and argumentation, 73
  data selection and restraint, 14–15
  neutrality of, 321, 325
  persuasion and rhetoric, 11–12, 14
  purpose, 12–16
  visual hierarchy, 15
maps as production systems, 338–339
memory and persistence systems, 262–264, 272, 275, 277–278
  dashboards, 262–264, 278
  panels, 262–264, 272, 275, 277–278
memory design, 28, 30–32, 35–42
  balancing and structuring, 35–36
  deliberate forgetting, 37–38, 41–42
  externalizing memory, 28, 31–32, 40–41
  over-retention, 37–38, 41
  workflow role, 39–41
metadata standards, 104
metrics, 83, 85–89, 91; *see also* design quality assurance process
  contrast, 85–87
  density, 85, 87
  performance, 85–87
  as signals, not verdicts, 86, 88, 91
  thresholds and standards, 85–88, 91
Middle-Earth dataset, 280, 283–284, 286–287, 297
Minard, Josef, 5, *6*
  Napoleon's march map, 5, *6*
monitoring, 309–310, 312–314
  drift thresholds, 310
motion and continuity systems, 253–254, 267–269, 272, 277, 279
  abrupt transitions, 268–269
  animation, 253–254, 268, 272, 277, 279
  transitions, 253–254, 267–269, 277
multivariate maps, 209, 231–237
  cognitive load, 231, 233–234, 237
  colored dots, 232–233
  layer combinations, 231
  small multiples, 231, 233, 237

N

narrative memory, *see* memory and persistence systems
natural breaks quantiles, *see* classification schemes
natural language, 5, 7–9, 342
  as design interface, 5, 7–9
natural language cartography, 280, 285, 292
natural language GIS (NLGIS), 104, 108
natural language interfaces, 17, 20, *250*, 252, 344; *see also* prompt cartography
natural language prompts, 186, 188, 191–193
  typographic conventions, 186, 188, 191–193
natural language specification, 76–78, 90–91

natural language-based GIS and cartography, 2
natural-language aesthetic design, 114, *120*, 131
neatlines, 144, 147
NLGIS, *see* natural Language GIS (NLGIS)
normalization, 322, 324, 328
novelty-first design, 272

O

observational data, *see* observational ecological data
observational ecological data, 329–335
  citizen science records, 330–331
  human access and attention bias, 330–332, 335
  temporal limits, 334–335
omission as specification, 195, 197, 207
  symbol choices, 195, 197, 207
omission in cartography, 326
opacity, 201, *202–203*, 207–208
  hierarchy, context, and uncertainty, 201, *202–203*, 207–208
overly literal icons, 198, 206, 208
over-specification, 271, 277

P

pacing, 252, 257–259, 261, 277
palette design, 169, 174–182
pan and zoom controls, 147–148, 150
panning, 143–144, 146–148, 166
  constraints, 143–144, 146–148, 166
path animation, 243, 247
performance (map), *see* metrics, performance
pictographic symbols, 198–199, 206–208
pipeline (design), 43–50, 52–56, 59–63, 69–71, 73–74, 281–282, 284, 296–297, 303–304
  archiving schemas, 62–63, 69
  behaviors pipeline, 45
  conversational, 281
  critique, 296–297
  critique pipeline, 45, 52–53, 56
  data engineering, 284
  data engineering pipeline, 45–48, 56
  definition and purpose, 43–45, 55–56
  ethical prompt pipeline, 69–70
  interface pipeline, 45, 49–50, 56
  judgment and traceability, 43–44, 55–56
  prompt-to-map pipeline, 71
  publish and maintain pipelines, 45, 54, 56
  visual planning pipeline, 45, 48–49, 53, 55–56
PNG, *see* PNG symbols
PNG symbols, 204, 208
point symbols, 200, 204
prefers-reduced-motion, *see* reduced-motion preferences
presence absence and uncertainty, *see* mapping presence, absence, and uncertainty
preserve hierarchy, *see* visual hierarchy
print cartography, 140, 144, 150, 155
privacy, 58, 65, *66–67*, 69–70; *see also* aggregation; differential privacy
  aggregation and masking, 65, *66–67*, 69–70
  differential privacy, 69
  doxxing, 65, 70
  sensitive themes, 65, 69
procedural mapmaking workflows, 7, 10
progressive disclosure, 255, 258, 261, 271, 277
prompt cartographer, 5, 8–10, 12, 14
  role and responsibilities, 5, 8–10, 12, 14
prompt cartography, 1–3, 5, 7, 9–10, 12–14, 20–21, 24, 26, 30, 36, 40, 42–60, 62, 64–65, 68–74, 76–78, 80, 82, 84–86, 88, 90–94, 96, 98–100, 102–104, 106–108, 110–116, 131–138, 140, 142, 144, 148, 150, 153, 155, 159, 163–169, 171, 173–174, *176*, *178*, 180, 182–184, 188, 191–192, 194, 209–210, 234–236, 238–240, 242, 244, 246, 248, 251, 269, 273–274, 276, 280–283, 285–288, 295–299,

301–304, 306–308, 311–314, 317–319, 322–328, 330–332, 334–336, 340–345; *see also* civic cartography; data engineering pipeline; interactive web maps; large language models (LLMs); natural language-based GIS and cartography; pipeline (design); prompting; web cartography
accountability in civic mapping, 322, 326, 328
aesthetic vocabulary in, 114–116, 134–135
animation in, 239–240, 242, 244, 246, 248
capstone project, 340, 342
collaboration with agents, 341–342
collaborative nature, 9–10, 12
creativity and judgment, 344
critique in, 288, 295–298, 303
critique upstream in, 76–77, 90–91
dataset description in, 93–94, 106, 111–113
definition, 2
definition and scope, 5
design language, 317
design responsibility in, 137–138, 164–167
early stage and future evolution, 343, 345
ethics in, 102–104, 106–108, 111–112, 286, 297–298, 303
ethics of, 58, 65, 69–72
explicit intent, 322–323, 328
fundamentals, 1, 3
guardrails, 331, 336
interactive web maps, 344–345
interactivity in, 251, 269, 273–274, 276
iteration in, 96, 98–99, 102, 111–112
language as design tool, 325–326, 328
maintenance and stewardship, 307–308, 312–314
making data limits explicit, 331–332, 336
map elements as language, 138, 140, 148, 150, 153, 163, 165
natural language and, 342
natural-language map design, 344–345
as newest iteration of cartography, 5, 7, 14
prompt literacy, 163, 165–167
promptable form variables, 131–132, 135
prompt-based critique, 325, 327
prompt-to-map workflow, 44–45, 50, 55, 57
publication and reuse, 299, 301–304
removing predictive language, 335
seminar scenarios, 318–319
stack of pipelines, 282, 303
style drift in, 132–133, 135
thematic mapping, 209–210, 234–236, 238
transfer across domains, 283, 286, 302, 304
transition to practice, 74
typography in, 183–184, 188, 191–192, 194
web infrastructure, 306
workflow, 10, 12–14, 43–57, 90, 281–282, 297, 302–303, 318
prompt cartography fundamentals, *see* prompt cartography, fundamentals
prompt cartography stack, 282
prompt cartography supports, *see* prompt cartography
prompt cartography workflow, *see* prompt cartography, workflow
prompt chaining, 235
complex thematic decisions, 235
prompt director, 52, 62, 65, 70, 236, 318
prompt director mindset, 9–11
prompt framing, 17, 24–26
prompt guardrails, *see* guardrails, prompting
prompt literacy, 25–26; *see also* design intent
prompt metadata, *see* prompts, metadata
prompt pipeline, *see* pipeline (design), prompt cartography
prompt specification, 238
thematic maps, 238

prompt templates, 16
promptable form variables, 114–115, 122–123, 125, 127, 130–133, 135
basemap design and selection, 131–132, 135
color palette, 114–115, 123, 127, 131, 135
layout, 130–131, 133, 135
symbology, 114, 125, 131, 133, 135
typography, 114, 123, 125, 127, 131–132, 135
promptable variables, *see* promptable form variables
prompt-based cartography, *see* prompt cartography
prompt-based workflows, 5, 7, 9–10
prompt-driven automation pipelines, *see* automation pipelines
prompt-driven cartography, *see* prompt cartography
prompt-driven design, 251, 253, 270–271
prompt-generated maps, 306, 312
prompting, 8–9, 11, 13–14, 16, 73, 183–184, 188, 191–194
as design language, 73
as design literacy, 11, 13–14
micro-prompt exercise, 16
structured language and specifications, 8–9, 11, 14
typographic intent, 183–184, 188, 191–194
prompting for color palettes, 172, 174–175, 177, 179–182
prompting for symbology, 197–200, 204–208; *see also* large language models (LLMs)
constraints and guardrails, 199–200, 205–206, 208
failure modes, 198, 206
prompt template, 197–198, 208
prompting patterns, *see* prompts
prompts, 2–3, 17, 20–27, 30–31, 39, 42, 58–62, 64–65, 68–72, 74, 93–96, 98–115, 119, 122–123, 125, 127, 130, 132–136, 239–241, 243–245, 248–249, 251, 254–255, 257–259, 261–271, 273–274, 276, 278–279, 317–319
aesthetic specificity in, 114–115, 122, 134
for animated maps, 239–241, 243–245, 248
for animation and motion, 254, 268–269
archiving, 61–62, 69
casual versus directive prompting, 17, 20–21
clarity and diagnosis, 21, 26–27
for commitment systems, 258–259, 261
conceptual exercise, 42
constraint specification, 244, 248
constraint-driven dataset queries, 94–96, 98, 112
for control systems, 265–267
definition as natural language directives, 17
as deliberate directives, 74
as design artifacts, 22–23, 26
as design specifications, 26
for disclosure systems, 255, 257
ethical clauses, 107–108, 111–112
ethical directives, 60, 65, 69
LLM prompts, 2
metadata, 59, 61, 69
natural-language intent, 318
for panels and persistence, 262–264
structure versus cleverness, 23–24, 26
structured output formats, 94–95, 98, 102, 106, 112–113
structured prompt exercise, 136
structures and transcripts, 318
system prompts, 3, 21, 26
system prompts and schemas, 123, 132, 135
templates, 72
tool prompts, 22, 26
types and roles, 21–22, 26
for user flow, 269–270, 279
user prompts, 21, 26, 30–31
prompt-to-map pipeline, *see* pipeline (design)
proportional symbol maps, 209, 214–217, *219*, 220, 229, 231, 233, 236
graduated symbols, 214–215, 217
legends and tooltips, 217

overlap and occlusion, 217
scaling by area, 216–217, 236
proportional symbology, 200–201, 206–207
area versus radius scaling, 200, 207
proportional symbols, *see* proportional symbology
provenance (cartographic), 58–59, 61, 63, 65, 69–71; *see also* attribution
chain of custody for data, 58–59, 71
prompt metadata fields, 59, 61, 69
reproducibility, 59
provenance ethics, *see* ethics
provenance ethics and privacy, *see* ethics
public risk data, 322–323, 326
publishing, 54, 56; *see also* responsibility
accountability, 54, 56
publishing civic maps, 327
publishing maps, 296, 299–304; *see also* quality assurance (QA)
as design decision, 299
responsibility after release, 299, 304

Q

QA toolkit, *see* design quality assurance process
quality assurance (QA), 299–301
fragility, 300
testing as intended audience, 300–301

R

RAG, *see* retrieval-augmented generation (RAG)
RAG systems, *see* retrieval-augmented generation (RAG)
raw counts, 211, 214–215, 217, 236, 238
mapping cautions, 211, 214–215, 217, 236, 238
readiness, 298, 301, 305
versus perfection, 298, 301, 305
reduced-motion preferences, 241–242, 244, 247
repeatable design processes, 43–44, 55
repositories, 306–307
map deployment, 306–307
representation ethics, 103, 107–108, 111–112
responsibility, 53–56
for LLM-assisted maps, 53–56
responsible framing, *see* framing (maps)
responsible map design, 341–342
retrieval (in RAG), *see* retrieval-augmented generation (RAG)
retrieval-augmented generation (RAG), 3, 29–32, *35*, 35–42
constraints, 37, 41
definition and basic workflow, 29–30
forgetting and maintenance, 37–38, 41–42
limitations, 39, 41
retrieval as editorial act, 31, 36–37, 40–41
training versus retrieval augmentation, 30–31
retrieval-augmented generation RAG, *see* retrieval-augmented generation (RAG)
reuse, 301–302, 304
prompt-directed maps, 301–302, 304
review, 45, 52
quality assurance, 45, 52
RGB color model, 173, 181
rhetoric (cartography), 28, 36, 39
rhetoric, in map critique, 75, 77, 79, 90–91

S

sans-serif fonts, 184–185, 188–189, 191, 193
scalable vector graphics (SVG), *see* SVG symbols
scope creep, 265, 267, 271–272
scripting and code, 17, *19*
self-critique, 133, 135
LLM design outputs, 133, 135
semantic categories, 255, 257, 275
filters and legends, 255, 257, 275
semantic color prompts, 177, 181; *see also* prompting for color palettes
semantic fingerprint, *see* semantic search, semantic fingerprints
semantic search, 95–99, 109, *110*, 112; *see also* data discovery
embeddings and vectors, 95–96, *97*

vs keyword search, 95–96, *97*, *110*, 112
semantic fingerprints, 96, 112
sequential color ramps, 324
sequential palette, *see* color schemes, sequential schemes
serif fonts, 184–185, 188–189, 191, 193
shape, *196*, 198–199, 207–208
categorical encoding, *196*, 198–199, 207–208
size, 198, 200–201, 207–208
quantitative encoding, 198, 200–201, 207–208
slippy web maps, 251
small multiples, 231, 233, 237
smooth transitions, 243–248
spatial grammar layout, *see* layout systems, spatial grammar
specifications (LLM-era), 50–53, 56; *see also* critique
as contracts, 50–52, 56
drift, 53
as shared contracts, 50
speculative datasets, 340–341
splash screens, 142–143
state, 254–256, 263, 271–272, 275, 277–278
as rhetorical design decision, 254–256, 263, 271–272, 275, 277–278
stewardship, 312–313
over control, 312–313
stop-frame transitions, 243–244, 246–248
structured language, 8–9, 14
structured prompting, 23, 26
structured prompts, *see* structured prompting
style drift (LLM), 132–133, 135, 206; *see also* aesthetic defaults (in LLMs)
across map layers, 206
causes of, 132–133
mitigation strategies, 132–133
style templates, 123, 132–133, 135–136
SVG, *see* SVG symbols
SVG symbols, 205–208
symbol maps, *see* proportional symbol maps
symbol rhetoric, 197–198, 206–208
symbology, 114, 125, 131, 133, 135, 159, 162–163, 280–281, 286–289, *290*, 295
meaning before beauty, 288, *290*
synchronization rules, 255, 257, 262–263, 268, 272, 275
system prompts, 3

T

tacit knowledge (in cartography), 28–29, 32, 40–41
team of agents, *see* AI agents
temporal animation, 241–243, 247–248
temporal controls, 241, 248
temporal legends, 242–243, 245, 247–248; *see also* temporal controls
bar scales, 242–243
circles, 242
clocks and calendars, 242
temporal thinking without prediction, 334–335
temporal variables, 239–242, 245, 247
duration, 239–241, 245, 247
order, 239, 241, 247
rate of change, 239, 241–242, 247
synchronization, 239–241, 245, 247
testing interactive prompts, 273–278
communicative testing, 273, 275–277
iteration by subtraction, 273–274
variation testing, 274
texture, 201, *203*, 207
as visual redundancy, 201, *203*, 207
thematic map, *see* thematic map methods
thematic map methods, 209–211, 214, 216–218, 220, 223–225, 228, 230, 234–238
default lies, 209, 211, 214, 216–218, 220, 223–225, 228, 230, 234–347
method selection framework, 234, 236, 238
thematic mapping, 200, 206; *see also* proportional symbology
symbol sizing, 200, 206

thematic representation, *see* thematic map methods
titles, 140–143, 166, 168
  human-facing and model-facing, 141–143
  map, 140–143, 166, 168
  rhetorical framing, 140–143
traceability, 39, 41, 55–56
  in prompt cartography, 55–56
  RAG as paper trail, 39, 41
training data, 137, 151
  and map-design defaults, 137, 151
transparency, 60–61, 70, 72
tweening, *see* smooth transitions
typeface, *see* font
typography, 183–194
  legibility, 183, 185–186, 192–193
  prompting for, 183–184, 186, 188, 191–194
  robustness in web maps, 183, 187, 189, 191, 193
  visual hierarchy, 183, 186, 188–189, 191–194
  as voice or tone of map, 183, 189–193

U

uncertainty as message, 330
under-specification, 256, 270–271, 277
user flows, 269–270, 277, 279
  pauses and rest, 270, 279
  transition conditions, 270

V

visual hierarchy, *see* visual hierarchy
validation bias, 272–273, 278
vector databases, 30
versioning, 310, 315
  map integrity, 310, 315
visual clutter, 160, 162
visual communication, 115–116, 119, *120*
visual direction, 287–289; *see also* symbology; visual hierarchy
  style, symbology, and hierarchy, 287–289
visual fluency, 118
visual form, *see* map form
visual hierarchy, 50, 53, 117, 127, 133, 137, 148, 150–153, *154*, 165, 167–168, 183, 186, 188–189, 191–194, 255, 257–259, 264, 266–267, 274–275, 277–278, 289, *290*, 292, 300, 304; *see also* font properties
  interactive behavior and, 255, 257–259, 264, 266–267, 274–275, 277–278
  linguistic priority, 151, 153, 165, 167
  mapped data as primary, 151–153, 165, 168
  preserving in typography, 183, 186, 188–189, 191–194
  templates, 153, *154*
visual language, 115, 123, 131, 135
visual literacy, 115–119, 122, 134, 285, 287; *see also* Dondis, Donis A.
  Dondis framework and, 115–116, 119, 122, 134
visual noise, 218, 221, 236
visual planning, 45, 48–49, 56; *see also* visual hierarchy
  designing before execution, 45, 48–49, 56
  hierarchy, emphasis, and tone, 48–49, 56
visual rhetoric, 246
  animation as, 246
visual style, 280–281, 283, 285, 287
visual variables, 65, 119, 122, 134, 174, 179, *196*, 201, 215, 225–226, 228, 231–232, 239–241, 247; *see also* flow maps; proportional symbol maps
  animation, 239–241, 247
  color, 174, 179
  empathy as, 65

W

warm and cool colors, 172–173, 181
web cartographer, *see* web cartography

web cartography, 5, 7, 10, 13–14, 60, 65, 140, 148, 153, 155, 159, 166, 249, 251–252, 254, 256, 269, 271, 342
  interactivity in, 249, 251–252, 254, 256, 269, 271
web cartography workflows, 75–77
web map design, 318; *see also* prompt cartography
web map director, *see* map director
web maps, 5, 14–16, 140, 142–143, 146–150, 153–155, 158–159, 164, 183–187, 191, 193, *250*, 251, 254, 256, 270–271, 278, 306–307, 313–314, 318–319; *see also* interactivity; maintenance pipelines
  interactive, *250*, 251, 254, 256, 270–271, 278
  interactivity and use, 5, 14–16
  maintenance after publishing, 306–307, 313–314
  over-inclusion of data, 15
  prompt-generated, 318–319
  typographic constraints, 183–187, 191, 193
web typography, 184, 187–189, 191, 193
  font substitution and rendering inconsistencies, 184, 187–189, 191, 193
WIMP interfaces, *250*
working memory, 245, 247
  animation limits and, 245, 247

Y

YAML, 32, **33**

Z

zoom animation, 243
zoom-dependent visibility, 183, 191, 193
  labels, 183, 191, 193
zooming, 143–148, 165–168
  infinite zoom, 143–146, 167–168
  limits and valid comparison, 144, 146–148, 165–166, 168

For Product Safety Concerns and Information please contact our EU representative GPSR@taylorandfrancis.com Taylor & Francis Verlag GmbH, Kaufingerstraße 24, 80331 München, Germany